T0395791

Electromagnetic Boundary Problems

Electromagnetics, Wireless, Radar, and Microwaves

Electromagnetic Boundary Problems

Edward F. Kuester
University of Colorado Boulder

David C. Chang
Polytechnic School of Engineering, New York University

CRC Press
Taylor & Francis Group
Boca Raton   London   New York

CRC Press is an imprint of the
Taylor & Francis Group, an **informa** business

CRC Press
Taylor & Francis Group
6000 Broken Sound Parkway NW, Suite 300
Boca Raton, FL 33487-2742

© 2016 by Taylor & Francis Group, LLC
CRC Press is an imprint of Taylor & Francis Group, an Informa business

No claim to original U.S. Government works

Printed on acid-free paper
Version Date: 20150813

International Standard Book Number-13: 978-1-4987-3026-6 (Hardback)

Library of Congress Cataloging-in-Publication Data

Kuester, Edward F.
  Electromagnetic boundary problems / Edward F. Kuester and David C. Chang.
      pages cm -- (Electromagnetics, wireless, radar, and microwaves)
  "A CRC title."
  Includes bibliographical references and index.
  ISBN 978-1-4987-3026-6 (hardcover : alk. paper) 1. Electromagnetic fields. 2.
  Boundary value problems. I. Chang, David C. II. Title.

  QC665.E4K77 2016
  530.14'101515625--dc23   2015008681

Visit the Taylor & Francis Web site at
http://www.taylorandfrancis.com

and the CRC Press Web site at
http://www.crcpress.com

Dedication

For Dan, Nikki, and Lisi

To Cecilia

However mysterious energy (and its flux) may be in some of its theoretical aspects, there must be something in it, because it is convertible into dollars, the ultimate official measure of value.

—Oliver Heaviside,
"Electromagnetic theory - L,"
The Electrician, vol. 31, 1893.

The interpretation just given [of Poynting's theorem] suggests several comments. In the first place, the present authors are not able to ascribe any significance whatever to the phrase "localized energy." They do not believe that "Where?" is a fair or sensible question to ask concerning energy. Energy is a function of configuration, just as the beauty of a certain black-and-white design is a function of configuration. The authors see no more reason or excuse for speaking of a spatial energy density than they would for saying, in the case of a design, that its beauty was distributed over it with a certain density. Such a view would lead one to assign to a perfectly blank square inch in one portion of the design a certain amount of beauty, and to an equally blank square inch in another portion a certain different amount of beauty.

—Max Mason and Warren Weaver,
The Electromagnetic Field.
Chicago: University of Chicago Press, 1929.

"Maxwell's equations, they're called. I can say them forward and backward, but I don't know what they are or what they mean."
"Neither do I, but you're not supposed to know."

—Walter M. Miller, Jr.,
Saint Leibowitz and the Wild Horse Woman.
New York: Bantam Books, 1997.

Contents

List of Figures

List of Tables

Series Preface

Electromagnetics is a fascinating discipline of science and engineering with endless technological and societal impacts. The purpose of the Electromagnetics, Wireless, Radar, and Microwaves Book Series is to present — in an innovative and reader-centered way — both the fundamentals and the state of the art of electromagnetics at all frequencies and a broad spectrum of its application areas. This includes antennas and wireless systems, RF and microwave engineering, computational electromagnetics, radar, remote sensing, high-speed electronics, power and energy, metamaterials, and bioelectromagnetics.

Electromagnetics is a foundation of electrical and computer engineering as a whole. As we celebrate 150 years of Maxwell's equations, we are still not done in our attempts to solve them in yet more efficient and accurate ways, to enable computer simulations of yet larger and more complex electromagnetic problems. Concepts, models, devices, and systems based on Maxwell's equations are at the forefront of discovery, advancement, and practice of cutting-edge wireless, radar, electronics, communication, and microwave technologies of today and tomorrow. They also have immediate impact on emerging interdisciplinary applications spread across most prominent disciplines of science and engineering.

The series features authored and edited titles, including textbooks, research monographs, guides, and handbooks. It provides versatile, comprehensive, and diverse resources for learning, understanding, using, and advancing the science and engineering of electromagnetic fields, waves, devices, and systems, as well as all related applications. The texts are aimed for the audiences spanning students, researchers, teachers, practicing engineers and scientists, and lifelong learners.

Preface

This book, which is the first in the series, Electromagnetics, Wireless, Radar, and Microwaves (edited by Branislav Notaroš), arose out of course notes for a one-semester second-year graduate course offered at the University of Colorado Boulder on the formulation and solution of the boundary problems of electromagnetics. The reader who consults other references on the subject will note that, while some of our coverage is quite standard, other parts of the treatment follow a rather idiosyncratic path, reflecting our desire to keep the presentation at a certain level or our conviction that some concepts have more importance than is often accorded them. Thus, for example, we have avoided any use of the apparatus of complex analysis in our treatment of radiation and scattering problems. Although this apparatus is undoubtedly of great value, we feel that students without this background should not have to wait to study, for example, an asymptotic far field approximation, and that an exposure to a more traditional approach later on will impart a deeper appreciation of such topics than if only a single approach has been seen. In addition, we have treated the following topics in more detail or from different viewpoints than is usual: jump conditions, finite propagation velocity of wavefronts, Bloch-wave representations in periodic media, and variational methods of different kinds. Potentials are presented as a consequence of the converse of Poincaré's lemma, the proper justification for them.

The problems at the ends of the chapters are designed in some cases to test the reader's understanding of a concept or technique by asking to modify a derivation or formulation from the notes so as to apply to a somewhat different problem. In other cases, a problem may be a minor research topic requiring a certain amount of preliminary thought or discussion before putting pencil to paper. In any event, many of these problems (in L. A. Vainshtein's marvelous phrase) require a certain perseverance on the part of the student.

Earlier versions of the material in this book have been used in teaching hundreds of our former students. Many inaccuracies have been corrected from previous versions by the careful attention of those students, and to them the authors are grateful. If you find any more errors during your study of them, please let us know: the authors will be grateful to you as well, and you will have the satisfaction of knowing that a true understanding of any subject can only come after many mistakes are made and corrected. Suggestions and contributions of some material by Dr. David A. Hill of the National Institute of Standards and Technology in Boulder are also gratefully acknowledged.

Though this book is meant to be a self-contained text, there are a number of books available that can be valuable in gaining additional insights or alternate viewpoints for certain topics covered herein. We list below the ones that we have found most valuable.

BIBLIOGRAPHY

C. A. Balanis, *Advanced Engineering Electromagnetics*. New York: Wiley, 1989.

L. Eyges, *The Classical Electromagnetic Field*. New York: Dover, 1980.

R. F. Harrington, *Time-Harmonic Electromagnetic Fields*. New York: McGraw-Hill, 1961.

A. Ishimaru, *Electromagnetic Wave Propagation, Radiation, and Scattering*. Englewood Cliffs, NJ: Prentice-Hall, 1991.

J. D. Jackson, *Classical Electrodynamics*, second edition. New York: Wiley, 1975.

J. A. Kong, *Electromagnetic Wave Theory*. New York: Wiley, 1986.

J. A. Stratton, *Electromagnetic Theory*. New York: McGraw-Hill, 1941.

G. Tyras, *Radiation and Propagation of Electromagnetic Waves*. New York: Academic Press, 1969.

J. Van Bladel, *Electromagnetic Fields*, second edition. Hoboken, NJ: Wiley, 2007.

W. L. Weeks, *Electromagnetic Theory for Engineering Applications*. New York: Wiley, 1964.

Authors

Edward F. Kuester received a B.S. degree from Michigan State University, East Lansing, in 1971, and M.S. and Ph.D. degrees from the University of Colorado at Boulder, in 1974 and 1976, respectively, all in electrical engineering.

Since 1976, he has been with the Department of Electrical, Computer, and Energy Engineering at the University of Colorado at Boulder, where he is currently a Professor. In 1979, he was a Summer Faculty Fellow at the Jet Propulsion Laboratory, Pasadena, CA. From 1981 to 1982, he was a Visiting Professor at the Technische Hogeschool, Delft, The Netherlands.

In 1992 and 1993, he was an Invited Professor at the École Polytechnique Fédérale de Lausanne, Switzerland. He has held the position of Visiting Scientist at the National Institute of Standards and Technology (NIST), Boulder, CO in 2002, 2004, and 2006. His research interests include the modeling of electromagnetic phenomena of guiding and radiating structures, metamaterials, applied mathematics, and applied physics. He is the coauthor of one book, author of chapters in two others, and has translated two books from Russian. He is co-holder of two U.S. patents, and author or coauthor of more than 90 papers in refereed technical journals.

Dr. Kuester is a fellow of the Institute of Electrical and Electronics Engineers (IEEE), and a member of the Society for Industrial and Applied Mathematics (SIAM) and Commissions B and D of the International Union of Radio Science (URSI).

 **David C. Chang** holds M.S. and Ph.D. degrees in applied physics from Harvard University and a bachelor's degree in electrical engineering from National Cheng-Kung University in Taiwan. His academic career began as an Assistant Professor of Electrical and Computer Engineering at the University of Colorado at Boulder, Colorado in 1967, Associate Professor in 1971, and Full Professor in 1975. He was Chair of that department for the period from 1981 to 1989. Dr. Chang helped establish the National Science Foundation (NSF) Industry/University Cooperative Research on Microwave and Millimeter Wave Computer-Aided Design in that department in 1988 and served as its Director until 1992, when he became Dean of Engineering and Applied Sciences at the Arizona State University in Tempe, AZ.

Dr. Chang was named President of Polytechnic University, Brooklyn, New York, in 1994, and was appointed as its Chancellor in July 2005. He retired from that position in the summer of 2013, and is now Professor Emeritus at the same university, which was merged into New York University in January 2014 and is now known as the Polytechnic School of Engineering of NYU. Dr. Chang was named a Fellow of the Institute of Electrical and Electronics Engineers (IEEE) in 1985 and Life Fellow in 2006. He is a past president of the IEEE Antennas and Propagation Society, 1990 and an Associate Editor of its Transactions from 1978 to 1980.

Dr. Chang was also active in the International Scientific Radio Union (URSI); he became chairman of the U.S. National Committee of 1993-96 and was the Editor of *Radio Science*, a joint publication with the American Geographical Union from 1990 to 1994. Dr. Chang has a distinguished record of academic achievements in electromagnetic theory, antennas analysis, and both microwave and optical integrated circuits. He was awarded the IEEE Third Millennium Medal in 2000, and has also been named an honorary professor at five major Chinese universities. He has served as Chairman of the International Board of Advisors at the Hong Kong Polytechnic University since 2009. He was also appointed in 2013 as Special Advisor to the President of Nanjing University in China.

1 Maxwell's Equations and Sources

1.1 MAXWELL EQUATIONS IN FREE SPACE

The Maxwell equations describing a time-dependent electromagnetic field $\mathbf{E}$, $\mathbf{B}$ existing in the presence of an electric charge density ρ_{tot}, and an electric current density $\mathbf{J}_{\text{tot}}$, are

$$\nabla \cdot (\epsilon_0 \mathbf{E}) = \rho_{\text{tot}} \tag{1.1}$$

$$\nabla \cdot \mathbf{B} = 0 \tag{1.2}$$

$$\nabla \times \frac{\mathbf{B}}{\mu_0} = \mathbf{J}_{\text{tot}} + \epsilon_0 \frac{\partial \mathbf{E}}{\partial t} \tag{1.3}$$

$$\nabla \times \mathbf{E} = -\frac{\partial \mathbf{B}}{\partial t} \tag{1.4}$$

where $\mathbf{E}$ is the electric field and $\mathbf{B}$ is the magnetic induction. The constants $\mu_0 = 4\pi \times 10^{-7}$ H/m and $\epsilon_0 = 8.854\ldots \times 10^{-12}$ F/m are the permeability and permittivity, respectively, of free space. Their presence is associated with the choice of SI (MKS) units for the fields, charges, and currents.

The fields $\mathbf{E}$ and $\mathbf{B}$ have direct *physical* meaning: the force on an infinitesimal (point) charged particle of charge q_{te} (often called a "test" charge) is given by the Lorentz[1] force

$$\mathbf{F} = q_{\text{te}}(\mathbf{E} + \mathbf{v} \times \mathbf{B}) \tag{1.5}$$

where $\mathbf{v}$ is the particle's velocity. The $\mathbf{E}$ and $\mathbf{B}$ fields in (1.5) are those produced by other charges and their motions, but *not* by the test charge itself. If we replace the infinitesimal test particle by a small volume element ΔV carrying test charge and current densities ρ_{te} and $\mathbf{J}_{\text{te}}$, then there will exist a corresponding Lorentz force $\Delta \mathbf{F}$ on this volume element given by:

$$\Delta \mathbf{F} = (\rho_{\text{te}} \mathbf{E} + \mathbf{J}_{\text{te}} \times \mathbf{B}) \, \Delta V \tag{1.6}$$

and again, the $\mathbf{E}$ and $\mathbf{B}$ fields in (1.7) are those produced by all sources except the test charge and current. We can then identify a force density (per unit volume) as

$$\mathbf{F}_v = \rho_{\text{te}} \mathbf{E} + \mathbf{J}_{\text{te}} \times \mathbf{B} \tag{1.7}$$

---

[1] After the Dutch physicist H. A. Lorentz—not to be confused with the Danish physicist L. Lorenz.

However, if we wish to obtain the force on a portion of the charge and current that produce the field, we must proceed a little more carefully.

Let us consider a small volume element ΔV carrying a portion of the total charge and current densities ρ_{tot} and $\mathbf{J}_{\text{tot}}$. The total field is then separated into a portion due only to the sources in ΔV (which we will call $\mathbf{E}_{\Delta V}$ and $\mathbf{B}_{\Delta V}$), and a portion due to all the remaining sources (which we will designate as the "local field" $\mathbf{E}_{\text{loc}}$ and $\mathbf{B}_{\text{loc}}$):

$$\mathbf{E} = \mathbf{E}_{\Delta V} + \mathbf{E}_{\text{loc}}; \quad \mathbf{B} = \mathbf{B}_{\Delta V} + \mathbf{B}_{\text{loc}} \tag{1.8}$$

The force on the portion of the sources in ΔV is now

$$\Delta \mathbf{F} = (\rho_{\text{tot}} \mathbf{E}_{\text{loc}} + \mathbf{J}_{\text{tot}} \times \mathbf{B}_{\text{loc}}) \Delta V \tag{1.9}$$

We must proceed in this way because the origin of the Lorentz force is traced to the Coulomb force between two point charges, and we do not include the field of the test charge (which is infinite at the location of the charge) when computing the force on it. This distinction is not important when dealing with continuous volume distributions of charge and current. Indeed, if we take the limit as $\Delta V \to 0$ we recover the force density (1.7), but if concentrated sources like point or line charges are present this will no longer be the case.

Note that taking the divergence of (1.3) and using (1.1) leads to

$$\nabla \cdot \mathbf{J}_{\text{tot}} + \frac{\partial \rho_{\text{tot}}}{\partial t} = 0 \tag{1.10}$$

because $\nabla \cdot \nabla \times \mathbf{B} \equiv 0$. Equation (1.10) is the law of conservation of charge. Likewise, taking the divergence of (1.4) gives

$$\frac{\partial}{\partial t} (\nabla \cdot \mathbf{B}) = 0$$

or

$$\nabla \cdot \mathbf{B} = \text{constant in time}$$

Equation (1.2) fixes this constant at zero, and thus serves as an initial condition for $\nabla \cdot \mathbf{B}$, as does Eqn. (1.1) for the quantity $[\nabla \cdot (\epsilon_0 \mathbf{E}) - \rho_{\text{tot}}]$.

Boundary conditions for (1.1)-(1.4) at a surface in space are traditionally obtained by integration of (1.1) and (1.2) over pillbox volumes cutting across the surface, or (1.3) and (1.4) over thin rectangular areas cutting through the surface (Figure 1.1), in the limit as the thickness of the pillbox or rectangle approaches zero. The results are

$$\begin{aligned}
\mathbf{u}_n \times (\mathbf{E}_1 - \mathbf{E}_2) &= 0 \\
\mathbf{u}_n \times (\mathbf{B}_1 - \mathbf{B}_2) &= \mu_0 \mathbf{J}_{S,\text{tot}} \\
\mathbf{u}_n \cdot (\mathbf{B}_1 - \mathbf{B}_2) &= 0 \\
\mathbf{u}_n \cdot (\mathbf{E}_1 - \mathbf{E}_2) &= \frac{\rho_{S,\text{tot}}}{\epsilon_0}
\end{aligned}$$

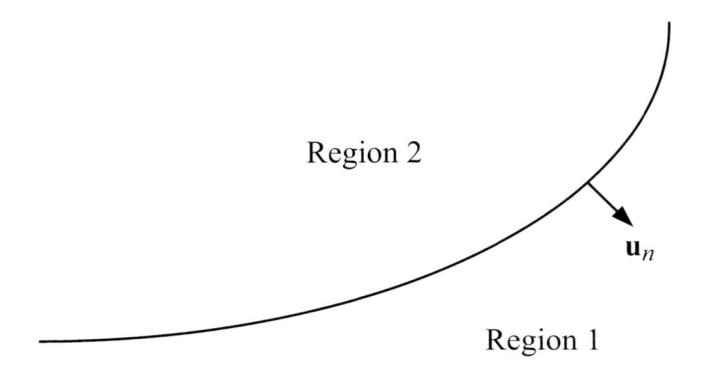

Figure 1.1 Interface between two media.

where $\mathbf{u}_n$ is the unit vector normal to the surface directed towards region 1 and $\rho_{S,\mathrm{tot}}$ and $\mathbf{J}_{S,\mathrm{tot}}$ are surface charge and current densities on the boundary surface. We will see later on how to obtain a more general set of such boundary conditions in a systematic fashion using generalized functions.

1.1.1   ENERGY TRANSFER AND POYNTING'S THEOREM

Motion of charges in the presence of the Lorentz force is associated with a conversion of energy between electromagnetic and non-electromagnetic forms. From basic physics, we know that the power being delivered to a single test charge (the rate of work being done *against* the Lorentz force to move the charge) is equal to

$$P = \mathbf{F} \cdot \mathbf{v} = q\mathbf{v} \cdot \mathbf{E} \tag{1.11}$$

inasmuch as $\mathbf{v} \cdot \mathbf{v} \times \mathbf{B} = 0$ since the cross product of $\mathbf{v}$ with any other vector will be perpendicular to $\mathbf{v}$. In the case of a continuous volume charge density, there is in the differential volume element dV a power $P_{v,\mathrm{tot}}\, dV$ delivered to the charge density, where the power density $P_{v,tot}$ is

$$P_{v,\mathrm{tot}} = \mathbf{F}_v \cdot \mathbf{v} = \rho_{\mathrm{tot}}\mathbf{v} \cdot \mathbf{E} = \mathbf{J}_{\mathrm{tot}} \cdot \mathbf{E} \tag{1.12}$$

since the current density $\mathbf{J}_{\mathrm{tot}}$ is equal to $\rho_{\mathrm{tot}}\mathbf{v}$. Thus, the power being delivered to maintain a current density $\mathbf{J}_{\mathrm{tot}}$ in a volume V of space (the so-called Joule power) is

$$P_{J,\mathrm{tot}} = \int_V P_v\, dV = \int_V \mathbf{J}_{\mathrm{tot}} \cdot \mathbf{E}\, dV \tag{1.13}$$

Again, this discussion is valid for continuous volume distributions of current, but if $\mathbf{J}_{\mathrm{tot}}$ contains singular terms such as line currents or point dipoles, the Joule power may be infinite. This results from the over-idealization of the current of the distribution of current in space, and will cause such things as uniqueness proofs to fail. We should keep this in mind when considering power and energy in the electromagnetic field, and be willing to consider line and point sources as limiting cases of more realistic situations.

If the power (1.13) is positive, energy is being delivered to the system, and dissipated (converted to some non-electromagnetic form, such as heat). If this power is negative, energy is coming from the system (electromagnetic energy is being generated from some non-electromagnetic form, such as chemical energy in the case of a battery).

We can use Maxwell's equations to express the Joule power in other forms. We take the dot product of (1.3) with $\mathbf{E}$, and the dot product of (1.4) with $\mathbf{B}/\mu_0$, and subtract the results. Using the vector identity (D.9), we finally obtain

$$-\nabla \cdot \left(\mathbf{E} \times \frac{\mathbf{B}}{\mu_0} \right) = \mathbf{J}_{\text{tot}} \cdot \mathbf{E} + \frac{\partial}{\partial t} \left[\frac{B^2}{2\mu_0} + \frac{\epsilon_0 E^2}{2} \right] \tag{1.14}$$

We can integrate this result over a fixed (independent of time) volume V (bounded by a surface S whose outward unit normal vector is $\mathbf{u}_n$) and apply the divergence theorem (D.20) to the integral of the divergence term to obtain

$$-\oint_S \left(\mathbf{E} \times \frac{\mathbf{B}}{\mu_0} \right) \cdot \mathbf{u}_n \, dS = P_{J,\text{tot}} + \frac{d}{dt} \int_V \left[\frac{B^2}{2\mu_0} + \frac{\epsilon_0 E^2}{2} \right] dV \tag{1.15}$$

This is Poynting's theorem. Its left side is conventionally interpreted as the rate of *electromagnetic* energy flow *into* the volume V: the vector direction of $d\mathbf{S} = \mathbf{u}_n \, dS$ is *out of* V. The first term on the right side is the Joule power converted into other forms (heat, motion, chemical energy, etc.) inside V, and the second term is interpreted as the time rate of change of the total electromagnetic (electric plus magnetic) stored energy

$$\int_V \left[\frac{B^2}{2\mu_0} + \frac{\epsilon_0 E^2}{2} \right] dV \tag{1.16}$$

inside V. Taken as a whole, Poynting's theorem expresses the conservation of energy for the volume V.

It should be noted that while the physical interpretation of the Joule power P_J is unambiguous, there is a certain arbitrariness in the interpretations of the left side of (1.15) and of (1.16). This is because the curl of any vector $\mathbf{Q}$ could be added to the electromagnetic *Poynting vector*

$$\mathbf{E} \times \frac{\mathbf{B}}{\mu_0}$$

without affecting the value of the left side of (1.15). A similar statement can be made about (1.16): only in the static case is it unequivocally identifiable with stored energy. Nevertheless, it does no harm to make the conventional interpretation of these terms, and we will see later that Poynting's theorem has important applications independent of these interpretations.

1.2   MACROSCOPIC MAXWELL EQUATIONS IN MATERIAL MEDIA

The densities ρ_{tot} and $\mathbf{J}_{\text{tot}}$ contain all charge and current distributions in space, whether *impressed* (that is, independently specified by some external agency—these are the analogs of ideal generators in circuit analysis) or *induced* in a material medium by the action of the fields on the atoms, ions, or molecules of the medium. The impressed charges and currents will be designated ρ_{ext} and $\mathbf{J}_{\text{ext}}$, are assumed to separately obey the charge conservation law (1.10), and will always be assumed to be known. The induced portions of $\mathbf{J}_{\text{tot}}$ and ρ_{tot} are often expressed in terms of the conduction current, polarization, and magnetization densities in the medium. Before we discuss these, we must pause for a brief discussion of multipoles.

1.2.1   MULTIPOLE EXPANSIONS FOR CHARGES AND CURRENTS

Assume that a charge density is concentrated near the origin. Then we can use the three-dimensional extension of the expansion (A.19) in the Appendix to express ρ as a series of *multipole sources*[2] concentrated *at* the origin:

$$\rho(\mathbf{r}, t) = Q(t)\delta(\mathbf{r}) - \mathbf{p}(t) \cdot \nabla\delta(\mathbf{r}) + \dots \tag{1.17}$$

where

$$Q(t) = \int \rho(\mathbf{r}, t)\, dV$$

is the total charge (monopole moment) of the distribution, and

$$\mathbf{p}(t) = \int \mathbf{r}\rho(\mathbf{r}, t)\, dV \tag{1.18}$$

is the dipole moment of the distribution. Further terms, not shown in (1.17), are quadrupoles and higher multipoles into which we will not delve here. The total charge Q now appears as a point charge concentrated at the origin.

To interpret the dipole moment term, consider the traditional charge dipole oriented along the x-axis, made up of a negative point charge $-q_d$ located at $\mathbf{r} = (-d/2, 0, 0)$ and a positive point charge q_d located at $\mathbf{r} = (d/2, 0, 0)$. The charge density of this system is (ignoring any time dependence for purposes of this discussion)

$$\rho(\mathbf{r}) = q_d\delta(y)\delta(z) \left[\delta(x - d/2) - \delta(x + d/2)\right]$$

Expanding $\delta(x \pm d/2)$ in powers of $d/2$ using Taylor's series (or alternatively using (A.19)), we can write this expression as

$$-q_d\delta(y)\delta(z) \left[d\delta'(x) + \frac{d^3}{24}\delta'''(x) + \dots\right]$$

---

[2]C. A. Kocher, *Amer. J. Phys.*, vol. 46, pp. 578-579, 1978.

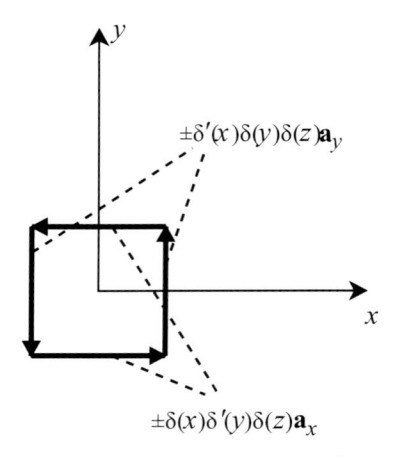

Figure 1.2 "Current loop" interpretation of a z-directed magnetic dipole.

If $d \to 0$ at the same time as $q_d \to \infty$ in such a way that $q_d d \to p_x$, this expression becomes $-p_x \delta'(x)\delta(y)\delta(z)$, in agreement with the second term of (1.17) when only an x-directed dipole moment is present.

The current density has a similar expansion:

$$\mathbf{J}(\mathbf{r}, t) = \frac{d\mathbf{p}}{dt}\delta(\mathbf{r}) - \mathbf{m}(t) \times \nabla\delta(\mathbf{r}) - \frac{d\overleftrightarrow{\mathbf{Q}}_s(t)}{dt} \cdot \nabla\delta(\mathbf{r}) + \dots \quad (1.19)$$

$$= \frac{d\mathbf{p}}{dt}\delta(\mathbf{r}) + \nabla \times [\mathbf{m}(t)\delta(\mathbf{r})] - \frac{d\overleftrightarrow{\mathbf{Q}}_s(t)}{dt} \cdot \nabla\delta(\mathbf{r}) + \dots \quad (1.20)$$

where

$$\mathbf{m}(t) = \frac{1}{2}\int \mathbf{r} \times \mathbf{J}(\mathbf{r}, t)\,dV \quad (1.21)$$

is the magnetic dipole moment of the current distribution, and the dyadic

$$\overleftrightarrow{\mathbf{Q}}_s(t) = \int \mathbf{r}\mathbf{r}\rho(\mathbf{r}, t)\,dV$$

is the symmetric quadrupole moment of the charge distribution. Physically, the magnetic dipole moment can be thought of as shown in Figure 1.2. For a z-directed magnetic dipole, the corresponding electric current density is

$$-m_z \mathbf{u}_z \times \nabla\delta(\mathbf{r}) = m_z\left[-\mathbf{u}_y \delta'(x)\delta(y) + \mathbf{u}_x \delta(x)\delta'(y)\right]\delta(z)$$

The derivatives of the delta functions represent approximately pairs of electric dipole currents displaced along the x or y direction as shown. The overall effect is that of a small loop of electric current lying in the plane $z = 0$.

The details of the derivation of (1.19) are as follows. To get the first term on the right side, we have used the following transformation to examine a typical component of $\mathbf{J}$ along the direction of an arbitrary constant unit vector $\mathbf{u}_c$:

$$
\begin{aligned}
\int \mathbf{u}_c \cdot \mathbf{J}(\mathbf{r})\, dV \;&=\; \int \nabla(\mathbf{u}_c \cdot \mathbf{r}) \cdot \mathbf{J}(\mathbf{r})\, dV \\
&=\; \int \left\{ \nabla \cdot [(\mathbf{u}_c \cdot \mathbf{r})\mathbf{J}(\mathbf{r})] - (\mathbf{u}_c \cdot \mathbf{r})\nabla \cdot \mathbf{J}(\mathbf{r}) \right\}\, dV \\
&=\; \frac{d\mathbf{p}}{dt} \cdot \mathbf{u}_c
\end{aligned}
\tag{1.22}
$$

The integral of the divergence converts to a surface integral over the boundary of the region; we assume that $\mathbf{J} = 0$ outside a bounded volume, so that this surface integral can be made to vanish.

To get the second term on the right side of (1.19) we required the following typical component of the integral that appears in scalar product with $\nabla\delta$:

$$
\begin{aligned}
-\int \mathbf{J}(\mathbf{r} \cdot \mathbf{u}_c)\, dV \;=\;\; & -\tfrac{1}{2}\int [\mathbf{J}(\mathbf{r} \cdot \mathbf{u}_c) - \mathbf{r}(\mathbf{J} \cdot \mathbf{u}_c)]\, dV \\
& -\tfrac{1}{2}\int [\mathbf{J}(\mathbf{r} \cdot \mathbf{u}_c) + \mathbf{r}(\mathbf{J} \cdot \mathbf{u}_c)]\, dV
\end{aligned}
\tag{1.23}
$$

The first integral on the right side of (1.23) is transformed using identity (D.2) from the Appendix:

$$
-\frac{1}{2}\int [\mathbf{J}(\mathbf{r} \cdot \mathbf{u}_c) - \mathbf{r}(\mathbf{J} \cdot \mathbf{u}_c)]\, dV = -\frac{1}{2}\int [\mathbf{r} \times \mathbf{J}(\mathbf{r})] \times \mathbf{u}_c\, dV = -\mathbf{m} \times \mathbf{u}_c
\tag{1.24}
$$

and leads to the magnetic dipole term in (1.19). The second integral on the right side of (1.23), of which we take a component along the direction of another typical constant vector $\mathbf{u}_d$, transforms as follows:

$$
\begin{aligned}
-\frac{1}{2}\int & [(\mathbf{u}_d \cdot \mathbf{J})(\mathbf{r} \cdot \mathbf{u}_c) + (\mathbf{u}_d \cdot \mathbf{r})(\mathbf{J} \cdot \mathbf{u}_c)]\, dV \\
&= -\frac{1}{2}\int \mathbf{J} \cdot [\mathbf{u}_d(\mathbf{r} \cdot \mathbf{u}_c) + \mathbf{u}_c(\mathbf{u}_d \cdot \mathbf{r})]\, dV \\
&= -\frac{1}{2}\int \mathbf{J} \cdot \nabla [(\mathbf{u}_c \cdot \mathbf{r})(\mathbf{u}_d \cdot \mathbf{r})]\, dV \\
&= -\frac{1}{2}\int \left\{ \nabla \cdot [\mathbf{J}(\mathbf{u}_c \cdot \mathbf{r})(\mathbf{u}_d \cdot \mathbf{r})] - (\mathbf{u}_c \cdot \mathbf{r})(\mathbf{u}_d \cdot \mathbf{r})\nabla \cdot \mathbf{J} \right\}\, dV \\
&= -\frac{1}{2}\frac{d}{dt}\int (\mathbf{u}_c \cdot \mathbf{r})(\mathbf{u}_d \cdot \mathbf{r})\rho(\mathbf{r},t)\, dV = -\frac{1}{2}\mathbf{u}_c \cdot \frac{d\overset{\leftrightarrow}{\mathbf{Q}}_s}{dt} \cdot \mathbf{u}_d
\end{aligned}
$$

and leads to the quadrupole term in (1.19).

The calculation of higher multipole terms is usually too cumbersome to be practical. However, when the spatial extent of the source distribution is small, only the first few terms of the multipole series are required for an accurate evaluation of the field.

1.2.2 AVERAGING OF CHARGE AND CURRENT DENSITIES

In a material medium, currents and charges can take any of several forms. A portion of the current density may, for example, be in the form of motion of charges (e.g., electrons or ions) that are not bound to particular locations in space, and is called a *free* current density. In conductive materials, such a current is often produced in response to the presence of an electric field, and takes the form of a *conduction current* $\mathbf{J}_c$. Other currents and charges may arise that are localized near the (relatively immobile) sites of the "entities" (atoms, ions, and molecules) of the medium. The action of a field or other electromotive force on these entities (which would normally be electrically neutral—that is, their positive and negative charges would overlap and cancel—if no field were present) causes *bound* charges and currents to exist near the sites of the entities.

Hypothetically, a sufficiently fine measuring instrument could determine the microscopic distribution of current and charge density (and the fields that result from them) that varies rapidly on the scale of atomic and molecular dimensions, as shown in a one-dimensional schematic form in Figure 1.3. Here, $f(x)$ represents a charge density or component of a current density that spikes sharply near an atom or molecule (whose characteristic dimension we call a), but is nearly zero over most of the distance between atoms or molecules (this distance we call d). In practice, we do not usually have access to or require a measuring instrument that can resolve variations on such microscopic scales,

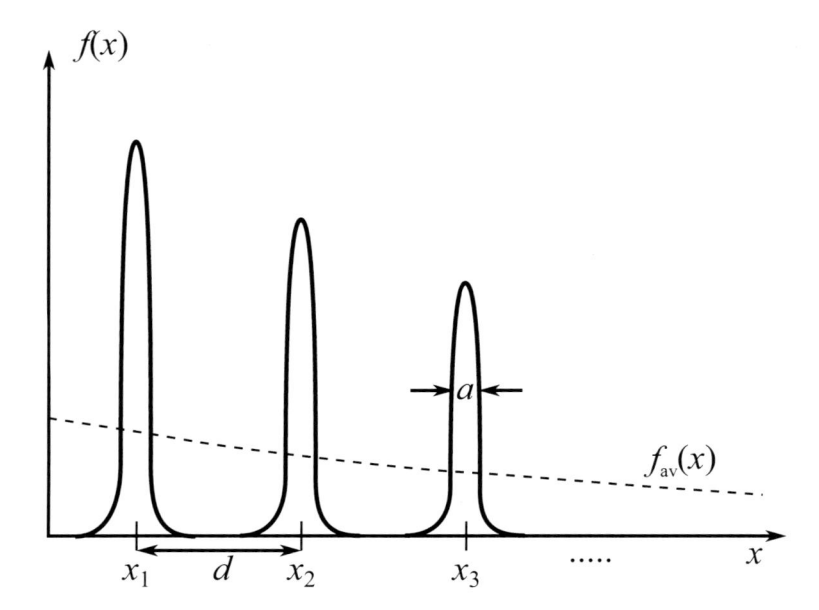

Figure 1.3 Microscopic variation of a charge or current density due to a distribution of atoms or molecules and its average (one-dimensional representation).

but instead will measure charges and fields averaged over distances that are large compared to d. In the terminology of generalized functions, the "good" weighting functions $g(\mathbf{r})$ of position that we allow to multiply the charge density, current density, or fields before integrating over all points of space as in (A.1) must vary slowly over the distance d.

More specifically, let $g(\mathbf{r})$ be a good function that is a maximum at $\mathbf{r} = 0$, decreases as $|\mathbf{r}|$ increases, and obeys

$$\int g(\mathbf{r}')\, dV' = 1 \qquad (1.25)$$

The condition that g is "slowly varying" is made concrete by stating that near any point $\mathbf{r}_0$ sufficiently close to the origin, $g(\mathbf{r}_0 + \Delta\mathbf{r}) \simeq g(\mathbf{r}_0)$ if $|\Delta\mathbf{r}| \leq d$, and roughly speaking this means that the second term of its Taylor series must be negligible compared to the first:

$$d\,|\nabla g(\mathbf{r}_0)| \ll |g(\mathbf{r}_0)| \qquad (1.26)$$

A one-dimensional representative of g is shown in Figure 1.4. We can now define an average $f_{\text{av}}(\mathbf{r})$ of a function $f(\mathbf{r})$ by

$$f_{\text{av}}(\mathbf{r}) \equiv \int g(\mathbf{r}')f(\mathbf{r} - \mathbf{r}')\, dV' \qquad (1.27)$$

where the integrations are taken over all space, $\mathbf{r}' = x'\mathbf{u}_x + y'\mathbf{u}_y + z'\mathbf{u}_z$ and $dV' = dx'dy'dz'$. A one-dimensional example of such an averaged function is

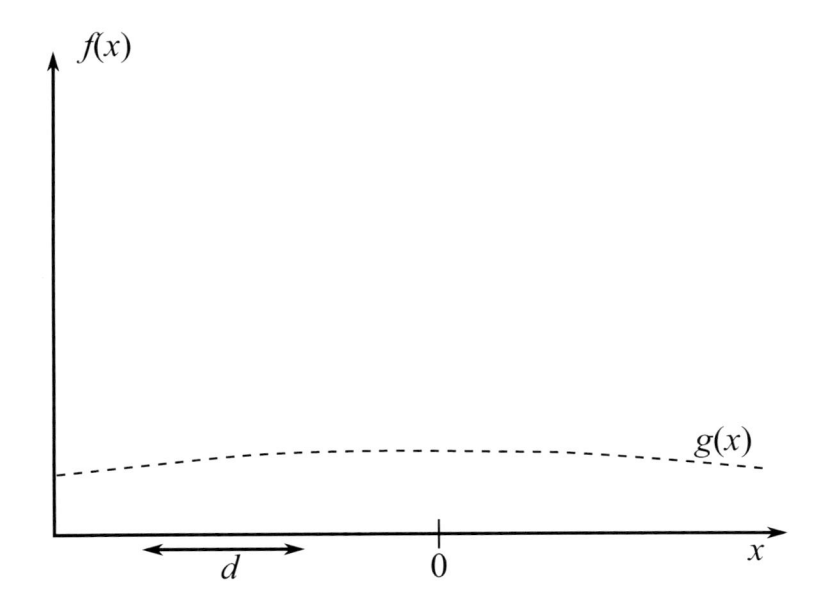

Figure 1.4 A typical weight function for averaging microscopic fields and sources.

shown in Figure 1.3. Note that many possible choices of g exist, and therefore the average of a function is not a uniquely defined quantity. However, the constraints we have placed on g mean that any average defined in this way will be close to any other one.[3]

Let us consider a bound charge density $\rho_b(\mathbf{r})$ that consists of many charge densities concentrated around the points $\mathbf{r}_i$, such that each one can be expressed in the form of (1.17):

$$\rho_b(\mathbf{r}) = \sum_i \left[-\mathbf{p}_i(t) \cdot \nabla \delta(\mathbf{r} - \mathbf{r}_i) + \ldots \right] \tag{1.28}$$

Here we have assumed that the total charge Q_i at each site $\mathbf{r}_i$ is equal to zero, as appropriate for a polarized but electrically neutral atom or molecule. Then the averaged bound charge density is given by

$$\rho_{b,\text{av}}(\mathbf{r}, t) = \sum_i \left[-\mathbf{p}_i(t) \cdot \nabla g(\mathbf{r} - \mathbf{r}_i) + \ldots \right] \tag{1.29}$$

In a similar way, an average bound current density corresponding to sums of terms of the form (1.20) is given by

$$\mathbf{J}_{b,\text{av}}(\mathbf{r}, t) = \sum_i \left[\frac{d\mathbf{p}_i}{dt} g(\mathbf{r} - \mathbf{r}_i) + \nabla \times [\mathbf{m}_i(t) g(\mathbf{r} - \mathbf{r}_i)] + \ldots \right] \tag{1.30}$$

From this averaged or macroscopic point of view, there appears to be a continuous density of electric dipoles, magnetic dipoles, electric quadrupoles, and so on, per unit volume of space. The fields of the higher multipoles can often be regarded as so highly localized that they are unimportant in macroscopic observations. The dipoles, however, form densities known as electric polarization $\mathcal{P}$ (**p** per unit volume) and magnetization $\mathcal{M}$ (**m** per unit volume):

$$\mathcal{P} = \sum_i \mathbf{p}_i(t) g(\mathbf{r} - \mathbf{r}_i), \qquad \mathcal{M} = \sum_i \mathbf{m}_i(t) g(\mathbf{r} - \mathbf{r}_i) \tag{1.31}$$

The part of the current density corresponding to these is:

$$\mathbf{J}_b = \frac{\partial \mathcal{P}}{\partial t} + \nabla \times \mathcal{M} \tag{1.32}$$

---

[3]Much more detail can be found in:

S. R. de Groot, *The Maxwell Equations: Non-Relativistic and Relativistic Derivations from Electron Theory.* Amsterdam: North-Holland, 1969.

G. Russakoff, *Amer. J. Phys.*, vol. 38, pp. 1188-1195, 1970.

F. N. H. Robinson, *Macroscopic Electromagnetism.* Oxford, UK: Pergamon Press, 1973.

A medium with free (e.g., conduction) charges at the points $\mathbf{r}_i$ has an average conduction charge density of

$$\rho_{c,\mathrm{av}} = \sum_i Q_{c,i}(t)g(\mathbf{r} - \mathbf{r}_i) \qquad (1.33)$$

that is obtained in the same way, where $Q_{c,i}$ is the free charge at the point $\mathbf{r}_i$. A corresponding average conduction current density $\mathbf{J}_{c,\mathrm{av}}$ can be defined similarly, and obeys charge conservation (1.10) together with $\rho_{c,\mathrm{av}}$. From here on, we will drop the subscript $_\mathrm{av}$ with the understanding that such averaging has been performed on all sources and fields.

1.2.3 CONDUCTION, POLARIZATION, AND MAGNETIZATION

The total current and charge densities are given by:

$$\mathbf{J}_{\mathrm{tot}} = \mathbf{J}_c + \frac{\partial \boldsymbol{\mathcal{P}}}{\partial t} + \nabla \times \boldsymbol{\mathcal{M}} \qquad (1.34)$$

and

$$\rho_{\mathrm{tot}} = \rho_c - \nabla \cdot \boldsymbol{\mathcal{P}} \qquad (1.35)$$

They have been split into portions associated with $\boldsymbol{\mathcal{P}}$ and $\boldsymbol{\mathcal{M}}$ that are localized near atomic sites, and the free charges and currents ρ_c and $\mathbf{J}_c$ that are at liberty to move across many atomic sites. It is customary at this point to introduce two auxiliary fields $\mathbf{H}$ and $\mathbf{D}$, called the magnetic field and the electric displacement, respectively, by means of the definitions

$$\mathbf{H} = \frac{\mathbf{B}}{\mu_0} - \boldsymbol{\mathcal{M}} \qquad (1.36)$$

$$\mathbf{D} = \epsilon_0 \mathbf{E} + \boldsymbol{\mathcal{P}} \qquad (1.37)$$

We then arrive at a modified form of Maxwell's equations:

$$\begin{aligned}
\nabla \cdot \mathbf{D} &= \rho_c & (1.38)\\
\nabla \cdot \mathbf{B} &= 0 & (1.39)\\
\nabla \times \mathbf{H} &= \mathbf{J}_c + \frac{\partial \mathbf{D}}{\partial t} & (1.40)\\
\nabla \times \mathbf{E} &= -\frac{\partial \mathbf{B}}{\partial t} & (1.41)
\end{aligned}$$

for description of the electromagnetic field in material media. However, the fields $\mathbf{D}$ and $\mathbf{H}$ have mainly been introduced for convenience, and different definitions for them may be more convenient depending on the intended application, as we will see below.

In a given medium, the presence of electric or magnetic fields can (by the action of the Lorentz force) *induce* a motion of the microscopic charges, which is called *induced current* and is denoted by $\mathbf{J}_{\text{ind}}$. On the other hand, sometimes the configuration of the medium itself or the action of non-electromagnetic forces can cause distributions of current in a material that are called *externally impressed*, or simply *impressed*. The first type of currents are analogous to those flowing in a resistor in response to an applied voltage, while those of the second type are analogous to ones produced by an ideal current generator. Permanent magnets, batteries, and electrets are examples where impressed currents are present; here the source densities are maintained by some process such as a chemical reaction, atomic or molecular forces, etc.

The conduction current, polarization, and magnetization densities are the sources of the electric and magnetic fields. Sometimes portions of these sources remain completely unchanged (or almost so) no matter what electric or magnetic fields are present, and can generally be regarded as impressed and assumed to be known. Induced sources on the other hand are functions of the electric and magnetic fields. Induced currents vanish when the fields are zero. In general, both types of source may exist at the same time. Denoting induced sources with a subscript $_{\text{ind}}$ and externally impressed sources with a subscript $_{\text{ext}}$, we write each one as the sum of two terms:

$$\mathbf{J}_c = \mathbf{J}_{c,\text{ind}} + \mathbf{J}_{c,\text{ext}} \tag{1.42}$$

$$\mathcal{P} = \mathcal{P}_{\text{ind}} + \mathcal{P}_{\text{ext}} \tag{1.43}$$

$$\mathcal{M} = \mathcal{M}_{\text{ind}} + \mathcal{M}_{\text{ext}} \tag{1.44}$$

A similar split of the charge density ρ_c corresponding to the conduction current is also made, in which the impressed conduction current and charge densities are assumed to separately fulfill conservation of charge (1.10). Impressed magnetization densities occur in permanent magnets, while impressed polarization densities are found in electrets; these are sometimes called remanent densities.

The impressed sources can be explicitly displayed in Maxwell's equations if the definitions of the auxiliary fields $\mathbf{H}$ and $\mathbf{D}$ are modified to

$$\mathbf{H}_{\text{ind}} = \frac{\mathbf{B}}{\mu_0} - \mathcal{M}_{\text{ind}} \tag{1.45}$$

$$\mathbf{D}_{\text{ind}}(\mathbf{r}, t) = \epsilon_0 \mathbf{E} + \mathcal{P}_{\text{ind}} + \int \mathbf{J}_{c,\text{ind}}\, dt \tag{1.46}$$

which include only the induced parts of the polarization and magnetization densities. The grouping (1.46) is motivated by the fact that in many common materials $\mathcal{P}_{\text{ind}}$ and $\mathbf{J}_{c,\text{ind}}$ are both proportional to $\mathbf{E}$ while $\mathcal{M}_{\text{ind}}$ is proportional to $\mathbf{B}$, and a simplification of the constitutive equations then results. Maxwell's equations in a material medium with induced currents now read

$$\nabla \cdot \mathbf{D}_{\mathrm{ind}}(\mathbf{r}, t) = \rho_{\mathrm{ext}} \tag{1.47}$$

$$\nabla \cdot \mathbf{B} = 0 \tag{1.48}$$

$$\nabla \times \mathbf{H}_{\mathrm{ind}} = \frac{\partial \mathbf{D}_{\mathrm{ind}}}{\partial t} + \mathbf{J}_{\mathrm{ext}} \tag{1.49}$$

$$\nabla \times \mathbf{E} = -\frac{\partial \mathbf{B}}{\partial t} \tag{1.50}$$

where the total impressed electric current density is given by

$$\mathbf{J}_{\mathrm{ext}} = \mathbf{J}_{c,\mathrm{ext}} + \frac{\partial \boldsymbol{\mathcal{P}}_{\mathrm{ext}}}{\partial t} + \nabla \times \boldsymbol{\mathcal{M}}_{\mathrm{ext}} \tag{1.51}$$

and the total impressed charge density is

$$\rho_{\mathrm{ext}} = \rho_{c,\mathrm{ext}} - \nabla \cdot \boldsymbol{\mathcal{P}}_{\mathrm{ext}} \tag{1.52}$$

If we provide the laws expressing how the induced sources depend on the electric and magnetic fields (the so-called constitutive equations discussed in Section 1.2.4 below), then (1.47)-(1.50) will be enough to completely solve for the fields if the impressed sources are given.

A derivation similar to that which led to (1.15) now gives the following modified version of Poynting's theorem:

$$-\oint_S (\mathbf{E} \times \mathbf{H}_{\mathrm{ind}}) \cdot \mathbf{u}_n \, dS = \int_V \mathbf{E} \cdot \mathbf{J}_{\mathrm{ext}} \, dV + \int_V \left(\mathbf{E} \cdot \frac{\partial \mathbf{D}_{\mathrm{ind}}}{\partial t} + \mathbf{H}_{\mathrm{ind}} \cdot \frac{\partial \mathbf{B}}{\partial t} \right) dV \tag{1.53}$$

Although formally similar to (1.15), the physical interpretation of this equation is different. The Joule power term

$$\int_V \mathbf{E} \cdot \mathbf{J}_{\mathrm{ext}} \, dV \tag{1.54}$$

is only that associated with the impressed sources. A portion of the term

$$\int_V \left(\mathbf{E} \cdot \frac{\partial \mathbf{D}_{\mathrm{ind}}}{\partial t} + \mathbf{H}_{\mathrm{ind}} \cdot \frac{\partial \mathbf{B}}{\partial t} \right) dV \tag{1.55}$$

will represent Joule energy lost to heating of conductors and dielectrics or to other phenomena, while the rest will correspond to the time derivative of stored energy. The stored energy in this case will include not only the electromagnetic energy, but also non-electromagnetic energy stored in the material medium associated with the induced currents (atomic oscillations, for example). In general, it is not possible to write explicit expressions for these separate parts of (1.55).[4]

---

[4]See

Yu. S. Barash and V. L. Ginzburg, *Sov. Phys. Uspekhi*, vol. 19, pp. 263-270, 1976.

V. L. Ginzburg, *Applications of Electrodynamics in Theoretical Physics and Astrophysics*. New York: Gordon and Breach, 1989, Chapter 13.

The foregoing is not the only way that impressed sources can be incorporated into Maxwell's equations. Instead of (1.45), let us define a modified $\mathbf{B}$ field:

$$\mathbf{B}_{\text{ind}} = \mathbf{B} - \mu_0 \boldsymbol{\mathcal{M}}_{\text{ext}} \tag{1.56}$$

so that the "original" $\mathbf{H}$ defined in (1.36) is now given by

$$\mathbf{H} = \frac{\mathbf{B}_{\text{ind}}}{\mu_0} - \boldsymbol{\mathcal{M}}_{\text{ind}}$$

Maxwell's equations using "induced" forms of $\mathbf{D}$ and $\mathbf{B}$ (rather than $\mathbf{D}$ and $\mathbf{H}$) now look like:

$$\nabla \cdot \mathbf{D}_{\text{ind}} = \rho_{\text{ext}} \tag{1.57}$$

$$\nabla \cdot \mathbf{B}_{\text{ind}} = \rho_{m,eq,\text{ext}} \tag{1.58}$$

$$\nabla \times \mathbf{H} = \frac{\partial \mathbf{D}_{\text{ind}}}{\partial t} + \mathbf{J}_{eq,\text{ext}} \tag{1.59}$$

$$\nabla \times \mathbf{E} = -\frac{\partial \mathbf{B}_{\text{ind}}}{\partial t} - \mathbf{M}_{eq,\text{ext}} \tag{1.60}$$

where the "equivalent" impressed electric current density is

$$\mathbf{J}_{eq,\text{ext}} = \mathbf{J}_{c,\text{ext}} + \frac{\partial \boldsymbol{\mathcal{P}}_{\text{ext}}}{\partial t}$$

while an equivalent impressed magnetic current density

$$\mathbf{M}_{eq,\text{ext}} = \mu_0 \frac{\partial \boldsymbol{\mathcal{M}}_{\text{ext}}}{\partial t}$$

now appears in Faraday's law, with a corresponding impressed equivalent magnetic charge density

$$\rho_{m,eq,\text{ext}} = -\mu_0 \nabla \cdot \boldsymbol{\mathcal{M}}_{\text{ext}}$$

appearing in the Gauss law for the magnetic field. The magnetic current $\mathbf{M}$ will sometimes be seen under the notation $\mathbf{J}_m$.[5] If magnetic monopoles (which would be to magnetic fields what point charges are to electric fields) existed, their volume density and flow density would appear exactly where ρ_m and $\mathbf{M}$ appear in (1.58) and (1.60). They (and $\mathbf{J}_{eq,\text{ext}}$) are denoted "equivalent" magnetic sources because solving (1.57)-(1.60) with the impressed sources given would yield not the true fields $\mathbf{E}$ and $\mathbf{B}$, but $\mathbf{E}$ and the modified magnetic

---

[5] Harrington's [1961] book has brought the notation $\mathbf{M}$ into popular use in the engineering literature. Traditionally, $\mathbf{M}$ had been used to denote the magnetization vector whose curl gives the equivalent volume electric current density resulting from alignment of microscopic magnetic dipoles in a magnetic medium. We have denoted this latter quantity as $\boldsymbol{\mathcal{M}}$.

field $\mathbf{B}_{\text{ind}}$. Since the true $\mathbf{B}$ is easily recovered using (1.56) if the impressed magnetization density is known, the alternative formulation (1.57)-(1.60) is in some sense "just as good" for finding the fields, and is said to be equivalent to the system (1.47)-(1.50) in this sense. This is an example of an equivalence principle; we will study more examples of this idea in Section 1.4, especially in connection with the use of stream potentials.

It is common practice in everyday usage of Maxwell's equations to ignore these subtle distinctions between the various definitions of the fields, dropping the subscripts $_{\text{ind}}$ on the fields, as well as on $\mathbf{J}_c$, $\boldsymbol{\mathcal{P}}$, and $\boldsymbol{\mathcal{M}}$, all of which will be understood from here on to mean induced sources only unless otherwise stated. The subscript $_{\text{ext}}$ will likewise be omitted from the impressed sources $\mathbf{J}$, $\mathbf{M}$, and the corresponding charge densities unless needed for emphasis; such quantities will henceforth be understood to be impressed unless stated otherwise. Maxwell's equations will thus henceforth be given in the form

$$\begin{aligned}
\nabla \cdot \mathbf{D} &= \rho \\
\nabla \cdot \mathbf{B} &= \rho_m \\
\nabla \times \mathbf{H} &= \mathbf{J} + \frac{\partial \mathbf{D}}{\partial t} \\
\nabla \times \mathbf{E} &= -\mathbf{M} - \frac{\partial \mathbf{B}}{\partial t}
\end{aligned} \tag{1.61}$$

unless special emphasis is needed. The reader should, however, keep the discussion of this subsection in mind if the "real" field is needed (e.g., if the true force on a charged particle is required).

As an example of the danger involved in ignoring this subtlety, consider the extension of Poynting's theorem to fields that obey (1.57)-(1.60). Again using a derivation like that which led to (1.15) and (1.53), we get

$$-\oint_S (\mathbf{E} \times \mathbf{H}) \cdot \mathbf{u}_n \, dS = \int_V (\mathbf{E} \cdot \mathbf{J}_{eq,\text{ext}} + \mathbf{H} \cdot \mathbf{M}_{eq,\text{ext}}) \, dV$$
$$+ \int_V \left(\mathbf{E} \cdot \frac{\partial \mathbf{D}_{\text{ind}}}{\partial t} + \mathbf{H} \cdot \frac{\partial \mathbf{B}_{\text{ind}}}{\partial t} \right) dV \tag{1.62}$$

Two things should be noted here. First, the first volume integral on the right side is not equal to the true impressed Joule power (1.54). There are differences due to the neglect of the Joule power associated with the magnetization current $\nabla \times \boldsymbol{\mathcal{M}}$. Moreover, the second volume integral on the right side is not, in general, expressible as the time derivative of a function we can identify as stored energy (this happens only in special cases, such as when $\boldsymbol{\mathcal{P}}_{\text{ind}}$ and $\mathbf{J}_{c,\text{ind}}$ are proportional to $\mathbf{E}$, and $\boldsymbol{\mathcal{M}}_{\text{ind}}$ to $\mathbf{B}$, with no dispersive or nonlinear effects present). So although (1.62) is certainly true as a mathematical identity, we must exercise extreme care in interpreting it physically.

1.2.4 MATERIAL PROPERTIES AND CONSTITUTIVE EQUATIONS

Maxwell's equations in any of the forms of the previous subsection are incomplete, until we specify the way in which the induced sources depend on the fields $\mathbf{E}$ and $\mathbf{B}$. These relationships are called *constitutive equations* and arise from the Lorentz forces, in accordance with the structure of the medium and the nature of its constituent entities.

The simplest such response is that of a linear isotropic medium, in which the conduction current obeys the spatially distributed form of Ohm's Law:

$$\mathbf{J}_c = \sigma \mathbf{E} \tag{1.63}$$

where the parameter σ is known as the conductivity of the medium. Likewise, in such a medium we expect that induced bound currents and charges should arise that are proportional to the fields. For a dielectric, we have

$$\mathcal{P} = \epsilon_0 \chi_e \mathbf{E} \tag{1.64}$$

where the dimensionless quantity χ_e is the *electric susceptibility* of the medium. This relation implies in turn the relationship

$$\mathbf{D} = \epsilon \mathbf{E} \tag{1.65}$$

where $\epsilon = \epsilon_0(1 + \chi_e)$ is the permittivity of the medium in question. The magnetic constitutive equation for a linear isotropic material is obtained similarly. We put

$$\mathcal{M} = \frac{\chi_m}{1 + \chi_m} \frac{\mathbf{B}}{\mu_0} \tag{1.66}$$

where χ_m is the *magnetic susceptibility* of the medium. The reason for the more complicated dependence of (1.66) on χ_m is to make the definition of the magnetic permeability of the medium from

$$\mathbf{B} = \mu \mathbf{H} \tag{1.67}$$

result in the relationship $\mu = \mu_0(1 + \chi_m)$, which is analogous in form to that of the the dielectric permittivity. Because the definitions of $\mathbf{B}$ and $\mathbf{H}$ may be chosen in different ways according to the previous subsection, we may need to keep in mind which definitions are being used in order to properly interpret these relations.

As an example of how induced and impressed sources may act simultaneously, consider a "battery" (voltaic cell) for which the induced conduction current is given by (1.63) and the total conduction current by (1.42). The chemical processes in the battery are responsible for an impressed conduction current $\mathbf{J}_{c,\text{ext}}$ (presumed known whatever the value of the electric field). When the battery is open-circuited, there is no current flow ($\mathbf{J}_c + \mathbf{J}_{c,\text{ext}} = 0$), but $\mathbf{J}_{c,\text{ext}} \neq 0$, so there must be an electric field

$$\mathbf{E}_{\text{oc}} = -\frac{\mathbf{J}_{c,\text{ext}}}{\sigma}$$

under these circumstances. The integration of this field between the terminals of the battery yields the open-circuit battery voltage.

In a real medium, the constitutive relations are not so simple as relations (1.63), (1.65), and (1.67) imply; there is often nonlinearity, spatial dispersion, or anisotropy in the response of an actual material (which we will not examine in any detail here) as well as temporal dispersion, which we will briefly describe below. The most general complexity of linear, nondispersive material constitutive response is that of *bianisotropy*, where both the polarization and magnetization density in a material depend on both the electric field $\mathbf{E}$ and on the magnetic field $\mathbf{B}$.[6] In such a case we have $\mathbf{D} = \overset{\leftrightarrow}{\epsilon} \cdot \mathbf{E} + \overset{\leftrightarrow}{\xi}_{em} \cdot \mathbf{B}$ and $\mathbf{H} = -\overset{\leftrightarrow}{\xi}_{me} \cdot \mathbf{E} + \overset{\leftrightarrow}{\nu} \cdot \mathbf{B}$, where $\overset{\leftrightarrow}{\epsilon}$, $\overset{\leftrightarrow}{\xi}_{em}$, $\overset{\leftrightarrow}{\xi}_{me}$, and $\overset{\leftrightarrow}{\nu}$ are *dyadic* constitutive parameters of the medium (note that the *magnetic reluctance* ν has been introduced here; it is the inverse of the permeability μ). Many interesting phenomena can arise in media with such general consitutive properties; the Hall effect is one well-known example.

In a linear, spatially nondispersive, and isotropic material, the electric displacement may depend not only on the current value of the field, but also on the field values at previous times, so that more generally than in (1.65), $\mathbf{D}_{\text{ind}}$ must be a temporal convolution of $\mathbf{E}$ with some susceptibility function that vanishes for $t < 0$:

$$\mathbf{D}_{\text{ind}}(\mathbf{r}, t) = \epsilon_0 \left(\mathbf{E}(\mathbf{r}, t) + \int_{-\infty}^{t} \mathbf{E}(\mathbf{r}, t') \chi_e(t - t') \, dt' \right) \tag{1.68}$$

Similar generalizations of (1.63) and (1.67) are also possible. Use of the induced electric displacement $\mathbf{D}_{\text{ind}}$ from (1.46):

$$\mathbf{D}_{\text{ind}}(\mathbf{r}, t) = \mathbf{D}(\mathbf{r}, t) + \sigma \int_{-\infty}^{t} \mathbf{E}(\mathbf{r}, t') \, dt'$$

gives a constitutive relation that differs from (1.68) only by the addition of a step function term $\sigma \vartheta(t)$ to $\chi_e(t)$. Although both σ and χ_e in principle affect the value of $\mathbf{D}$ via the entire time history of $\mathbf{E}$, the susceptibility is usually associated with "short-term memory" of the medium (χ_e is significantly different from zero only when t is small in some sense), while the conductivity is a phenomenon involving "long-term memory." Because in the frequency domain the addition of the term involving σ turns out to have the effect simply of making ϵ into a complex frequency-dependent quantity, it is customary

---

[6]See for example:

T. H. O'Dell, *The Electrodynamics of Magneto-Electric Media*. Amsterdam: North-Holland, 1970.

A. Lakhtakia, V.K. Varadan and V.V. Varadan, *Time-Harmonic Electromagnetic Fields in Chiral Media*. Berlin: Springer-Verlag, 1989.

I. V. Lindell et al., *Electromagnetic Waves and Chiral and Bi-Isotropic Media*. Boston: Artech House, 1994.

A. Lakhtakia, *Beltrami Fields in Chiral Media*. Singapore: World Scientific, 1994.

E. J. Post, *Formal Structure of Electromagnetics*. New York: Dover, 1997.

to deal with $\mathbf{D}_{\text{ind}}$ exclusively, rather than to separate conduction current effects from those of polarization currents. Thus, we will henceforth omit the subscript $_{\text{ind}}$ in $\mathbf{D}$, lumping consideration of conductivity and permittivity together. Time dispersion for magnetic, anisotropic, and chiral effects can be taken into account in a similar fashion.

Whatever the dependence of the induced currents in a material on the electric and magnetic fields, it is often the case that the material is *passive*— that is, it is incapable of delivering energy to the field. We state the passivity property precisely as follows. Suppose that no sources act before a certain time (which we choose to be $t = 0$ without loss of generality). Then all fields are likewise zero for $t < 0$ (this indicates that the material response is *causal*—it cannot occur before the excitation is present). Induced currents of a material are said to be passive if the Joule energy of these currents (the time integral of the Joule power) is nonnegative for all times $t_1 > 0$:

$$\int_0^{t_1} \int \mathbf{E} \cdot \mathbf{J}_{\text{ind}} \, dV \, dt \geq 0 \tag{1.69}$$

no matter in what way the fields inducing the currents depend on space and time. This notion of passivity generalizes one commonly used in circuit theory.[7] It is important to recognize that passivity does not require that the Joule *power* itself is always nonnegative. The medium is to some extent free to trade energy back and forth from the field. What passivity does require is that the medium cannot supply any more energy to the field than it has previously been given by externally impressed sources.

As an example, consider the case of ohmic conduction currents (1.63). The integral in (1.69) will be nonnegative (that is, a conductor is passive) only if $\sigma \geq 0$. Other more complicated constitutive relations will be constrained in various ways by the passivity criterion.[8]

Note that there are examples of active (i.e., nonpassive) media for which (1.69) does not hold: laser and maser materials, amplifying materials, etc. All such really turn out to be nonlinear materials to which an external source of energy is being supplied, and could be viewed as combinations of passive media and impressed sources. We will not consider active media in this book.

---

[7]See

D. C. Youla, L. J. Castriota, and H. J. Carlin, *IRE Trans. Circ. Theory*, vol. 6, pp. 102-124, 1959.

M. R. Wohlers, *Lumped and Distributed Passive Networks*. New York: Academic Press, 1969, Chapter 2.

V. S. Vladimirov, *Generalized Functions in Mathematical Physics*. Moscow: Mir, 1979, Chapter 19.

[8]M. Fabrizio and A. Morro, *Electromagnetism of Continuous Media*. Oxford, UK: Oxford University Press, 2003, Chapters 4-6.

1.3 TIME-HARMONIC PROBLEMS

A function of time (and, as we have here, possibly of space as well) can be expressed in terms of its frequency spectrum as a Fourier transform. For some functions this may need to be interpreted in the sense of generalized functions (see the Appendix), but for ordinary functions we have the Fourier transform pair

$$\hat{F}(\mathbf{r}, \omega) = \int_{-\infty}^{\infty} F(\mathbf{r}, t) e^{-j\omega t} \, dt \tag{1.70}$$

$$F(\mathbf{r}, t) = \frac{1}{2\pi} \int_{-\infty}^{\infty} \hat{F}(\mathbf{r}, \omega) e^{j\omega t} \, d\omega \tag{1.71}$$

If (as happens for physical quantities) $F(\mathbf{r}, t)$ is real, then from (1.70):

$$\begin{aligned}
\hat{F}^*(\mathbf{r}, \omega) &= \int_{-\infty}^{\infty} F(\mathbf{r}, t) e^{j\omega^* t} \, dt \\
&= \hat{F}(\mathbf{r}, -\omega^*)
\end{aligned} \tag{1.72}$$

where $*$ denotes complex conjugate. Thus the spectrum $\hat{F}(\mathbf{r}, \omega)$ for negative ω is not independent of that for positive ω. Using (1.72) in (1.71) for real ω, we have that

$$\begin{aligned}
F(\mathbf{r}, t) &= \frac{1}{2\pi} \left\{ \int_0^{\infty} \hat{F}(\mathbf{r}, \omega) e^{j\omega t} \, d\omega + \int_0^{\infty} \hat{F}^*(\mathbf{r}, \omega) e^{-j\omega t} \, d\omega \right\} \\
&= \operatorname{Re} \left\{ \frac{1}{\pi} \int_0^{\infty} \hat{F}(\mathbf{r}, \omega) e^{j\omega t} \, d\omega \right\}
\end{aligned} \tag{1.73}$$

so that only the positive frequency spectrum is needed to determine $F(\mathbf{r}, t)$. The function

$$F_A(\mathbf{r}, t) = \frac{1}{\pi} \int_0^{\infty} \hat{F}(\mathbf{r}, \omega) e^{j\omega t} \, d\omega \tag{1.74}$$

associated with $F(\mathbf{r}, t)$ is called the *analytic signal*, and is used extensively in communication theory.

If each component of the fields and sources in Maxwell's equations (1.61) is represented according to (1.71), then these spectral components (the time-harmonic field) are found to obey

$$\nabla \cdot \hat{\mathbf{D}} = \hat{\rho} \tag{1.75}$$

$$\nabla \cdot \hat{\mathbf{B}} = \hat{\rho}_m \tag{1.76}$$

$$\nabla \times \hat{\mathbf{H}} = \hat{\mathbf{J}} + j\omega \hat{\mathbf{D}} \tag{1.77}$$

$$\nabla \times \hat{\mathbf{E}} = -\hat{\mathbf{M}} - j\omega \hat{\mathbf{B}} \tag{1.78}$$

Equations (1.75)-(1.76) are redundant (they follow by taking the divergences of (1.77)-(1.78)) unless $\omega = 0$. All four of the Maxwell equations are still required for the complete specification of a static field.

The constitutive relations for a time-harmonic problem in a linear isotropic (but possibly temporally dispersive) medium follow by taking the Fourier transform of (1.65), (1.67) and (1.63), or of the dispersive forms such as (1.68) or (1.46).

$$\hat{\mathbf{D}}(\mathbf{r}, \omega) = \epsilon(\omega)\hat{\mathbf{E}}(\mathbf{r}, \omega) \tag{1.79}$$

$$\hat{\mathbf{B}}(\mathbf{r}, \omega) = \mu(\omega)\hat{\mathbf{H}}(\mathbf{r}, \omega) \tag{1.80}$$

where, for example, $\epsilon(\omega) = \epsilon_0[1 + \hat{\chi}_e(\omega)]$, and the Fourier transform $\hat{\chi}_e(\omega)$ of $\chi_e(t)$ is the frequency-dependent electric susceptibility of the medium. Because of the causality principle (there can be no response before an input) $\chi_e(t) = 0$ for $t < 0$, and this implies a certain relationship between the real and imaginary parts of $\hat{\chi}_e(\omega)$. Similar considerations apply in principle for the permeability $\mu(\omega)$, but for real magnetic materials nonlinearity is as often an issue as dispersion. It is common to use the following notations to indicate the real and imaginary parts of ϵ and μ:

$$\epsilon(\omega) = \epsilon' - j\epsilon''; \qquad \mu(\omega) = \mu' - j\mu''$$

As a simple example, consider the Debye relaxation model, often used to model the behavior of a lossy medium:

$$\chi_e(t) = \frac{C}{\tau} e^{-t/\tau} \vartheta(t) \tag{1.81}$$

where C is a dimensionless real constant, $\vartheta(t)$ is a Heaviside unit step function, and τ is the relaxation time characteristic of the medium. The Fourier transform of (1.81) yields

$$\hat{\chi}_e(\omega) = \frac{C}{1 + j\omega\tau} \tag{1.82}$$

Since $\hat{\chi}_e(\omega)$ is complex, the medium is lossy as well as dispersive. The real and imaginary parts of the permittivity are

$$\operatorname{Re}\left[\epsilon(\omega)\right] = \epsilon_0 \left[1 + \frac{C}{1 + \omega^2\tau^2}\right] \tag{1.83}$$

and

$$\operatorname{Im}\left[\epsilon(\omega)\right] = -\epsilon_0 \left[\frac{C\omega\tau}{1 + \omega^2\tau^2}\right] \tag{1.84}$$

The time-harmonic fields obey a form of Poynting's theorem known as the complex Poynting theorem. To obtain it, we form the divergence $\nabla \cdot (\hat{\mathbf{E}} \times \hat{\mathbf{H}}^*)$ and apply the same sequence of vector identities as in the time-domain case. The result is

$$-\oint_S (\hat{\mathbf{E}} \times \hat{\mathbf{H}}^*) \cdot \mathbf{u}_n \, dS = \int_V \left(\hat{\mathbf{E}} \cdot \hat{\mathbf{J}}^* + \hat{\mathbf{H}}^* \cdot \hat{\mathbf{M}}\right) dV$$

$$+ j\omega \int_V \left(\hat{\mathbf{H}}^* \cdot \hat{\mathbf{B}} - \hat{\mathbf{E}} \cdot \hat{\mathbf{D}}^*\right) dV \tag{1.85}$$

Again, this is a mathematical identity satisfied by any field that obeys Maxwell's equations, but its physical interpretation must be done with care. One half the real part of this equation expresses the conservation of time-average power in the electromagnetic system. For the case of linear isotropic media this becomes:

$$-\frac{1}{2}\text{Re}\oint_S (\hat{\mathbf{E}} \times \hat{\mathbf{H}}^*) \cdot \mathbf{u}_n\, dS \;=\; \frac{1}{2}\text{Re}\int_V \left(\hat{\mathbf{E}} \cdot \hat{\mathbf{J}}^* + \hat{\mathbf{H}}^* \cdot \hat{\mathbf{M}}\right) dV$$
$$+ \frac{\omega}{2}\int_V \left(\mu''|\hat{\mathbf{H}}|^2 + \epsilon''|\hat{\mathbf{E}}|^2\right) dV \quad (1.86)$$

The left side is the time-average electromagnetic power flowing into the surface S, the first volume integral on the right side is the time-average power delivered *to* the impressed sources, and the second term is the time-average power dissipated in the material (Joule losses). The imaginary part of (1.85) has *no* direct physical meaning, although that of the second volume integral on the right side appears to be proportional to the excess of stored magnetic energy over electric (this interpretation is problematic in the case of dispersive and lossy media).

Except for a few places (such as the discussion of uniqueness theorems in Section 3.2), we will from here on consider exclusively time-harmonic quantities rather than the corresponding functions of time (1.73). It will be convenient in such cases to drop the " ˆ " superscript, but it should be remembered that $\mathbf{E}(\mathbf{r}, \omega)$ is still complex, and a function of ω, and that the real-time function $\mathbf{E}(\mathbf{r}, t)$ must still be recovered from (1.73).

1.4 DUALITY; EQUIVALENCE; SURFACE SOURCES

1.4.1 DUALITY AND MAGNETIC SOURCES

Next, we look at the idea of *duality*. Our fields are required to satisfy the modified Maxwell's equations (1.61) with magnetic as well as electric sources present (corresponding to some choice of definitions for $\mathbf{B}$, $\mathbf{D}$, and $\mathbf{H}$ as in the previous section). Boundary conditions for the fields when magnetic surface currents and charges are present must also be modified, but we defer discussion of this topic to the next section. In a linear isotropic medium, these equations become

$$\nabla \cdot (\epsilon\mathbf{E}) = \rho$$
$$\nabla \cdot (\mu\mathbf{H}) = \rho_m$$
$$\nabla \times \mathbf{H} = \mathbf{J} + \frac{\partial(\epsilon\mathbf{E})}{\partial t}$$
$$\nabla \times \mathbf{E} = -\mathbf{M} - \frac{\partial(\mu\mathbf{H})}{\partial t} \quad (1.87)$$

If $\mathbf{E}$ and $\mathbf{H}$ satisfy (1.87) for some given ϵ, μ and source densities, then the *dual* fields

$$\mathbf{E}' = -\mathbf{H} \quad \text{and} \quad \mathbf{H}' = \mathbf{E}$$

will obey a dual set of Maxwell's equations as follows:

$$\nabla \cdot (\epsilon' \mathbf{E'}) = \rho'$$
$$\nabla \cdot (\mu' \mathbf{H'}) = \rho'_m$$
$$\nabla \times \mathbf{H'} = \mathbf{J'} + \frac{\partial(\epsilon' \mathbf{E'})}{\partial t}$$
$$\nabla \times \mathbf{E'} = -\mathbf{M'} - \frac{\partial(\mu' \mathbf{H'})}{\partial t}$$

in which the dual material parameters

$$\epsilon' = \mu \qquad \mu' = \epsilon$$

are obtained by interchanging the original dielectric and magnetic material parameters. The dual sources must be:

$$\mathbf{J'} = -\mathbf{M} \qquad \rho' = -\rho_m \tag{1.88}$$

$$\mathbf{M'} = \mathbf{J} \qquad \rho'_m = \rho \tag{1.89}$$

If boundaries are present, then dual boundary conditions will also apply. Notably, suppose that a perfect electric conductor (PEC) is located at the surface S, on which the boundary condition

$$\mathbf{u}_n \times \mathbf{E}|_S = 0$$

holds (where $\mathbf{u}_n$ is the outward unit vector normal to S). Then in the dual problem, S must be a *perfect magnetic conductor (PMC)*, on which the boundary condition

$$\mathbf{u}_n \times \mathbf{H}|_S = 0$$

holds. Because it had not often been encountered even approximately in practice, the PMC was traditionally regarded as a fictitious concept. However, recent work on artificial impedance surfaces (particularly, the so-called "mushroom" surface) has brought the concept into more common use in applications.[9]

This is the duality principle, which allows us (for example) to find the fields of a magnetic current distribution when those of a corresponding electric current distribution are known, with no extra labor. We have already seen one example of the convenience of using equivalent magnetic sources to find fields due to more complicated electric current distributions. Duality allows us to make use of already known solutions to Maxwell's equations to find those

---

[9]See, for example,

S. Tretyakov, *Analytical Modeling in Applied Electromagnetics*. Boston: Artech House, 2003.

of such equivalent problems without "reinventing the wheel," so to speak. Fictitious magnetic charges and currents will be useful not only when invoking duality to generate new classes of solutions to Maxwell's equations, but also, as we shall see, in a number of other contexts, especially those involving applications of the equivalence theorem.

1.4.2 STREAM POTENTIALS

We can further explore the usefulness of equivalent magnetic sources by first introducing a set of potentials for the electric and magnetic current densities called *stream potentials*.[10] To begin with, the electric (and magnetic) charge and current densities are related by the conservation laws

$$\nabla \cdot \mathbf{J} + \frac{\partial \rho}{\partial t} = 0$$

$$\nabla \cdot \mathbf{M} + \frac{\partial \rho_m}{\partial t} = 0 \qquad (1.90)$$

By Poincaré Lemma 1 of Appendix D, there is an infinite number of *stream potentials* $\mathbf{J}_e$ such that

$$\frac{\partial \rho}{\partial t} = -\nabla \cdot \mathbf{J}_e$$

In fact, any such $\mathbf{J}_e$ must differ from the actual $\mathbf{J}$ by the curl of some other vector (Poincaré Lemma 3), so we introduce a second stream potential $\mathbf{T}_e$ as:

$$\mathbf{J} = \mathbf{J}_e + \nabla \times \mathbf{T}_e \qquad (1.91)$$

Likewise, we introduce the stream potentials $\mathbf{T}_m$ and $\mathbf{M}_m$ by

$$\frac{\partial \rho_m}{\partial t} = -\nabla \cdot \mathbf{M}_m$$

and

$$\mathbf{M} = \mathbf{M}_m - \nabla \times \mathbf{T}_m \qquad (1.92)$$

Maxwell's equations are now

$$\nabla \cdot \left(\mathbf{B} + \int \mathbf{M}_m \, dt \right) = 0$$

$$\nabla \cdot \left(\mathbf{D} + \int \mathbf{J}_e \, dt \right) = 0 \qquad (1.93)$$

$$\nabla \times (\mathbf{E} - \mathbf{T}_m) = -\frac{\partial \mathbf{B}}{\partial t} - \mathbf{M}_m$$

$$\nabla \times (\mathbf{H} - \mathbf{T}_e) = \frac{\partial \mathbf{D}}{\partial t} + \mathbf{J}_e$$

---

[10]See:

A. Nisbet, *Proc. Roy. Soc. London A*, vol. 231, pp. 250-263, 1955.

—, *ibid.*, vol. 240, pp. 375-381, 1957.

Now let us define the hypothetical fields

$$\begin{aligned}
\mathbf{E}_0 &= \mathbf{E} - \mathbf{T}_m \\
\mathbf{H}_0 &= \mathbf{H} - \mathbf{T}_e
\end{aligned}$$

as well as

$$\begin{aligned}
\mathbf{B}_0 &= \mu_0 \left(\mathbf{H}_0 + \boldsymbol{\mathcal{M}}_0\right) \\
\mathbf{D}_0 &= \epsilon_0 \mathbf{E}_0 + \boldsymbol{\mathcal{P}}_0
\end{aligned}$$

analogous to the fields $\mathbf{H}_{\text{ind}}$, $\mathbf{D}_{\text{ind}}$, and $\mathbf{B}_{\text{ind}}$ defined in (1.45), (1.46), and (1.56). However, in this case we are extending these ideas without reference to any specific physical models for polarization, conduction, or magnetization currents. The hypothetical polarization and magnetization densities $\boldsymbol{\mathcal{P}}_0$ and $\boldsymbol{\mathcal{M}}_0$ are chosen so that the hypothetical fields obey the same constitutive relations as do the original fields. For example, for a linear isotropic medium where $\mathbf{D} = \epsilon\mathbf{E}$ and $\mathbf{B} = \mu\mathbf{H}$, we demand also that $\mathbf{D}_0 = \epsilon\mathbf{E}_0$ and $\mathbf{B}_0 = \mu\mathbf{H}_0$, so that

$$\begin{aligned}
\mathbf{B}_0 &= \mathbf{B} - \mu\mathbf{T}_e \\
\mathbf{D}_0 &= \mathbf{D} - \epsilon\mathbf{T}_m
\end{aligned}$$

From (1.93), these fields will satisfy the following version of Maxwell's equations:

$$\begin{aligned}
\nabla \cdot \mathbf{B}_0 &= \rho_{m,\text{eq}} \\
\nabla \cdot \mathbf{D}_0 &= \rho_{\text{eq}} \\
\nabla \times \mathbf{E}_0 &= -\frac{\partial \mathbf{B}_0}{\partial t} - \mathbf{M}_{\text{eq}} \\
\nabla \times \mathbf{H}_0 &= \frac{\partial \mathbf{D}_0}{\partial t} + \mathbf{J}_{\text{eq}}
\end{aligned} \tag{1.94}$$

with the equivalent sources

$$\mathbf{J}_{\text{eq}} = \mathbf{J} + \epsilon_0\frac{\partial \mathbf{T}_m}{\partial t} - \nabla \times \mathbf{T}_e + \frac{\partial \left(\boldsymbol{\mathcal{P}} - \boldsymbol{\mathcal{P}}_0\right)}{\partial t} \tag{1.95}$$

$$\mathbf{M}_{\text{eq}} = \mathbf{M} + \mu_0\frac{\partial \mathbf{T}_e}{\partial t} + \nabla \times \mathbf{T}_m + \mu_0\frac{\partial \left(\boldsymbol{\mathcal{M}} - \boldsymbol{\mathcal{M}}_0\right)}{\partial t} \tag{1.96}$$

$$\rho_{\text{eq}} = \rho - \epsilon_0\nabla \cdot \mathbf{T}_m - \nabla \cdot \left(\boldsymbol{\mathcal{P}} - \boldsymbol{\mathcal{P}}_0\right) \tag{1.97}$$

$$\rho_{m,\text{eq}} = \rho_m - \mu_0\nabla \cdot \mathbf{T}_e - \mu_0\nabla \cdot \left(\boldsymbol{\mathcal{M}} - \boldsymbol{\mathcal{M}}_0\right) \tag{1.98}$$

(we have used (1.91) and (1.92) to eliminate the stream potentials $\mathbf{J}_e$ and $\mathbf{M}_m$). For the example case of a linear isotropic medium, the equivalent current densities are:

$$\mathbf{J}_{\mathrm{eq}} = \mathbf{J} + \frac{\partial(\epsilon\mathbf{T}_m)}{\partial t} - \nabla \times \mathbf{T}_e \tag{1.99}$$

$$\mathbf{M}_{\mathrm{eq}} = \mathbf{M} + \frac{\partial(\mu\mathbf{T}_e)}{\partial t} + \nabla \times \mathbf{T}_m \tag{1.100}$$

$$\rho_{\mathrm{eq}} = \rho - \nabla \cdot (\epsilon\mathbf{T}_m) \tag{1.101}$$

$$\rho_{m,\mathrm{eq}} = \rho_m - \nabla \cdot (\mu\mathbf{T}_e) \tag{1.102}$$

The hypothetical field $\mathbf{E}_0$, $\mathbf{H}_0$ produced by these equivalent sources differs from the actual field $\mathbf{E}$, $\mathbf{H}$ by the terms involving $\mathbf{T}_m$ and $\mathbf{T}_e$, which can be chosen in an essentially arbitrary manner and can be regarded as known.

To summarize: Given an original set of sources $\mathbf{J}$ and $\mathbf{M}$, we may calculate the field $\mathbf{E}$, $\mathbf{H}$ produced by these sources from

$$\mathbf{E} = \mathbf{E}_0 + \mathbf{T}_m \tag{1.103}$$

$$\mathbf{H} = \mathbf{H}_0 + \mathbf{T}_e \tag{1.104}$$

where the stream potentials $\mathbf{T}_e$ and $\mathbf{T}_m$ are arbitrary functions, and the field $(\mathbf{E}_0, \mathbf{H}_0)$ is the solution of (1.94) with equivalent sources given by (1.95)-(1.98) or (1.99)-(1.102). The functions $\mathbf{T}_e$ and $\mathbf{T}_m$ may be chosen in any convenient way according to the application at hand.

1.4.3 EQUIVALENCE PRINCIPLES

As an example of the application of stream potentials, we will derive what is known as Love's equivalence principle. Consider the situation shown in Figure 1.5(a). In this case, there are sources both inside and outside some surface S which produce a certain field $\mathbf{E}$, $\mathbf{H}$. We seek a second configuration that will have identical fields and sources outside of S, but zero fields and sources

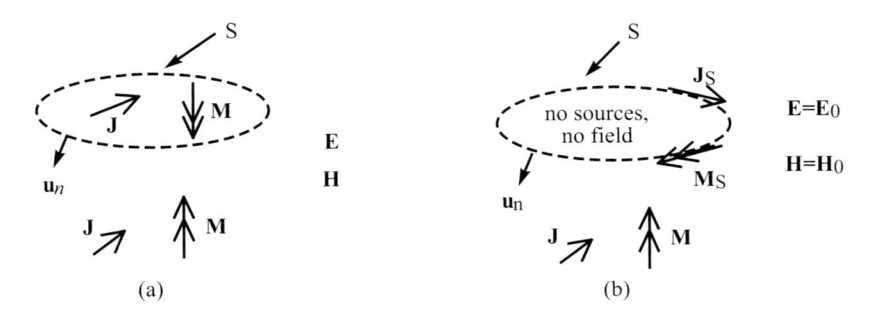

(a)   (b)

Figure 1.5 Equivalence principle based on tangential surface sources: (a) original problem; (b) equivalent problem.

inside S. If we exclude the possibility of fields containing delta functions or their derivatives of any order, the equivalent field $\mathbf{E}_0$, $\mathbf{H}_0$ must then be

$$\mathbf{E}_0 = \mathbf{E}\vartheta_S(n) \qquad \mathbf{H}_0 = \mathbf{H}\vartheta_S(n) \qquad (1.105)$$

where $\vartheta_S(n)$ is a Heaviside unit step function associated with S defined as:

$$\vartheta_S(n) = \left\{ \begin{array}{ll} 1 & (n > 0) \\ 0 & (n < 0) \end{array} \right\} \qquad (1.106)$$

Here n is the normal distance to the surface S, taken as positive for points outside S. The field $\mathbf{E}_0$, $\mathbf{H}_0$ conforms to the conditions imposed on the desired equivalent field above, and corresponds to the configuration shown in Figure 1.5(b).

We achieve the desired equivalence in a straightforward manner through the use of stream potentials. In order to have the equivalent field possess identical sources outside S, we must choose all stream potentials to vanish there, so they must contain a factor $[1 - \vartheta_S(n)]$. Inside S, we want the equivalent sources (1.99)-(1.102) to vanish, so we choose

$$\mathbf{T}_m = \mathbf{E} - \mathbf{E}_0 = \mathbf{E}\left[1 - \vartheta_S(n)\right] \qquad (1.107)$$

and likewise

$$\mathbf{T}_e = \mathbf{H} - \mathbf{H}_0 = \mathbf{H}\left[1 - \vartheta_S(n)\right] \qquad (1.108)$$

because (1.99)-(1.102) reduce to the original Maxwell equations in $n < 0$. When (1.107) and (1.108) are substituted into (1.99)-(1.100) and Maxwell's equations for the original configuration of Figure 1.5(a) are used, we obtain

$$\mathbf{J}_{\text{eq}} = \mathbf{J}\vartheta_S(n) + (\mathbf{u}_n \times \mathbf{H})\delta_S(n) \qquad (1.109)$$

$$\mathbf{M}_{\text{eq}} = \mathbf{M}\vartheta_S(n) + (\mathbf{E} \times \mathbf{u}_n)\delta_S(n) \qquad (1.110)$$

whence the vanishing of the equivalent sources inside S is verified. We have computed the derivatives of terms involving the step function by means of Dirac delta functions concentrated at the surface S:

$$\delta_S(n) \equiv \vartheta'_S(n) \qquad (1.111)$$

The subscripts S on ϑ_S and δ_S are, strictly speaking, redundant, since the definition of n is related to S, but serve to remind us of the geometry involved. We have in particular:

$$\begin{array}{rcl} \nabla\vartheta_S(n) & = & \mathbf{u}_n\delta_S(n) \\ \nabla\delta_S(n) & = & \mathbf{u}_n\delta'_S(n) \end{array} \qquad (1.112)$$

and so on, where $\mathbf{u}_n$ is the normal unit vector to S pointing to the outside of S.

We thus have *Love's equivalence principle*, which states that the equivalent sources (1.109) and (1.110) produce exactly the same field outside S as did the original ones. These sources contain, in addition to the surviving impressed volume sources outside S, tangentially directed surface sources $\mathbf{J}_S$ and $\mathbf{M}_S$ concentrated at S, as indicated by the delta functions δ_S, where

$$\begin{aligned} \mathbf{J}_S &= \mathbf{u}_n \times \mathbf{H}|_S \\ \mathbf{M}_S &= \mathbf{E} \times \mathbf{u}_n|_S \end{aligned} \qquad (1.113)$$

The effect of the surface sources replaces that of the sources interior to S that have been removed. We will take a more careful look at the effect of surface-concentrated currents and charges later on.

A second example of an equivalence principle is furnished by the multipole expansion of a current distribution discussed in Section 1.2.1. Consider the electric current density $\mathbf{J} = \nabla \times [\mathbf{m}(t)\delta(\mathbf{r})]$ that represents a magnetic dipole, with no magnetic currents present ($\mathbf{M} = 0$). Calculated explicitly, this electric current density has a relatively complicated form. However, if we take $\mathbf{T}_e = \mathbf{m}\delta(\mathbf{r})$ while $\mathbf{T}_m = 0$ in the equivalent electric current density given by (1.99), we can obtain equivalent sources of a much simpler form. The reason for the identification of the term involving $\mathbf{m}$ in (1.19) with a magnetic dipole can now be seen: The equivalent sources for a magnetic dipole involve no electric current density at all, but instead a magnetic current density $\mathbf{M}_{\mathrm{eq}} = \partial[\mu\mathbf{m}\delta(\mathbf{r})]/\partial t$ by (1.100), which is the dual of an electric Hertz dipole of similar mathematical form.

As a final example, suppose that there are no magnetic sources ($\mathbf{M} = 0$ and $\rho_m = 0$). Choose the stream potentials $\mathbf{T}_e = 0$ and

$$\mathbf{T}_m = -\frac{1}{\epsilon}\int \mathbf{J}\,dt$$

and use the equivalent sources (1.99)-(1.102), of which only

$$\mathbf{M}_{\mathrm{eq}} = -\frac{1}{\epsilon}\nabla \times \int \mathbf{J}\,dt$$

is different from zero. We see that a field caused exclusively by electric current sources can always be replaced by one produced exclusively by equivalent magnetic currents.

1.5 JUMP CONDITIONS

The classical boundary conditions at an interface, as has been implied by the foregoing, must be generalized when magnetic sources are present. The derivation can in fact be considerably simplified and extended by the use of the apparatus of generalized functions (see Appendix A), whence we arrive at so-called *jump conditions*.[11]

---

[11] Cf.:

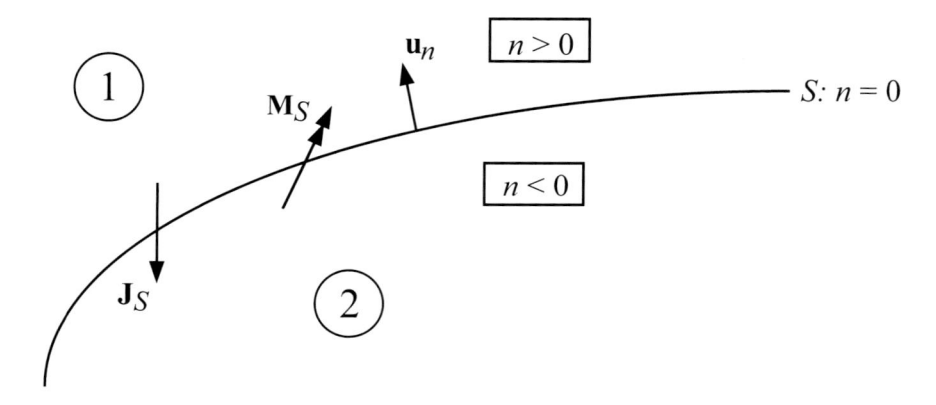

Figure 1.6 Fields and sources at an interface.

Suppose that all the impressed sources in a certain configuration are concentrated on a fixed (time-invariant) surface S in space (Figure 1.6). A coordinate system is set up that includes the normal distance n to the surface S as shown. A unit vector $\mathbf{u}_n$ is chosen normal to each point of S, directed towards the side of S where n is positive. The portion of space on this side is denoted region 1, and that on the other side region 2. As before, we define a Heaviside unit step function and a corresponding Dirac δ-function associated with S.

The volume source distributions $\mathbf{J}$ and $\mathbf{M}$ representing the surface source densities $\mathbf{J}_S$ and $\mathbf{M}_S$ (which are, without loss of generality, assumed to be independent of n) are thus expressed as:

$$\mathbf{J} = \mathbf{J}_S \delta_{S(e)}(n)$$
$$\mathbf{M} = \mathbf{M}_S \delta_{S(m)}(n) \qquad (1.114)$$

where the subscripts (e) and (m) refer to the possibility that these surface densities may be limiting cases of different approximate delta functions (see Appendix A; when this distinction is not necessary, we will omit these additional subscripts). If we do *not* make the assumption that $\mathbf{J}_S$ and $\mathbf{M}_S$ are

N. Marcuvitz and J. Schwinger, *J. Appl. Phys.*, vol. 22, pp. 806-819, 1951.

D. Bedeaux and J. Vlieger, *Physica*, vol. 67, pp. 55-73, 1973.

S. F. Timashev and M. A. Krykin, *Phys. Stat. Sol. (b)*, vol. 76, pp. 67-76, 1976.

C. Vassallo, *Théorie des Guides d'Ondes Electromagnetiques*, tome 1. Paris: Eyrolles, 1985, pp. 1-4.

M. Idemen and A. H. Serbest, *Electron. Lett.*, vol. 23, pp. 704-705, 1987.

J. Walsh and R. Donnelly, *Phys. Rev. A*, vol. 36, pp. 4474-4485, 1987.

M. Idemen, *Electron. Lett.*, vol. 24, pp. 663-665, 1988.

M. Idemen, *J. Phys. Soc. Japan*, vol. 59, pp. 71-80, 1990.

necessarily perpendicular to $\mathbf{u}_n$, then the conservation of electric and magnetic charge (1.90) implies the existence of both single-layer and double-layer surface charge densities, with

$$
\frac{\partial \rho}{\partial t} = -\nabla \cdot \mathbf{J}_S \delta_{S(e)}(n) - \mathbf{u}_n \cdot \mathbf{J}_S \delta'_{S(e)}(n)
$$

$$
\frac{\partial \rho_m}{\partial t} = -\nabla \cdot \mathbf{M}_S \delta_{S(m)}(n) - \mathbf{u}_n \cdot \mathbf{M}_S \delta'_{S(m)}(n) \qquad (1.115)
$$

The components of $\mathbf{J}_S$ and $\mathbf{M}_S$ perpendicular to S produce double-layer terms proportional to $\delta'_S(n)$ in (1.115); physically these are associated with layers of dipoles oriented perpendicular to S.[12]

1.5.1 GENERAL JUMP CONDITIONS AT A STATIONARY SURFACE

Let us consider an isotropic, homogeneous region where μ and ϵ are constant and scalar. Maxwell's equations (1.61) suggest that each of the fields $\mathbf{E}$, $\mathbf{H}$, $\mathbf{D}$, and $\mathbf{B}$ can be expressed as the sum of a continuous function, a step-function term, and a δ-function term as follows:

$$
\begin{aligned}
\mathbf{E} &= \mathbf{E}^c + \mathbf{E}^\vartheta \vartheta_S(n) + \mathbf{E}^\delta \delta_S(n) \\
\mathbf{H} &= \mathbf{H}^c + \mathbf{H}^\vartheta \vartheta_S(n) + \mathbf{H}^\delta \delta_S(n)
\end{aligned} \qquad (1.116)
$$

and so on. Here, $\mathbf{E}^c$, $\mathbf{E}^\vartheta$, $\mathbf{E}^\delta$, $\mathbf{H}^c$, $\mathbf{H}^\vartheta$, $\mathbf{H}^\delta$ etc., are to be treated as ordinary functions, continuous across S. We may, without loss of generality, take the terms $\mathbf{E}^\delta$, $\mathbf{H}^\delta$ and so forth, to be independent of n. From (1.114), (1.116), and the Maxwell curl equations, we get:

$$
\nabla \times \mathbf{E}^c + (\nabla \times \mathbf{E}^\vartheta)\vartheta_S(n) + [\mathbf{u}_n \times \mathbf{E}^\vartheta + \nabla \times \mathbf{E}^\delta]\delta_S(n) + (\mathbf{u}_n \times \mathbf{E}^\delta)\delta'_S(n)
$$

$$
= -\frac{\partial \mathbf{B}^c}{\partial t} - \frac{\partial \mathbf{B}^\vartheta}{\partial t}\vartheta_S(n) + [-\frac{\partial \mathbf{B}^\delta}{\partial t} - \mathbf{M}_S]\delta_S(n) \quad (1.117)
$$

$$
\nabla \times \mathbf{H}^c + (\nabla \times \mathbf{H}^\vartheta)\vartheta_S(n) + [\mathbf{u}_n \times \mathbf{H}^\vartheta + \nabla \times \mathbf{H}^\delta]\delta_S(n) + (\mathbf{u}_n \times \mathbf{H}^\delta)\delta'_S(n)
$$

$$
= \frac{\partial \mathbf{D}^c}{\partial t} + \frac{\partial \mathbf{D}^\vartheta}{\partial t}\vartheta_S(n) + [\frac{\partial \mathbf{D}^\delta}{\partial t} + \mathbf{J}_S]\delta_S(n) \quad (1.118)
$$

Although it can be argued that the singular parts of (1.117) and (1.118) containing delta functions are not observable (at least with macroscopic measurements), the discontinuous parts are observable, and as we will see below,

---

[12] It should be emphasized that if the surface S is not plane, then the terms in $\delta'_S(n)$ are not themselves double layers in the traditional sense, because they may contain nonzero total charge. For details, see

J. Van Bladel, *Singular Electromagnetic Fields and Sources*. Oxford: Clarendon Press, 1991.

J. Van Bladel, *Radio Science*, vol. 28, pp. 841-845, 1993.

are affected by the singular parts, which therefore cannot be ignored in our analysis.

The singular, stepwise discontinuous and continuous parts of (1.117) and (1.118) must all be separately equal. The terms therein containing $\delta'_S(n)$ imply that

$$\begin{aligned}
\mathbf{u}_n \times \mathbf{E}^\delta &= 0 \\
\mathbf{u}_n \times \mathbf{H}^\delta &= 0
\end{aligned} \tag{1.119}$$

which is to say that

$$\begin{aligned}
\mathbf{E}^\delta &= \mathbf{u}_n E_n^\delta \\
\mathbf{H}^\delta &= \mathbf{u}_n H_n^\delta
\end{aligned} \tag{1.120}$$

From (1.120) it follows that in this isotropic homogeneous medium:

$$\mathbf{B}_t^\delta = 0; \qquad \mathbf{D}_t^\delta = 0 \tag{1.121}$$

Using this result, and gathering the terms in $\delta_S(n)$ from (1.117) and (1.118) gives

$$\begin{aligned}
\mathbf{u}_n \times \mathbf{E}^\vartheta\big|_S - \mathbf{u}_n \times \nabla E_n^\delta &= -\mathbf{u}_n \frac{\partial B_n^\delta}{\partial t} - \mathbf{M}_S \\
\mathbf{u}_n \times \mathbf{H}^\vartheta\big|_S - \mathbf{u}_n \times \nabla H_n^\delta &= \mathbf{u}_n \frac{\partial D_n^\delta}{\partial t} + \mathbf{J}_S
\end{aligned} \tag{1.122}$$

since $\nabla \times \mathbf{u}_n = 0$.

The normal components of (1.122) imply that

$$\begin{aligned}
D_n^\delta &= -\int J_{Sn}\, dt \\
B_n^\delta &= -\int M_{Sn}\, dt
\end{aligned} \tag{1.123}$$

so that the δ-function portions of the fields $\mathbf{D}$ and $\mathbf{B}$ are present only if J_{Sn} or M_{Sn} is nonzero (i.e., there is a double layer of electric or magnetic charge on S). Since the medium is isotropic, we have further that

$$\begin{aligned}
E_n^\delta &= -\frac{\int J_{Sn}\, dt}{\epsilon} \\
H_n^\delta &= -\frac{\int M_{Sn}\, dt}{\mu}
\end{aligned} \tag{1.124}$$

Taking the tangential components of (1.122) and using (1.124) and (1.121) gives:

$$\begin{aligned}
\mathbf{E}^\vartheta \times \mathbf{u}_n\big|_S &= \mathbf{M}_{St} - \nabla\left(\frac{\int J_{Sn}\, dt}{\epsilon}\right) \times \mathbf{u}_n \\
\mathbf{u}_n \times \mathbf{H}^\vartheta\big|_S &= \mathbf{J}_{St} - \mathbf{u}_n \times \nabla\left(\frac{\int M_{Sn}\, dt}{\mu}\right)
\end{aligned} \tag{1.125}$$

where the subscript t denotes the components of a vector tangential to S. Since the left sides of (1.125) denote the jumps in tangential $\mathbf{E}$ and $\mathbf{H}$ across S, these equations represent the generalizations of the jump conditions on the tangential field components when double-layer sources are present at S.

The remaining terms in (1.117) and (1.118) that do not contain delta functions or their derivatives give us:

$$\nabla \times \mathbf{E}^c + (\nabla \times \mathbf{E}^\vartheta)\vartheta_S(n) = -\frac{\partial[\mathbf{B}^c + \mathbf{B}^\vartheta \vartheta_S(n)]}{\partial t}$$

$$\nabla \times \mathbf{H}^c + (\nabla \times \mathbf{H}^\vartheta)\vartheta_S(n) = \frac{\partial[\mathbf{D}^c + \mathbf{D}^\vartheta \vartheta_S(n)]}{\partial t} \tag{1.126}$$

It is possible that $\mathbf{E}^c$ and $\mathbf{H}^c$, while continuous themselves, have discontinuous derivatives that may result in further terms involving $\vartheta_S(n)$ in (1.126), which do not appear explicitly. However, such terms can only show up in the tangential components of this equation, and not in the normal component. Examining the discontinuities in the normal components of (1.126) then gives

$$\mathbf{u}_n \cdot \nabla \times \mathbf{E}^\vartheta \big|_S = -\mathbf{u}_n \cdot \frac{\partial \mathbf{B}^\vartheta}{\partial t}\bigg|_S$$

$$\mathbf{u}_n \cdot \nabla \times \mathbf{H}^\vartheta \big|_S = \mathbf{u}_n \cdot \frac{\partial \mathbf{D}^\vartheta}{\partial t}\bigg|_S \tag{1.127}$$

We next take the divergence of (1.125), use (1.127) and then some vector identities. The result is the jump conditions on the normal components of $\mathbf{D}$ and $\mathbf{B}$:

$$D_n^\vartheta \big|_S = -\int \nabla \cdot \mathbf{J}_{St}\, dt$$

$$B_n^\vartheta \big|_S = -\int \nabla \cdot \mathbf{M}_{St}\, dt \tag{1.128}$$

If a double layer is not present, the right sides of (1.128) are simply ρ_S and ρ_{mS}, and we recover the ordinary jump conditions for the normal field components across a single layer of surface charge.

Alternate forms of these jump conditions for media that are not isotropic (or even linear) can be obtained by introducing the magnetization and polarization densities $\mathcal{M}$ and $\mathcal{P}$ from (1.36) and (1.37). The derivation is unchanged up to (1.120), but (1.121) no longer holds. In place of (1.123), we now have

$$D_n^\delta = -\left[\int J_{Sn}\, dt + \mathcal{P}_n^\delta\right]$$

$$B_n^\delta = -\left[\int M_{Sn}\, dt + \mu_0\mathcal{M}_n^\delta\right] \tag{1.129}$$

Instead of (1.125), there is:

$$
\mathbf{E}^{\vartheta} \times \mathbf{u}_n\big|_S = \left(\mathbf{M}_{St} + \mu_0 \frac{\partial \boldsymbol{\mathcal{M}}_t^{\delta}}{\partial t} \right) - \nabla \left(\frac{\int J_{Sn}\, dt + \mathcal{P}_n^{\delta}}{\epsilon_0} \right) \times \mathbf{u}_n
$$

$$
\mathbf{u}_n \times \mathbf{H}^{\vartheta}\big|_S = \left(\mathbf{J}_{St} + \frac{\partial \boldsymbol{\mathcal{P}}_t^{\delta}}{\partial t} \right) - \mathbf{u}_n \times \nabla \left(\frac{\int M_{Sn}\, dt}{\mu_0} + \boldsymbol{\mathcal{M}}_n^{\delta} \right) \quad (1.130)
$$

and finally, the jump conditions (1.128) become:

$$
D_n^{\vartheta}\big|_S = -\nabla \cdot \left(\int \mathbf{J}_{St}\, dt + \boldsymbol{\mathcal{P}}_t^{\delta} \right)
$$

$$
B_n^{\vartheta}\big|_S = -\nabla \cdot \left(\int \mathbf{M}_{St}\, dt + \mu_0 \boldsymbol{\mathcal{M}}_t^{\delta} \right) \quad (1.131)
$$

It can be seen that the effect of surface densities of magnetization and polarization is very similar to that of surface densities of impressed current sources. The only difference is that $\boldsymbol{\mathcal{M}}^{\delta}$ and $\boldsymbol{\mathcal{P}}^{\delta}$ depend on the field values through some constitutive relations not specified here, so that (1.129)-(1.131) are not a complete set of jump conditions until those constitutive relations are provided.

1.5.2 EXAMPLE: THIN-SHEET BOUNDARY CONDITIONS

These results can be applied to the derivation of approximate boundary conditions for thin material sheets.[13] Suppose that a thin material sheet of permittivity ϵ, permeability μ, and thickness h in the z-direction is centered about the plane $z = 0$. The remainder of space has the parameters of vacuum, and no impressed currents are assumed to be present at the sheet. Because the sheet is thin, the polarization and magnetization densities are concentrated near $z = 0$, and the first term of the moment expansion (A.19) can be used to approximate them: $\boldsymbol{\mathcal{P}} \simeq \boldsymbol{\mathcal{P}}^{\delta}\delta(z)$ and $\boldsymbol{\mathcal{M}} \simeq \boldsymbol{\mathcal{M}}^{\delta}\delta(z)$, where

$$
\boldsymbol{\mathcal{P}}^{\delta}(x, y) = \int_{-h/2}^{h/2} \boldsymbol{\mathcal{P}}(x, y, z)\, dz \quad (1.132)
$$

and similarly for $\boldsymbol{\mathcal{M}}^{\delta}$.

---

[13]See:

B. Z. Katsenelenbaum, *Izv. Akad. Nauk SSSR, Otdel. Tekh. Nauk*, no. 7, pp. 9-22, 1955.

T. B. A. Senior and J. L. Volakis, *Radio Science*, vol. 22, pp. 1261-1272, 1987.

M. Idemen, *Electron. Lett.*, vol. 24, pp. 663-665, 1988.

M. Idemen, *J. Phys. Soc. Japan*, vol. 59, pp. 71-80, 1990.

T. B. A. Senior and J. L. Volakis, *Approximate Boundary Conditions in Electromagnetics*. London: IEE, 1995.

C. L. Holloway et al., *IEEE Ant. Prop. Mag.*, vol. 54, no. 2, pp. 10-35, 2012.

Since $\mathbf{J}_S$ and $\mathbf{M}_S$ are zero, we have from (1.129) that

$$H_z^\delta = -\mathcal{M}_z^\delta; \qquad E_z^\delta = -\frac{\mathcal{P}_z^\delta}{\epsilon_0}$$

and consequently from (1.130) that

$$\mathbf{E}^\vartheta \times \mathbf{u}_z\big|_{z=0} = -\nabla\left(\frac{\mathcal{P}_z^\delta}{\epsilon_0}\right) \times \mathbf{a}_z + \mu_0\frac{\partial \mathcal{M}_t^\delta}{\partial t}$$

$$\mathbf{u}_z \times \mathbf{H}^\vartheta\big|_{z=0} = -\mathbf{u}_z \times \nabla\mathcal{M}_z^\delta + \frac{\partial \mathcal{P}_t^\delta}{\partial t} \qquad (1.133)$$

The components of $\mathcal{M}^\delta$ and $\mathcal{P}^\delta$ can be calculated to a good approximation using the following argument. Consider $\mathcal{P}_t = (\epsilon - \epsilon_0)\mathbf{E}_t$ within the sheet. Let $\mathbf{E}^+$ denote the electric field just above the top face of the sheet in free space— that is, at $z = (h/2)^+$, and likewise denote the electric field at $z = (-h/2)^-$ as $\mathbf{E}^-$. Since the tangential electric field is continuous as it passes through the free space/material interface, and since the sheet is thin, it should be sufficient to approximate $\mathbf{E}_t$ by the average of $\mathbf{E}_t^+$ and $\mathbf{E}_t^-$. We then have:

$$\mathcal{P}_t^\delta \simeq \frac{h}{2}(\epsilon - \epsilon_0)(\mathbf{E}_t^+ + \mathbf{E}_t^-) \qquad (1.134)$$

The other components of the polarization and magnetization surface densities are found similarly, the only changes being to note that it is D_z, $\mathbf{H}_t$, and B_z that are continuous. The results, accurate to order $O(h)$, are:

$$\mathcal{P}_z^\delta \simeq \frac{h}{2}\left(\frac{1}{\epsilon_0} - \frac{1}{\epsilon}\right)(D_z^+ + D_z^-) \qquad (1.135)$$

$$\mathcal{M}_t^\delta \simeq \frac{h}{2}\frac{\mu - \mu_0}{\mu_0}(\mathbf{H}_t^+ + \mathbf{H}_t^-) \qquad (1.136)$$

$$\mathcal{M}_z^\delta \simeq \frac{h}{2}\left(\frac{1}{\mu_0} - \frac{1}{\mu}\right)(B_z^+ + B_z^-) \qquad (1.137)$$

where the superscripts $\pm$ now denote evaluation of the fields extrapolated to $z = 0^\pm$ through free space, instead of at $(h/2)^+$ and $(-h/2)^-$. This is permissible because the change makes only a difference of $O(h^2)$ on the right-hand sides of (1.134)-(1.137).

If these expressions are inserted into (1.133), we obtain the so-called generalized approximate boundary conditions (or sheet transition conditions) for the description of the sheet *as a zero-thickness object*:

$$(\mathbf{E}^+ - \mathbf{E}^-) \times \mathbf{u}_z = -\frac{1}{2\epsilon_0}\nabla\left[\chi_{es}^z(D_z^+ + D_z^-)\right] \times \mathbf{u}_z + \frac{\mu_0}{2}\chi_{ms}^t\frac{\partial(\mathbf{H}_t^+ + \mathbf{H}_t^-)}{\partial t}$$

$$\mathbf{u}_z \times (\mathbf{H}^+ - \mathbf{H}^-) = -\frac{1}{2\mu_0}\mathbf{u}_z \times \nabla\left[\chi_{ms}^z(B_z^+ + B_z^-)\right] + \frac{\epsilon_0}{2}\chi_{es}^t\frac{\partial(\mathbf{E}_t^+ + \mathbf{E}_t^-)}{\partial t}$$

$$(1.138)$$

where the quantities

$$\chi_{es}^{z} = h\left(1 - \frac{\epsilon_0}{\epsilon}\right)$$

$$\chi_{ms}^{t} = h\left(\frac{\mu}{\mu_0} - 1\right)$$

$$\chi_{ms}^{z} = \left(1 - \frac{\mu_0}{\mu}\right)$$

$$\chi_{es}^{t} = h\left(\frac{\epsilon}{\epsilon_0} - 1\right)$$

are called surface susceptibilities of the sheet, analogous to the bulk suscepti-bilities χ_e and χ_m of a material medium. In contrast with the latter, however, surface susceptibilities have units of length, and represent polarizability per unit area rather than per unit volume. These boundary conditions are gen-eralized impedance and admittance conditions on the surface, through the parameters χ_{es}^{t} and χ_{ms}^{t}, together with further nonlocal (spatially dispersive) effects associated with the parameters χ_{es}^{z} and χ_{ms}^{z}.

By being more careful during the approximation process, more accurate boundary conditions of higher order can be derived. Other kinds of thin layers such as gratings and thin granular films can also be described approximately by boundary conditions of the form (1.138), using different expressions for the surface susceptibilities.

1.5.3   JUMP CONDITIONS AT A MOVING SURFACE

Suppose now that the surface at which jumps are to be considered is moving, so that it is now denoted by $S(t)$, whose unit normal vector directed towards the region $n > 0$ is $\mathbf{u}_n$ as before. To be more precise, let $S(t)$ be defined by the equation

$$f(\mathbf{r}, t) = 0 \tag{1.139}$$

where f is a real function, chosen to be equal to n for points near enough to $S(t)$, where n is the normal coordinate (distance) of a point to $S(t)$. If $\mathbf{u}_n$ is the unit normal vector pointing to the "outside" of $S(t)$ where $f > 0$, then from (D.66) in the Appendix we have that the normal velocity of a point on $S(t)$ is:

$$v_n = \mathbf{v} \cdot \mathbf{u}_n = -\frac{\partial f}{\partial t} \tag{1.140}$$

and furthermore, $\nabla f = \mathbf{u}_n$ near $S(t)$.

Let us assume the presence of surface charge and surface current densities on this surface,

$$\rho = \rho_S \delta\left[f(\mathbf{r}, t)\right]; \quad \rho_m = \rho_{mS} \delta\left[f(\mathbf{r}, t)\right]; \quad \mathbf{J} = \mathbf{J}_S \delta\left[f(\mathbf{r}, t)\right]; \quad \mathbf{M} = \mathbf{M}_S \delta\left[f(\mathbf{r}, t)\right] \tag{1.141}$$

acting in free space. In place of (1.116), we now postulate

$$\mathbf{E} = \mathbf{E}^\vartheta \vartheta[f(\mathbf{r}, t)] + \text{smoother terms} \tag{1.142}$$

$$\mathbf{H} = \mathbf{H}^\vartheta \vartheta[f(\mathbf{r}, t)] + \text{smoother terms} \tag{1.143}$$

supposing that no delta-function fields are present (we will discover below under what conditions this supposition is true). We insert (1.142)-(1.143) into (1.61), together with $\mathbf{D} = \epsilon_0 \mathbf{E}$ and $\mathbf{B} = \mu_0 \mathbf{H}$. Note that

$$\frac{\partial \vartheta[f(\mathbf{r}, t)]}{\partial t} = \delta[f(\mathbf{r}, t)] \frac{\partial f(\mathbf{r}, t)}{\partial t} = -(\mathbf{v} \cdot \mathbf{u}_n) \, \delta[f(\mathbf{r}, t)]$$

and

$$\nabla \vartheta[f(\mathbf{r}, t)] = \nabla f(\mathbf{r}, t) \delta[f(\mathbf{r}, t)] = \mathbf{u}_n \delta[f(\mathbf{r}, t)]$$

so if we equate the terms involving delta functions as before, we find from the Gauss laws:

$$\mathbf{u}_n \cdot \mathbf{E}^\vartheta = \frac{\rho_S}{\epsilon_0}; \quad \mathbf{u}_n \cdot \mathbf{H}^\vartheta = \frac{\rho_{mS}}{\mu_0} \tag{1.144}$$

while from Ampère's law we have

$$\mathbf{u}_n \times \mathbf{H}^\vartheta = -\epsilon_0 \left(\mathbf{v} \cdot \mathbf{u}_n\right) \mathbf{E}^\vartheta + \mathbf{J}_S \tag{1.145}$$

and from Faraday's law,

$$\mathbf{u}_n \times \mathbf{E}^\vartheta = \mu_0 \left(\mathbf{v} \cdot \mathbf{u}_n\right) \mathbf{H}^\vartheta - \mathbf{M}_S \tag{1.146}$$

Taking the normal components of (1.145)-(1.146) and using (1.144), we obtain

$$\mathbf{u}_n \cdot (\rho_S \mathbf{v} - \mathbf{J}_S) = 0; \quad \mathbf{u}_n \cdot (\rho_{mS} \mathbf{v} - \mathbf{M}_S) = 0$$

meaning that normal components of the surface current densities appear only due to the normal velocities of the surface charge densities if no delta functions are allowed in the fields. The tangential components of (1.145)-(1.146) give

$$\mathbf{u}_n \times \mathbf{H}^\vartheta + \frac{v_n}{c} \frac{\mathbf{E}_t^\vartheta}{\zeta_0} = \mathbf{J}_{St} \tag{1.147}$$

and

$$\mathbf{E}^\vartheta \times \mathbf{u}_n + \frac{v_n}{c} \zeta_0 \mathbf{H}_t^\vartheta = \mathbf{M}_{St} \tag{1.148}$$

where $v_n = \mathbf{v} \cdot \mathbf{u}_n$ is the normal velocity of the surface, $c = 1/\sqrt{\mu_0 \epsilon_0}$ is the speed of light in vacuum, and $\zeta_0 = \sqrt{\mu_0/\epsilon_0}$ is the free-space wave impedance. Equations (1.147)-(1.148) can be solved for the tangential field discontinuities as follows:

$$\mathbf{u}_n \times \mathbf{H}^\vartheta = \frac{\mathbf{J}_{St} - \frac{v_n}{c} \frac{\mathbf{u}_n \times \mathbf{M}_{St}}{\zeta_0}}{1 - \frac{v_n^2}{c^2}} \tag{1.149}$$

and

$$\mathbf{E}^{\vartheta} \times \mathbf{u}_n = \frac{\mathbf{M}_{St} + \frac{v_n}{c}\zeta_0 \mathbf{u}_n \times \mathbf{J}_{St}}{1 - \frac{v_n^2}{c^2}} \tag{1.150}$$

These conditions correspond to (1.125) for a stationary surface (at least for the case when no delta-function fields are present).

It is to be noted that our derivation has made no explicit use of the theory of relativity,[14] although an independent deduction of these and related results can be made based on relativistic considerations.[15] This is a consequence of the fact that Maxwell's theory of electromagnetism is in fact compatible with relativity theory without the need for additional modifications.

1.5.4 FORCE ON SURFACE SOURCES

To illustrate the use of generalized functions in the analysis of problems with surface-concentrated sources, consider the following example. Let ρ_{Sf} and $\mathbf{J}_{Sf}$ be surface densities of free charge and current (in a thin conducting shell, for example) acting in free space so that

$$\rho_f = \rho_{Sf}\delta_S(n) \tag{1.151}$$

and

$$\mathbf{J}_f = \mathbf{J}_{Sf}\delta_S(n) \tag{1.152}$$

concentrated at a surface S. By conservation of charge,

$$(\nabla \cdot \mathbf{J}_{Sf})\delta_S(n) + \mathbf{J}_{Sf} \cdot \mathbf{u}_n \delta_S'(n) + \frac{\partial \rho_{Sf}}{\partial t}\delta_S(n) = 0 \tag{1.153}$$

Because the terms involving δ_S and those involving δ_S' must separately equate to zero in (1.153), we see that $\mathbf{J}_{Sf} \cdot \mathbf{u}_n = 0$ (the current must be tangential to S) and

$$\nabla \cdot \mathbf{J}_{Sf} + \frac{\partial \rho_{Sf}}{\partial t} = 0$$

For $\mathbf{E}$ and $\mathbf{H}$ to be compatible with these sources, Maxwell's equations demand that these fields have the form:

$$\begin{aligned} \mathbf{E} &= \mathbf{E}_2 + (\mathbf{E}_1 - \mathbf{E}_2)\vartheta_S(n) \\ \mathbf{H} &= \mathbf{H}_2 + (\mathbf{H}_1 - \mathbf{H}_2)\vartheta_S(n) \end{aligned}$$

---

[14] A similar approach to other moving medium problems can be found in:

B. M. Bolotovskii and S. N. Stolyarov, *Sov. Phys. Uspekhi*, vol. 32, pp. 813-827, 1989.

If the moving surface is a material interface, similar considerations to those of Section 1.5.5 about the microscopic properties of the interface may arise; see

L. A. Ostrovskii, *Sov. Phys. Uspekhi*, vol. 18, pp. 452-458, 1975.

[15] J. Van Bladel, *Relativity and Engineering*. Berlin: Springer-Verlag, 1984.

where the unit vector points normal to S from side 2 to side 1, which the subscripts of the fields in the equations above serve to denote.

If we substitute these expressions for the fields into Ampere's law, we get

$$\nabla \times \mathbf{H}_2 + \nabla \times (\mathbf{H}_1 - \mathbf{H}_2)\vartheta_S + \mathbf{u}_n \times (\mathbf{H}_1 - \mathbf{H}_2)\delta_S =$$
$$\epsilon_0 \frac{\partial}{\partial t} [\mathbf{E}_2 + (\mathbf{E}_1 - \mathbf{E}_2)\vartheta_S] + \mathbf{J}_{Sf}\delta_S(n)$$

or, grouping together all the terms involving delta functions that must separately cancel each other, we find the traditional boundary (or jump) condition

$$\mathbf{u}_n \times (\mathbf{H}_1 - \mathbf{H}_2)|_S = \mathbf{J}_{Sf}$$

A similar approach to Faraday's law gives that tangential $\mathbf{E}$ is continuous, while the divergence equations yield the continuity of normal $\mathbf{B}$ at S, and the jump condition

$$\mathbf{u}_n \cdot (\mathbf{E}_1 - \mathbf{E}_2)|_S = \frac{\rho_{Sf}}{\epsilon_0}$$

The force density on these surface densities can be computed from these results.[16] Since the delta function appearing in the surface charge and current densities is strongly equal to the derivative of the step function appearing in the expressions for the electric and magnetic fields, we have from (A.27) that

$$\delta_S \vartheta_S = \frac{1}{2}\delta$$

and therefore the total volume force density on the surface source charge and current densities is:

$$\begin{aligned}\mathbf{F}_v &= \frac{1}{2}\delta_S [\rho_{Sf}(\mathbf{E}_1 + \mathbf{E}_2) + \mathbf{J}_{Sf} \times (\mathbf{B}_1 + \mathbf{B}_2)]\\ &= \delta_S(n) \left[\rho_{Sf}\mathbf{E}_t + \rho_{Sf}\mathbf{u}_n \frac{E_{1n} + E_{2n}}{2} + \mathbf{J}_{Sf} \times \mathbf{u}_n B_n + \mathbf{J}_{Sf} \times \frac{\mathbf{B}_{1t} + \mathbf{B}_{2t}}{2}\right]\end{aligned}$$

where the subscript t indicates the components of a vector tangential to the surface S and n the component normal to S. If region 2 is a solid perfect conductor so that $\mathbf{E}_2 \equiv 0$ and $\mathbf{B}_2 \equiv 0$, then in addition $\mathbf{E}_t = 0$ and $B_n = 0$ at S, and so the total force is

$$\mathbf{F} = \frac{1}{2} \int_S \mathbf{u}_n \left(\frac{\rho_{Sf}^2}{\epsilon_0} - \mu_0 |\mathbf{J}_{Sf}|^2\right) dS$$

using the boundary conditions on $\mathbf{E}_1$ and $\mathbf{B}_1$ at S. This well-known result says that the presence of charge density on a conducting body tends to push the surface of the body outward, while the forces on the surface current tend to push it inwards (towards the conductor).

---

[16]This is a bit more rigorous approach to obtaining the result of:
S. Ragusa, *Amer. J. Phys.*, vol. 58, pp. 364-366, 1990.

1.5.5 EXAMPLE: CHARGE DIPOLE SHEET AT A DIELECTRIC INTERFACE

An additional complication arises when surface source densities are positioned at an interface between two media. In such a case, the material parameters (μ, ϵ, etc.) contain a step-function variation with the normal coordinate n, and if the field contains delta-function terms, the product of this "material" step-function with a Dirac delta function from the field expression will appear. Such a product must be calculated with care (see Section A.2 in the Appendix). We will illustrate this point using a specific electrostatic example.

We consider the situation of a sheet of normally directed electric dipoles in the plane $z = 0$, described by the "excess" free charge density

$$\rho_e = -\mathcal{P}_{Sez}(x,y)\delta'_{(e)}(z) = -\nabla \cdot \left[\boldsymbol{\mathcal{P}}_{eS}(x,y)\delta_{(e)}(z)\right] \tag{1.154}$$

where $\boldsymbol{\mathcal{P}}_{eS}(x,y) = \mathbf{u}_z \mathcal{P}_{Sez}(x,y)$ is a z-directed surface polarization density (dipole moment per unit area) in the plane $z = 0$. A single charge dipole located at $x = y = 0$ would have $\mathcal{P}_{Sez} = p\delta(x)\delta(y)$, where p is the dipole moment of the (z-directed) dipole. We make no assumption about how this charge distribution arises, but we do assume that it does not include any bound charges arising from the response of a medium to the field present in it (this is what we mean by "excess" charge).

We will attack this problem in several stages. We first assume that the dipole sheet acts in free space. We can analyze the delta function and step function behaviors of the $\mathbf{E}$ and $\mathbf{D}$ fields in the same way we did in Section 1.5.1. From the electrostatic field equations

$$\nabla \cdot \mathbf{D} = \rho_e \tag{1.155}$$

$$\nabla \times \mathbf{E} = 0 \tag{1.156}$$

we obtain:

$$\mathbf{D}^\delta = \mathbf{u}_z D_z^\delta \tag{1.157}$$

$$\mathbf{E}_t^\vartheta\big|_{z=0} = \nabla_t E_z^\delta \tag{1.158}$$

$$D_z^\delta = -\mathcal{P}_{Sez}(x,y) \tag{1.159}$$

Equations (1.157)-(1.159) do not depend on what medium the dipole sheet is located in, and will be valid for all cases considered below, not just for free space. Since in this case the dipole sheet does act in free space, we have $\mathbf{D} = \epsilon_0 \mathbf{E}$, and therefore

$$E_z^\delta = -\frac{\mathcal{P}_{Sez}(x,y)}{\epsilon_0} \tag{1.160}$$

and consequently there will be a discontinuity of tangential $\mathbf{E}$ at $z = 0$:

$$\mathbf{E}_t^\vartheta\big|_{z=0} = -\frac{1}{\epsilon_0}\nabla_t \mathcal{P}_{Sez}(x,y) \tag{1.161}$$

Next, we suppose that this dipole sheet acts in an infinite homogeneous dielectric whose permittivity is ϵ. Equations (1.157)-(1.159) remain valid, but (1.160)-(1.161) must be replaced by

$$E_z^\delta = -\frac{\mathcal{P}_{Sez}(x,y)}{\epsilon} \tag{1.162}$$

$$\mathbf{E}_t^\vartheta\big|_{z=0} = -\frac{1}{\epsilon}\nabla_t \mathcal{P}_{Sez}(x,y) \tag{1.163}$$

Now, the excess charge density ρ_e has produced an electric field, which in turn has caused a bound polarization density response

$$\mathcal{P}_b = (\epsilon - \epsilon_0)\mathbf{E} = \left(1 - \frac{\epsilon_0}{\epsilon}\right)\mathbf{D} \tag{1.164}$$

corresponding to a bound charge density

$$\rho_b = -\nabla \cdot \mathcal{P}_b = -\left(1 - \frac{\epsilon_0}{\epsilon}\right)\nabla \cdot \mathbf{D} = -\left(1 - \frac{\epsilon_0}{\epsilon}\right)\rho_e \tag{1.165}$$

in the dielectric. The field could just as well be regarded as having been produced by the net charge density

$$\rho_n \equiv \rho_e + \rho_b = \frac{\epsilon_0}{\epsilon}\rho_e \tag{1.166}$$

acting in free space. Defining a net surface polarization density

$$\mathcal{P}_{Snz}(x,y) = \frac{\epsilon_0}{\epsilon}\mathcal{P}_{Sez}(x,y) \tag{1.167}$$

we can re-express (1.162)-(1.163) as

$$E_z^\delta = -\frac{\mathcal{P}_{Snz}(x,y)}{\epsilon_0} \tag{1.168}$$

$$\mathbf{E}_t^\vartheta\big|_{z=0} = -\frac{1}{\epsilon_0}\nabla_t \mathcal{P}_{Snz}(x,y) \tag{1.169}$$

which are the same as (1.160)-(1.161), except that the excess surface polarization has been replaced by the net surface polarization.

We finally assume that the excess dipole moment sheet acts in a dielectric medium whose permittivity has a jump discontinuity at the plane $z = 0$. Again, Equations (1.157)-(1.159) still hold, but attempting to imitate the procedure above to determine E_z^δ will encounter the difficulty that the delta function associated with $\mathcal{P}_{Sez}$ will be multiplied by a step function associated with $\frac{1}{\epsilon}$:

$$\frac{1}{\epsilon(z)} = \frac{1}{\epsilon_2} + \left(\frac{1}{\epsilon_1} - \frac{1}{\epsilon_2}\right)\vartheta_{(\mathrm{mat})}(z) \tag{1.170}$$

where the step-function $\vartheta_{(\mathrm{mat})}$ describes the microscopic behavior of a material property (the dielectric constant), as opposed to that which characterizes

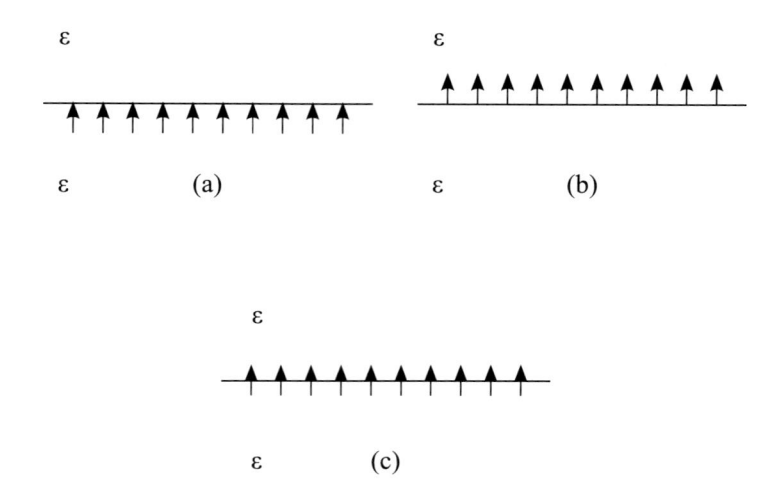

Figure 1.7 Dipole sheet at an interface: (a) sheet at $z = 0^-$; (b) sheet at $z = 0^+$; (c) sheet partially on each side of the interface.

the location of the dipole sheet. To proceed further, we must consider what we mean when we write Maxwell's equations in the sense of generalized functions. As seen above, the presence of the z-directed dipole at $z = 0$ implies a delta-function portion of **E**. The step-function behavior of the material properties at $z = 0$ is not necessarily related in any way to the location of the excess dipole sheet. The presence of step functions or delta functions in a field or source density indicates a limiting case of a physical situation where the physical quantity varies rapidly (or is highly concentrated) in the vicinity of a surface (or line, or point) in space, but is not discontinuous or singular there. In a physically more realistic problem, there would be a continuous (though rapid) transition between ϵ_2 for $z < 0$ and ϵ_1 for $z > 0$ as z passes through 0, and **E** will have portions that are highly concentrated near $z = 0$. Depending on the details of these transitions, the concentrated (δ-function) parts of the fields might be at a location where the permittivity is close to ϵ_2 (Figure 1.7(a)), or to ϵ_1 (Figure 1.7(b)), or perhaps some value in between (Figure 1.7(c)). Colloquially speaking, we have to say "where" the dipole is relative to the surface of dielectric discontinuity in order to make the problem precisely stated.

In any event, Maxwell's equations can be expected to hold "microscopically" throughout the rapid transition region around $z = 0$, and not just in a limiting sense as $z \to 0$ from either side. According to Section A.2 in the Appendix, this means that we should understand Maxwell's equations as *strong* equalities in the sense of generalized functions, meaning that we have to consider the microscopic detail of the step functions and delta functions at $z = 0$ in our calculations. Let us note that the delta function implied by the representation

$$\mathbf{D} = \mathbf{D}^\delta \delta_{(e)}(z) + \text{smoother terms} \tag{1.171}$$

must therefore be the same as the one appearing in the excess surface polarization above, because of the strong equality implied by Gauss' law (1.155). However, the delta function appearing in $\mathbf{E}$ is not necessarily the same (microscopically) as $\delta_{(e)}(z)$. The product of $\frac{1}{\epsilon}$ and the $\mathbf{D}$ field that appears in the constitutive relation $\mathbf{E} = \frac{\mathbf{D}}{\epsilon}$ must then be calculated according to the rules of Section A.2, but the equation itself need only be taken to be a weak equality or association. According to this theory, we must specify the pertinent step and delta functions by retaining certain information about how the sequences that converge to them behave during the limiting process. That is, their *microscopic* variations near $z = 0$ before the limit is taken must be "remembered" by the generalized functions that describe the fields, sources, and material parameter variations. Since the step function in $\frac{1}{\epsilon}$ is related to the microscopic position of the dielectric interface, we have indicated this in (1.170) through the use of a specific unit step function $\vartheta_{(\mathrm{mat})}$. Since the microscopic position of the surface dipole layer is generally independent of the dielectric surface, it has been expressed in terms of the delta function $\delta_{(e)}$, generally distinct from $\delta_{(\mathrm{mat})}$.

Following Section A.2 of the Appendix, we let A be the constant obeying

$$\vartheta_{(\mathrm{mat})}(z)\delta_{(e)}(z) \approx A\delta_{(r)}(z) \tag{1.172}$$

for *some* delta function $\delta_{(r)}(z)$, where the sign $\approx$ denotes weak equality of generalized functions. The delta-function terms of $\mathbf{E}$ are then found to be

$$E_z^\delta = \left[\frac{1}{\epsilon_2} + A \left(\frac{1}{\epsilon_1} - \frac{1}{\epsilon_2} \right) \right] D_z^\delta \tag{1.173}$$

Then from (1.158)-(1.159) we obtain

$$E_z^\delta = -\frac{\mathcal{P}_{Snz}(x,y)}{\epsilon_0} \tag{1.174}$$

where the net surface polarization density is now given by

$$\mathcal{P}_{Snz}(x,y) = \frac{\epsilon_0}{\epsilon_{\mathrm{eff}}}\mathcal{P}_{Sez}(x,y) \tag{1.175}$$

and ϵ_{eff} is an effective permittivity at the (microscopic) location of the dipole sheet,

$$\frac{1}{\epsilon_{\mathrm{eff}}} \equiv \frac{1}{\epsilon_2} + A \left(\frac{1}{\epsilon_1} - \frac{1}{\epsilon_2} \right) \tag{1.176}$$

For a specific sequence of continuous transition functions describing $\vartheta_{(e)}$ and $\delta_{(p)}$, we could compute (numerically if by no other means) the constant A, and therefore the effective permittivity ϵ_{eff}. Now from (1.158), we have that

$$\mathbf{E}_t^\vartheta \big|_{z=0} = -\nabla_t \left(\frac{\mathcal{P}_{Sez}(x,y)}{\epsilon_{\mathrm{eff}}} \right) = -\nabla_t \left(\frac{\mathcal{P}_{Snz}(x,y)}{\epsilon_0} \right) \tag{1.177}$$

or in other words, there is a jump in tangential $\mathbf{E}$ across the dipole sheet equal to:

$$\mathbf{E}_t\big|_{z=0^-}^{0^+} = -\nabla_t\left(\frac{\mathcal{P}_{Sez}(x,y)}{\epsilon_{\text{eff}}}\right) = -\nabla_t\left(\frac{\mathcal{P}_{Snz}(x,y)}{\epsilon_0}\right) \tag{1.178}$$

We see that, even without fictitious magnetic currents present, there is the possibility of a discontinuity in tangential $\mathbf{E}$ in certain idealized situations, and that we may, in principle, compute explicitly what this discontinuity is in terms of known quantities. Moreover, exactly what this discontinuity is will depend on a microscopic relation between the location of the dipole sheet and that of the material discontinuity. This can be important when modeling electromagnetic field behavior numerically or when examining the properties of thin films.[17]

---

[17]See, e.g.,

D. Bedeaux and J. Vlieger, *Optical Properties of Surfaces*. London: Imperial College Press, 2002.

M. A. Mohamed et al., *PIER B*, vol. 16, pp. 1-20, 2009.

1.6   PROBLEMS

1–1 A static electric current I flows along a path C (not necessarily lying in a plane) in the direction of the unit vector $\mathbf{u}_l$ tangent to this curve. In other words, the current density is given by:

$$\mathbf{J} = I\mathbf{u}_l\delta_C(\mathbf{r})$$

where δ_C represents a delta function concentrated on C in the sense that

$$\int_V g(\mathbf{r})\delta_C(\mathbf{r})\,dV = \int_C g(\mathbf{r})\,dl$$

for any good function g. Show that the magnetic dipole moment of this current loop is $\mathbf{m} = I\mathbf{S}$, where $\mathbf{S} = \int_S \mathbf{u}_n\,dS$ is the "vector area" of any surface S that spans (is bounded by) C, and $\mathbf{u}_n$ is the unit vector normal to S related to $\mathbf{u}_l$ by the right hand rule. If the loop lies in a plane, then $\mathbf{S} = \mathbf{u}_n S$.

1–2 A long rectangular loop of thin wire (dimensions $a \times b$, where a is the length of the loop along the z-axis, b the dimension along the x-axis, and $a \gg b$) carries a current I. Show that the magnetic dipole moment $\mathbf{m}$ of this loop is $Iab\mathbf{u}_y$, and that the contribution to $\mathbf{m}$ from the long sides is $Iab\mathbf{u}_y/2$, exactly the same as that from the short sides. Thus, if we were to attempt to define a magnetic dipole moment per unit length $\mathbf{m}_l$ for a two-dimensional magnetic field $\mathbf{H}_t = \mathbf{u}_x H_x + \mathbf{u}_y H_y$ transverse to z, use of only the z-directed currents corresponding to $\mathbf{H}_t$ to compute $\mathbf{m}_l$ would give only one half the value obtained by considering a long but finite loop and taking the limit as the length of the loop became infinite. As a result of this, justify the use of

$$\mathbf{m}_l = \int\int \boldsymbol{\rho} \times \mathbf{J}\,dxdy$$

to compute the magnetic dipole moment per unit length in a general two-dimensional ($\partial/\partial z = 0$ and $\mathbf{H} = \mathbf{H}_t$) magnetostatic problem. Here $\boldsymbol{\rho} = x\mathbf{u}_x + y\mathbf{u}_y$ denotes the two-dimensional position vector in the xy-plane.

1–3 A circular wire loop of radius a lies in the plane $z = 0$, centered at the origin. A line charge density

$$\rho_l = \frac{Q}{2a}\cos\omega t\cos\phi$$

exists on this loop, where ϕ is the angular coordinate in the circular cylindrical coordinate system, being zero along the x-axis.

(a)   Show that Q is the value of the total charge on the half of the loop lying in $-\pi/2 < \phi < \pi/2$, at the time $t = 0$.

(b) Obtain an expression for the electric dipole moment $\mathbf{p}$ of the loop.

(c) Assume that the current $I(\phi, t)$ on the wire is equal to zero at $\phi = 0$ for all t. Obtain an expression for this current at any values of ϕ and t.

(d) Evaluate the magnetic dipole moment $\mathbf{m}$ of the loop.

1–4 (a) If the origin of coordinates is re-designated as $\mathbf{r}_0$, the definition of the electric dipole moment will change from (1.18) to

$$\mathbf{p}(t) = \int (\mathbf{r} - \mathbf{r}_0)\rho(\mathbf{r}, t)\, dV$$

Under what condition on the charge density ρ will the value of $\mathbf{p}$ be independent of $\mathbf{r}_0$?

(b) A similar change of origin modifies the magnetic dipole moment from (1.21) to

$$\mathbf{m}(t) = \frac{1}{2} \int (\mathbf{r} - \mathbf{r}_0) \times \mathbf{J}(\mathbf{r}, t)\, dV$$

Under what condition on the dipole moment $\mathbf{p}$ will the value of $\mathbf{m}$ be independent of $\mathbf{r}_0$? Comment on the likelihood of this condition being satisfied.

1–5 Suppose that the equivalence principle has been used (as in Section 1.4.3) to re-express the radiation problem of given sources inside a fictitious closed surface S as one involving only equivalent tangential surface currents $\mathbf{J}_{S,tan}$ and $\mathbf{M}_{S,tan}$ impressed at S, with zero fields and sources inside S, as shown in Figure 1.5.

(a) If we also wish to impress normally directed surface currents J_{Sn} and M_{Sn} at S (but without terms in $\mathbf{J}_{\text{eq}}$ or $\mathbf{M}_{\text{eq}}$, which involve $\delta_s'(n)$ or higher derivatives of the delta function), and yet still maintain the original fields $\mathbf{E}$ and $\mathbf{H}$ exterior to S, what modifications must we make to $\mathbf{J}_{S,tan}$ and $\mathbf{M}_{S,tan}$?

(b) If a perfectly conducting surface is placed immediately behind the surface sources at S, which of the components of $\mathbf{J}_S$ and $\mathbf{M}_S$ as found in part (a) may be removed without affecting the fields exterior to S? [See also, e.g., Section 3.3.]

1–6 The various equivalence theorems imply that different distributions of electric and magnetic current densities inside a certain region of space can produce the same electromagnetic field outside that source region. This in turn implies that the difference between these two source distributions (which is nonzero only within the source region) produces a field that is *identically zero* everywhere outside the source region. Such a distribution of sources is called a nonradiating source distribution, and is notable for being undetectable from outside the source region.[18] This would have obvious implications for "stealth" technology, electromagnetic interference, and remote sensing applications.

Guided by the equivalence theorems we have studied so far, we can only construct nonradiating source distributions using both electric and magnetic sources together. Show that, if $\mathbf{Q}(\mathbf{r}, t)$ is any vector function that vanishes outside a bounded region of space, then the electric current density given by

$$\mathbf{J}(\mathbf{r}, t) = \nabla \times \left(\frac{1}{\mu} \nabla \times \mathbf{Q} \right) + \epsilon \frac{\partial^2 \mathbf{Q}}{\partial t^2}$$

is, by itself, a nonradiating source distribution in a medium of permittivity ϵ and permittivity μ. Put another way, this result means that such a source distribution could be added to any other without changing the external electromagnetic field: the problem of determining the source distribution of an electromagnetic field cannot be solved uniquely using only information about the field exterior to the source region.

1–7 A thin material sheet of permittivity ϵ, permeability μ, and thickness h lies on top of a perfectly conducting plane located at $z = 0$. For $z > h$, there is only free space. Similar to the example in Section 1.5.2, derive equivalent boundary conditions for the electric and magnetic fields $\mathbf{E}$ and $\mathbf{H}$ extrapolated to the surface at $z = 0^+$. Your result should reduce to the expected form if the sheet is absent ($h = 0$ or $\epsilon = \epsilon_0$ and $\mu = \mu_0$). In other words, one of your conditions should read $\mathbf{u}_n \times \mathbf{E} = 0$ at $z = 0^+$ in that case.

1–8 Consider a perpendicularly polarized plane wave obliquely incident from free space onto a thin magneto-dielectric slab of thickness h, permittivity ϵ, and permeability μ as shown in Figure 1.8. Its electric field has only the y-component

$$E_y^i = e^{-jk_0(x \sin \theta - z \cos \theta)}$$

where θ is the angle of incidence and $k_0 = \omega \sqrt{\mu_0 \epsilon_0}$ is the plane-wave wavenumber in free space.

---

[18]See:

G. H. Goedecke, *Phys. Rev.*, vol. 135, pp. B281-B288, 1964.

A. J. Devaney and E. Wolf, *Phys. Rev. D*, vol. 8, pp. 1044-1047, 1973.

M. A. Miller, *Radiophys. Quantum Electron.*, vol. 29, pp. 747-760, 1986.

A. Sihvola, G. Kristensson and I. V. Lindell, *IEEE Trans. Micr. Theory Tech.*, vol. 45, pp. 2155-2159, 1997.

N. K. Nikolova and Y. S. Rickard, *Phys. Rev. E*, vol. 71, art. 016617, 2005.

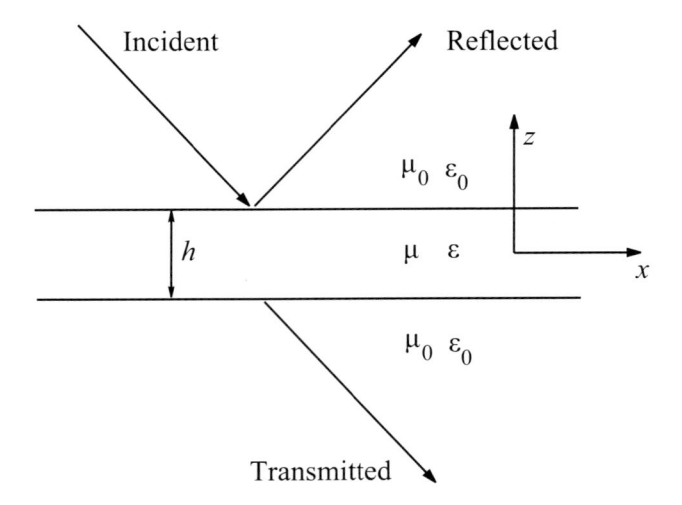

Figure 1.8

(a) Obtain an expression for the exact reflection and transmission coefficients Γ and T of this incident wave. Assume that the electric fields are

$$E_y = e^{-jk_0(x\sin\theta - z\cos\theta)} + \Gamma e^{-jk_0(x\sin\theta + z\cos\theta)}$$

in $z > h/2$,

$$E_y = Ae^{-jk(x\sin\theta_d - z\cos\theta_d)} + Be^{-jk(x\sin\theta_d + z\cos\theta_d)}$$

in $-h/2 < z < h/2$, and

$$E_y = Te^{-jk_0(x\sin\theta - z\cos\theta)}$$

in $z < -h/2$. Here $k = \omega\sqrt{\mu\epsilon}$ is the wavenumber in the layer, θ_d is the angle of the plane wave propagation in the layer, and A and B are complex constants to be determined.

(b) Assume the same forms for the field in free space as used in part (a), but employ instead (1.138) to determine Γ and T. Demonstrate that if h is "small" in some sense, these results approach those of part (a).

1–9 Repeat the example from Section 1.5.5 for the magnetostatic case. Specifically, let a sheet of tangentially directed magnetic dipoles of excess surface magnetization density $\mathcal{M}_{Set}(x, y)$ be located in the plane $z = 0$, so that the excess magnetization density $\mathcal{M} = \mathcal{M}_{Set}(x, y)\delta_{(e)}(z)$. The permeability is given by the generalized function $\mu(z) = \mu_2 + (\mu_1 - \mu_2)\vartheta_{(mat)}(z)$. Find the delta and step portions

$\mathbf{H}^{\delta}$, $\mathbf{H}^{\vartheta}$, $\mathbf{B}^{\delta}$, and $\mathbf{B}^{\vartheta}$ of the magnetic fields in terms of $\mathcal{M}_{Set}$. What assumption do you have to make about $\delta_{(e)}(z)$ and $\vartheta_{(\text{mat})}(z)$ to obtain your answer?

1–10 Suppose that the surface S at which the sources are concentrated in the derivation of the jump conditions in Section 1.5 is also a surface of discontinuity for ϵ and μ. In particular, let

$$\epsilon|_{n=0-} = \epsilon_a \qquad \epsilon|_{n=0+} = \epsilon_b$$

$$\mu|_{n=0-} = \mu_a \qquad \mu|_{n=0+} = \mu_b$$

and the jumps in ϵ and μ are described (using strong equality) by a certain step function $\vartheta_{S(\text{mat})}(n)$ associated with the surface of material discontinuity as

$$\frac{1}{\epsilon(z)} = \frac{1}{\epsilon_a} + \left(\frac{1}{\epsilon_b} - \frac{1}{\epsilon_a}\right)\vartheta_{(\text{mat})}(z); \quad \frac{1}{\mu(z)} = \frac{1}{\mu_a} + \left(\frac{1}{\mu_b} - \frac{1}{\mu_a}\right)\vartheta_{(\text{mat})}(z)$$

Let the delta function that describes the surface concentration of the sources $\mathbf{J}_S$ and $\mathbf{M}_S$ be given by $\delta_S(n)$, where we do not assume that $\delta_S(n) = \vartheta'_{S(\text{mat})}(n)$, but that

$$\vartheta_{S(\text{mat})}(n)\delta_S(n) \approx A\delta(n)$$

for some constant A, which is assumed to be known. Show that Equations (1.119)-(1.123) are still true, and that (1.124) and (1.125) remain valid if in them we replace ϵ and μ by ϵ_{av} and μ_{av}, respectively, giving expressions for ϵ_{av} and μ_{av}. Show that Equation (1.128) is still true.

1–11 A perfectly conducting surface $S(t)$ with outward unit normal vector $\mathbf{u}_n$ is moving through free space with velocity $\mathbf{v}$. It may support an induced surface electric current density $\mathbf{J}_S$, but no magnetic surface currents or charges. Obtain the boundary conditions for tangential $\mathbf{E}$ and $\mathbf{H}$ at this surface: $\mathbf{H}_t$ related to $\mathbf{J}_S$, and $\mathbf{E}_t$ related to $\mathbf{H}_t$. Give a physical interpretation of the boundary condition for tangential $\mathbf{E}$.

1–12 A plane wave

$$\mathbf{E}^i = \mathbf{u}_x E_0 \cos\left[\omega\left(t - \frac{z}{c}\right)\right]$$

$$\mathbf{H}^i = \mathbf{u}_y \frac{E_0}{\zeta_0} \cos\left[\omega\left(t - \frac{z}{c}\right)\right]$$

is normally incident onto a perfectly conducting plane $z = vt$ moving in the z-direction with a velocity v. Postulate a reflected plane wave

$$\mathbf{E}^r = \mathbf{u}_x E_1 \cos\left[\omega_1\left(t + \frac{z}{c}\right)\right]$$

$$\mathbf{H}^r = -\mathbf{u}_y \frac{E_1}{\zeta_0} \cos\left[\omega_1\left(t + \frac{z}{c}\right)\right]$$

and use the boundary condition found in Problem 1-11 to determine the reflection coefficient

$$\Gamma \equiv \frac{E_1}{E_0}$$

and the frequency ω_1 of the reflected wave. Obtain an expression for the induced surface current $\mathbf{J}_S$ on the conducting plane in terms of the amplitude E_0 of the incident wave. How does the frequency of the induced surface current compare with those of the incident and reflected plane waves? The resulting changes in frequency are due to the *Doppler shift*, and that of the reflected wave has important applications in radar systems.

2 Potential Representations of the Electromagnetic Field

2.1 LORENZ POTENTIALS AND THEIR DUALS (A, Φ, F, Ψ)

Consider for the moment a field in a linear isotropic medium produced solely by electric currents $\mathbf{J}$ and their associated charges ρ. Maxwell's equations (1.61) apply, with $\mathbf{M}$ and ρ_m equal to zero. This set of equations can be inconvenient to deal with directly, due to its complexity (the curl equations alone form a system of 6 coupled scalar equations in 6 scalar unknowns). For this reason, vector and scalar potentials are often used to simplify the solution procedure. In this section we derive the Lorenz[1] potentials.

Note first Gauss' law for the magnetic field $\nabla \cdot \mathbf{B} = 0$. Poincaré's Lemma 3 (Appendix D) states that a vector $\mathbf{B}$ whose divergence is zero is always expressible as the curl of some vector field $\mathbf{A}$, so long as certain conditions are satisfied by $\mathbf{B}$. If, therefore, we put

$$\mathbf{B} = \mu\mathbf{H} = \nabla \times \mathbf{A} \tag{2.1}$$

where $\mathbf{A}$ is called the (Lorenz) vector potential, then Faraday's law

$$\nabla \times \mathbf{E} = -\frac{\partial(\mu\mathbf{H})}{\partial t} \tag{2.2}$$

will be satisfied identically if

$$\nabla \times (\mathbf{E} + \frac{\partial \mathbf{A}}{\partial t}) = 0$$

from which, by Poincaré's Lemma 2 (Appendix D), it follows that

$$\mathbf{E} + \frac{\partial \mathbf{A}}{\partial t} = -\nabla\Phi \tag{2.3}$$

for some scalar potential Φ. There remains only Ampère's law

$$\nabla \times \mathbf{H} = \frac{\partial(\epsilon\mathbf{E})}{\partial t} + \mathbf{J} \tag{2.4}$$

and its related divergence equation

$$\nabla \cdot \mathbf{D} = \nabla \cdot (\epsilon\mathbf{E}) = -\int \nabla \cdot \mathbf{J}\, dt = \rho \tag{2.5}$$

---

[1] After the Danish physicist L. Lorenz—not to be confused with the Dutch physicist H. A. Lorentz.

to be satisfied. By (2.1)-(2.4), we have

$$-\nabla \times (\frac{1}{\mu}\nabla \times \mathbf{A}) - \frac{\partial^2 (\epsilon \mathbf{A})}{\partial t^2} - \frac{\partial(\epsilon \nabla \Phi)}{\partial t} = -\mathbf{J} \tag{2.6}$$

while (2.3) and (2.5) imply

$$\nabla \cdot (\epsilon \nabla \Phi) + \frac{\partial}{\partial t}\nabla \cdot (\epsilon \mathbf{A}) = -\rho \tag{2.7}$$

Now, what we have done so far is not likely to be much help unless further simplification and decoupling occur. We note that $\mathbf{A}$ has not been determined uniquely because only its curl has been specified (by connecting it to the physical quantity $\mathbf{B}$) and not its divergence. Once again, Poincaré's lemma is called upon, and we see that the gradient of any scalar function χ may be added to a valid vector potential $\mathbf{A}$, and the result will be a potential $\mathbf{A}'$ that obeys (2.1):

$$\mathbf{A}' = \mathbf{A} + \nabla \chi \tag{2.8}$$

If $\mathbf{A}$ is modified in this way, we must also modify Φ in order to ensure that (2.3) still holds for the new potentials. Thus we put

$$\Phi' = \Phi - \frac{\partial \chi}{\partial t} \tag{2.9}$$

and we have

$$\begin{aligned} \mathbf{B} &= \nabla \times \mathbf{A}' \\ \mathbf{E} &= -\frac{\partial \mathbf{A}'}{\partial t} - \nabla \Phi' \end{aligned} \tag{2.10}$$

for any choice of χ. Equations (2.6) and (2.7) are also valid with $\mathbf{A}$ and Φ replaced by $\mathbf{A}'$ and Φ'.

In a homogeneous isotropic region (one where ϵ and μ are scalars that are independent of position), we may decouple (2.6) and (2.7) by imposing the Lorenz gauge condition

$$\nabla \cdot \mathbf{A} = -\mu\epsilon\frac{\partial \Phi}{\partial t} \tag{2.11}$$

which amounts to a limitation on the choice of the potential Φ (or gauge function χ) once $\mathbf{A}$ is specified. Then (2.6) can be written (using the gradient of (2.11)) as

$$-\nabla \times \nabla \times \mathbf{A} + \nabla\nabla \cdot \mathbf{A} - \mu\epsilon\frac{\partial^2 \mathbf{A}}{\partial t^2} = \nabla^2 \mathbf{A} - \mu\epsilon\frac{\partial^2 \mathbf{A}}{\partial t^2} = -\mu\mathbf{J} \tag{2.12}$$

where the second equality follows from identity (D.13). Likewise, (2.7) can be expressed as

$$\nabla^2 \Phi - \mu\epsilon\frac{\partial^2 \Phi}{\partial t^2} = -\rho/\epsilon \tag{2.13}$$

although it is now almost redundant to solve (2.13) for Φ because (2.11) is available. Even if we introduce the arbitrary scalar function χ as above, we may still impose the Lorenz gauge on the new potentials $\mathbf{A}'$, Φ' provided that

$$\nabla^2 \chi - \mu\epsilon \frac{\partial^2 \chi}{\partial t^2} = 0 \tag{2.14}$$

This amounts to saying that a solution Φ of (2.13) is arbitrary to within an additive solution of the corresponding homogeneous equation. A unique solution to (2.13)—known as the inhomogeneous wave equation—is only obtained with the imposition of suitable boundary conditions on the potential Φ as required for specific problems.

Using the duality concept from Section 1.4, we can also construct fields from dual vector and scalar potentials (also called Fitzgerald potentials) $\mathbf{F}$ and Ψ when only magnetic current sources $\mathbf{M}$ are present. We put

$$-\epsilon\mathbf{E} = \nabla \times \mathbf{F} \tag{2.15}$$

$$\mathbf{H} = -\frac{\partial \mathbf{F}}{\partial t} - \nabla\Psi \tag{2.16}$$

and enforce the dual Lorenz gauge condition

$$\nabla \cdot \mathbf{F} = -\mu\epsilon \frac{\partial \Psi}{\partial t} \tag{2.17}$$

for homogeneous regions. The resulting Helmholtz equations analogous to (2.12) and (2.13) are then

$$-\nabla \times \nabla \times \mathbf{F} + \nabla\nabla \cdot \mathbf{F} - \mu\epsilon \frac{\partial^2 \mathbf{F}}{\partial t^2} = \nabla^2\mathbf{F} - \mu\epsilon \frac{\partial^2 \mathbf{F}}{\partial t^2} = -\epsilon\mathbf{M} \tag{2.18}$$

$$\nabla^2\Psi - \mu\epsilon \frac{\partial^2 \Psi}{\partial t^2} = \frac{\int \nabla \cdot \mathbf{M}\, dt}{\mu} = -\rho_m/\mu \tag{2.19}$$

In the most general case when both $\mathbf{J}$ and $\mathbf{M}$ are present at once, the field can be written as a superposition of the two kinds of potential representation (in the Lorenz gauge):

$$\mathbf{E} = -\frac{\partial \mathbf{A}}{\partial t} + \frac{\nabla\nabla \cdot \int \mathbf{A}\, dt}{\mu\epsilon} - \frac{1}{\epsilon}\nabla \times \mathbf{F} \tag{2.20}$$

$$\mathbf{H} = -\frac{\partial \mathbf{F}}{\partial t} + \frac{\nabla\nabla \cdot \int \mathbf{F}\, dt}{\mu\epsilon} + \frac{1}{\mu}\nabla \times \mathbf{A} \tag{2.21}$$

For time-harmonic fields, this representation takes the form

$$\mathbf{E} = -j\omega\mathbf{A} + \frac{\nabla\nabla \cdot \mathbf{A}}{j\omega\mu\epsilon} - \frac{1}{\epsilon}\nabla \times \mathbf{F} \tag{2.22}$$

$$\mathbf{H} = -j\omega\mathbf{F} + \frac{\nabla\nabla \cdot \mathbf{F}}{j\omega\mu\epsilon} + \frac{1}{\mu}\nabla \times \mathbf{A} \tag{2.23}$$

2.2 HERTZ VECTOR POTENTIALS

Consider a homogeneous region of space (ϵ and μ are independent of position). There is some redundancy in (2.20) and (2.21), since according to the equivalence principle it is possible that a given field in some region of space might be produced by a variety of different equivalent source distributions, consisting of either pure electric currents, pure magnetic currents, or some mixture of the two. Thus, either $\mathbf{A}$ alone, or $\mathbf{F}$ alone, or a mixture of the two might suffice to describe the field in a region free of sources, for example.

It is customary to explore the flexibility of this representation in terms of a set of related potentials known as *Hertz vectors*,[2] denoted by $\mathbf{\Pi}_e$ and $\mathbf{\Pi}_m$ and defined as follows. Suppose some set of stream potentials has been introduced to allow the representation of a field in terms of some equivalent field $\mathbf{E}_0$ and $\mathbf{H}_0$ as in Section 1.4. We now introduce Lorenz potentials $\mathbf{A}_0$ and $\mathbf{F}_0$ as in the previous section, such that $\mathbf{A}_0$ is used to express the part of $\mathbf{E}_0$ and $\mathbf{H}_0$ due to $\mathbf{J}_{eq}$, and $\mathbf{F}_0$ the part due to $\mathbf{M}_{eq}$. We define the Hertz vectors in terms of $\mathbf{A}_0$ and $\mathbf{F}_0$ as

$$\mathbf{\Pi}_e = \frac{\int \mathbf{A}_0 \, dt}{\mu\epsilon} \tag{2.24}$$

$$\mathbf{\Pi}_m = \frac{\int \mathbf{F}_0 \, dt}{\mu\epsilon} \tag{2.25}$$

after which, using (2.12), (2.18), (2.20), and (2.21), we have

$$\mathbf{E} = \mathbf{T}_m - \mu\epsilon \frac{\partial^2 \mathbf{\Pi}_e}{\partial t^2} + \nabla\nabla \cdot \mathbf{\Pi}_e - \mu \frac{\partial}{\partial t} \nabla \times \mathbf{\Pi}_m \tag{2.26}$$

$$\mathbf{H} = \mathbf{T}_e - \mu\epsilon \frac{\partial^2 \mathbf{\Pi}_m}{\partial t^2} + \nabla\nabla \cdot \mathbf{\Pi}_m + \epsilon \frac{\partial}{\partial t} \nabla \times \mathbf{\Pi}_e \tag{2.27}$$

where $\mathbf{\Pi}_e$ and $\mathbf{\Pi}_m$ satisfy

$$\begin{aligned}
\left(\nabla^2 - \mu\epsilon \frac{\partial^2}{\partial t^2} \right) \mathbf{\Pi}_e &= -\frac{\int \mathbf{J}_{eq} \, dt}{\epsilon} \\
\left(\nabla^2 - \mu\epsilon \frac{\partial^2}{\partial t^2} \right) \mathbf{\Pi}_m &= -\frac{\int \mathbf{M}_{eq} \, dt}{\mu}
\end{aligned} \tag{2.28}$$

where $\mathbf{J}_{eq}$ and $\mathbf{M}_{eq}$ are given by (1.95)-(1.96) or (1.99)-(1.100).

---

[2] See:

A. Nisbet, *Proc. Roy. Soc. London A*, vol. 231, pp. 250-263, 1955.

—, *ibid.*, vol. 240, pp. 375-381, 1957.

A. Mohsen, *Appl. Phys.*, vol. 10, pp. 53-55, 1976.

2.2.1 JUMP CONDITIONS FOR HERTZ POTENTIALS

It is possible to use Hertz potentials in two adjacent regions to represent the fields. In order to determine the potentials uniquely, we must then be able to apply appropriate boundary conditions at the interface S between the regions.

Two ways of addressing these boundary conditions are possible. An indirect way, which is useful in situations where the potential representation (2.26)-(2.27) does not hold (in the sense of generalized functions) at S due to discontinuities in ϵ and μ, even though it does throughout the adjoining volumes, is to take the limits of the field expressions (2.26)-(2.27) as we approach S from either side ($n \to 0^{\pm}$) and substitute them into the field jump conditions (1.125)-(1.128).[3] The resulting conditions lead to the desired constraints on $\mathbf{\Pi}_{e,m}$ at S without explicitly obtaining boundary or jump conditions on them.

If on the other hand (2.26)-(2.27) do hold at S, we can follow the procedure of Section 1.4 to obtain the jump conditions on $\mathbf{\Pi}_{e,m}$ at S in a much more direct fashion. Let S contain surface current densities as given by (1.114), but suppose that the arbitrary functions $\mathbf{T}_m$ and $\mathbf{T}_e$ contain no terms in either δ_S or ϑ_S. Expanding $\mathbf{\Pi}_e$ and $\mathbf{\Pi}_m$ in a form like (1.116), we substitute into (2.26)-(2.27), and make use of (1.116) itself. We readily find that

$$\mathbf{\Pi}_{e,m}^{\delta} \equiv 0$$

$$\mathbf{\Pi}_{e,m}^{\vartheta}\Big|_S \equiv 0 \tag{2.29}$$

or in other words, the Hertz vectors are continuous across these surface sources under these conditions on $\mathbf{T}_m$ and $\mathbf{T}_e$.

It is therefore only derivatives of $\mathbf{\Pi}_{e,m}$ that may be discontinuous at S. We thus postulate that

$$\mathbf{\Pi}_{e,m} = \mathbf{\Pi}_{e,m}^{R} R_S(n) + \mathbf{\Pi}_{e,m}^{s} \tag{2.30}$$

where R_S is the ramp function defined in (A.22) (with the subscript S included to emphasize the reference surface for the normal coordinate n) and $\mathbf{\Pi}_{e,m}^{s}$ represents "smoother terms" than $R_S(n)$ in the expansion whose normal derivatives are continuous at S. Substituting (2.30) into (2.28) and noting that

$$\nabla^2 \mathbf{\Pi}_{e,m} = \mathbf{\Pi}_{e,m}^{R} \delta_S(n) + \text{smoother terms}$$

we obtain

$$\mathbf{\Pi}_e^{R}\Big|_S = -\frac{\int \mathbf{J}_S \, dt}{\epsilon}$$

$$\mathbf{\Pi}_m^{R}\Big|_S = -\frac{\int \mathbf{M}_S \, dt}{\mu} \tag{2.31}$$

---

[3]Or the more general conditions in Problem 1–10.

Put another way, we have the jump conditions

$$\left.\frac{\partial \mathbf{\Pi}_e}{\partial n}\right|_{n=0^-}^{n=0^+} = -\frac{\int \mathbf{J}_S\,dt}{\epsilon}$$

$$\left.\frac{\partial \mathbf{\Pi}_m}{\partial n}\right|_{n=0^-}^{n=0^+} = -\frac{\int \mathbf{M}_S\,dt}{\mu} \tag{2.32}$$

If we allow $\mathbf{T}_m$ and $\mathbf{T}_e$ to possess step function or delta function terms, these results have to be modified somewhat.

2.2.2 TIME-HARMONIC HERTZ POTENTIALS

In the time-harmonic case, the fields are expressed in terms of Hertz potentials as:

$$\mathbf{E} = \mathbf{T}_m + k^2\mathbf{\Pi}_e + \nabla\nabla\cdot\mathbf{\Pi}_e - j\omega\mu\nabla\times\mathbf{\Pi}_m \tag{2.33}$$

$$\mathbf{H} = \mathbf{T}_e + k^2\mathbf{\Pi}_m + \nabla\nabla\cdot\mathbf{\Pi}_m + j\omega\epsilon\nabla\times\mathbf{\Pi}_e \tag{2.34}$$

where $k = \omega\sqrt{\mu\epsilon}$ and $\mathbf{\Pi}_e$ and $\mathbf{\Pi}_m$ satisfy

$$(\nabla^2 + k^2)\mathbf{\Pi}_e = -\frac{\mathbf{J}_{\text{eq}}}{j\omega\epsilon}$$

$$(\nabla^2 + k^2)\mathbf{\Pi}_m = -\frac{\mathbf{M}_{\text{eq}}}{j\omega\mu} \tag{2.35}$$

The use of the vector identity $\nabla\nabla\cdot\mathbf{\Pi} = \nabla^2\mathbf{\Pi} + \nabla\times\nabla\times\mathbf{\Pi}$ together with (2.35), (1.99), and (1.100) in (2.33) and (2.34) yields the alternate form

$$\mathbf{E} = -\frac{\mathbf{J}}{j\omega\epsilon} + \frac{1}{j\omega\epsilon}\nabla\times\mathbf{T}_e + \nabla\times\nabla\times\mathbf{\Pi}_e - j\omega\mu\nabla\times\mathbf{\Pi}_m \tag{2.36}$$

$$\mathbf{H} = -\frac{\mathbf{M}}{j\omega\mu} - \frac{1}{j\omega\mu}\nabla\times\mathbf{T}_m + \nabla\times\nabla\times\mathbf{\Pi}_m + j\omega\epsilon\nabla\times\mathbf{\Pi}_e \tag{2.37}$$

of the Hertz potential representation.

2.3 SPECIAL HERTZ POTENTIALS

Most engineering applications of the Hertz vector potentials occur in time-harmonic problems. There are several possible specializations of the Hertz potentials that permit only two scalar potential functions to model an essentially arbitrary field, and we will present two of them here.

2.3.1 WHITTAKER POTENTIALS

The arbitrariness of the stream potentials $\mathbf{T}_m$ and $\mathbf{T}_e$ allows considerable freedom in determining which components of $\mathbf{J}_{\mathrm{eq}}$ and $\mathbf{M}_{\mathrm{eq}}$ (and therefore, which components of $\mathbf{\Pi}_e$ and $\mathbf{\Pi}_m$) can be chosen to be identically zero. A widely encountered choice is that of Whittaker potentials[4] —Hertz vectors with only z-components:

$$\mathbf{\Pi}_e = \mathbf{u}_z V$$

$$\mathbf{\Pi}_m = \mathbf{u}_z U \tag{2.38}$$

To demonstrate that this is possible, we need to show that the components of $\mathbf{J}_{\mathrm{eq}}$ and $\mathbf{M}_{\mathrm{eq}}$ transverse to z (which we denote by a subscript t) can be set equal to zero. From (1.99)-(1.100), this means that we need to find vectors $\mathbf{T}_e$ and $\mathbf{T}_m$ such that

$$0 = \mathbf{J}_{\mathrm{eq},t} = \mathbf{J}_t + j\omega\epsilon\mathbf{T}_{mt} - \mathbf{u}_z \times \left[\frac{\partial \mathbf{T}_{et}}{\partial z} - \nabla_t T_{ez}\right]$$

$$0 = \mathbf{M}_{\mathrm{eq},t} = \mathbf{M}_t + j\omega\mu\mathbf{T}_{et} + \mathbf{u}_z \times \left[\frac{\partial \mathbf{T}_{mt}}{\partial z} - \nabla_t T_{mz}\right] \tag{2.39}$$

One way of accomplishing this would be to take $T_{ez} = T_{mz} = 0$, which results in a differential equation for $\mathbf{T}_{mt}$:

$$\frac{\partial^2 \mathbf{T}_{mt}}{\partial z^2} + k^2 \mathbf{T}_{mt} = \mathbf{u}_z \times \frac{\partial \mathbf{M}_t}{\partial z} + j\omega\mu\mathbf{J}_t \tag{2.40}$$

whose solution is well known:

$$\mathbf{T}_{mt} = \mathbf{T}_0(x,y)\cos kz + \mathbf{T}_1(x,y)\sin kz$$
$$+ \frac{1}{k}\int_0^z \left[\mathbf{u}_z \times \frac{\partial \mathbf{M}_t(x,y,z')}{\partial z'} + j\omega\mu\mathbf{J}_t(x,y,z')\right]\sin k(z-z')\,dz' \tag{2.41}$$

Here $\mathbf{T}_0$ and $\mathbf{T}_1$ are any transverse vector functions of (x,y) only. $\mathbf{T}_{et}$ is then obtained from

$$\mathbf{T}_{et} = -\frac{1}{j\omega\mu}\left[\mathbf{M}_t + \mathbf{u}_z \times \frac{\partial \mathbf{T}_{mt}}{\partial z}\right] \tag{2.42}$$

The importance of this result is not so much that we would actually use it to calculate $\mathbf{T}_{mt}$ and $\mathbf{T}_{et}$, but instead to demonstrate that such functions exist, thereby justifying the use of (2.38). For an interval of values of z (i.e., a layer) throughout which $\mathbf{M}_t$ and $\mathbf{J}_t$ are identically zero, we can choose $\mathbf{T}_0$ and $\mathbf{T}_1$ so that $\mathbf{T}_{mt}$ and $\mathbf{T}_{et}$ also vanish in that layer. Therefore, in such a region the Whittaker potentials can directly represent the *actual* field $\mathbf{E}$, $\mathbf{H}$ as well as the hypothetical field $\mathbf{E}_0$, $\mathbf{H}_0$.

---

[4]E. T. Whittaker, *Proc. Lond. Math. Soc.*, ser. 2, vol. 1, pp. 367-372, 1903.

This choice of $\mathbf{T}_e$ and $\mathbf{T}_m$ is not the only way to obtain a Whittaker potential representation. Let

$$\begin{aligned}
\mathbf{T}_e &= \mathbf{u}_z f_e + \nabla g_e \\
\mathbf{T}_m &= \mathbf{u}_z f_m + \nabla g_m
\end{aligned}$$

where $f_{e,m}$ and $g_{e,m}$ are scalar functions. For a Whittaker potential representation to be valid we now need

$$\begin{aligned}
\mathbf{J}_t &= -j\omega\epsilon\nabla_t g_m - \mathbf{u}_z \times \nabla_t f_e \\
\mathbf{M}_t &= -j\omega\mu\nabla_t g_e + \mathbf{u}_z \times \nabla_t f_m
\end{aligned}$$

The two-dimensional Helmholtz decomposition theorem (Appendix D) assures the existence of $f_{e,m}$ and $g_{e,m}$ that satisfy these conditions. Moreover, these functions can be chosen so that $\mathbf{T}_e$ and $\mathbf{T}_m$ vanish for all z where the conditions

$$\begin{aligned}
\mathbf{J}_t(x, y, z) &= 0 \\
\mathbf{M}_t(x, y, z) &= 0
\end{aligned}$$

hold for all (x, y) (whether or not this is a single interval in z). In particular, if $\mathbf{J}_t$ and $\mathbf{M}_t$ are proportional to $\delta(z - z_0)$, then so are $\mathbf{T}_{et}$ and $\mathbf{T}_{mt}$, while T_{ez} and T_{mz} also contain terms in $\delta'(z - z_0)$.

However the stream potentials are chosen, the fields are expressed in terms of Whittaker potentials as

$$\mathbf{E} = \mathbf{T}_m + k^2\mathbf{u}_z V + \nabla\frac{\partial V}{\partial z} - j\omega\mu\nabla \times (\mathbf{u}_z U) \tag{2.43}$$

$$\mathbf{H} = \mathbf{T}_e + k^2\mathbf{u}_z U + \nabla\frac{\partial U}{\partial z} + j\omega\epsilon\nabla \times (\mathbf{u}_z V) \tag{2.44}$$

Most often, we will want to choose $\mathbf{T}_m$ and $\mathbf{T}_e$ such that they vanish throughout regions in which we wish to compute the fields, so that the first terms in the right sides of this representation will not be present at those points in space.

2.3.2 DEBYE POTENTIALS

In certain coordinate systems, use of the Lorenz gauge conditions (2.11) and (2.17) may not be the most convenient way by which to carry out our calculations. This situation most often arises when we wish to use Hertz vectors that have only one component along a non-Cartesian coordinate direction, because the sources have components only in that direction. These components of the Hertz vectors will not themselves satisfy the scalar Helmholtz equation, but a more complicated one instead. We will illustrate how to circumvent this difficulty for geometries whose boundaries are coordinate surfaces in a spherical coordinate system.

We return to the Lorenz potentials $\mathbf{A}$ and Φ as appearing in (2.6). Let $\mathbf{J} = \mathbf{u}_r J_r$, and assume also that $\mathbf{A} = \mathbf{u}_r A_r$ has only a single component in the radial direction, but do not assume that the Lorenz gauge condition holds. We assume that μ and ϵ are independent of position and consider the time-harmonic case. Then (2.6) can be written as

$$\mathbf{u}_r \left[\nabla^2 A_r - \frac{1}{r^2} \frac{\partial}{\partial r} \left(r^2 \frac{\partial A_r}{\partial r} \right) + k^2 A_r - j\omega\mu\epsilon \frac{\partial \Phi}{\partial r} \right]$$
$$- \nabla_t \left(\frac{\partial A_r}{\partial r} \right) - j\omega\mu\epsilon \nabla_t \Phi = -\mathbf{u}_r \mu J_r \qquad (2.45)$$

where we have used the result of Problem D-2, and

$$\nabla_t f = -\mathbf{u}_r \times (\mathbf{u}_r \times \nabla f) \qquad (2.46)$$

denotes the portion of the gradient perpendicular to the r-direction. We see that the transverse components of (2.45) can be eliminated if we impose the gauge condition

$$\frac{\partial A_r}{\partial r} = -j\omega\mu\epsilon\Phi \qquad (2.47)$$

which we will call the Debye gauge [note that this is different than the Lorenz gauge (2.11) since $\nabla \cdot (\mathbf{u}_r A_r) \neq \partial A_r / \partial r$]. The remaining r-component of (2.45) becomes

$$\nabla^2 A_r - \frac{1}{r^2} \frac{\partial}{\partial r} \left(r^2 \frac{\partial A_r}{\partial r} \right) + k^2 A_r + \frac{\partial^2 A_r}{\partial r^2} = -\mu J_r \qquad (2.48)$$

which can be shown to be equivalent to

$$\left(\nabla^2 + k^2 \right) \left(\frac{A_r}{r} \right) = -\mu \frac{J_r}{r} \qquad (2.49)$$

The radial component of the electric Hertz potential is now introduced as

$$A_r = j\omega\mu\epsilon \Pi_{er} \qquad (2.50)$$

A similar development can be applied to the dual potentials $\mathbf{F} = \mathbf{u}_r F_r$ and Π_{mr}, with a radial magnetic current source M_r.

We see that the radial components of the Lorenz and Hertz potentials do not themselves obey the Helmholtz equation, in contrast to their Cartesian components. This motivates the introduction of two scalar potentials v and w called *Debye potentials* by means of the relations

$$\Pi_e = rv$$
$$\Pi_m = rw \qquad (2.51)$$

where $\mathbf{r} = x\mathbf{u}_x + y\mathbf{u}_y + z\mathbf{u}_z = r\mathbf{u}_r$ as before. We find that the Debye potentials do satisfy the Helmholtz equations

$$\left(\nabla^2 + k^2 \right) \left\{ \begin{array}{c} v \\ w \end{array} \right\} = \left\{ \begin{array}{c} -\frac{J_r}{j\omega\epsilon r} \\ -\frac{M_r}{j\omega\mu r} \end{array} \right\} \qquad (2.52)$$

The fields are then found from

$$\mathbf{E} = k^2 \mathbf{u}_r(rv) + \nabla \left[\frac{\partial(rv)}{\partial r} \right] + j\omega\mu\mathbf{u}_r \times \nabla(rw) \qquad (2.53)$$

$$\mathbf{H} = k^2 \mathbf{u}_r(rw) + \nabla \left[\frac{\partial(rw)}{\partial r} \right] - j\omega\epsilon\mathbf{u}_r \times \nabla(rv) \qquad (2.54)$$

The proof that this representation is valid in general for regions bounded by spherical surfaces (or, for that matter, for regions bounded by surfaces of constant θ or ϕ) parallels that for the Whittaker potentials, and is left as an exercise.

2.4 PROBLEMS

2–1 Instead of the Lorenz gauge (2.11), suppose we instead require the satisfaction of the *Coulomb* gauge condition:
$$\nabla \cdot \mathbf{A} = 0$$

(a) Find the equations that must be satisfied by the vector potential $\mathbf{A}$ and scalar potential Φ. Comment on the independence (or not) of these equations, and in what order they must be solved.

(b) What condition on the gauge function χ must be imposed if the fields are to be unchanged by its introduction into the potentials via (2.8) and (2.9) (compare with (2.14))?

(c) How is the significance of the scalar potential Φ different in this case from what it is in the case of the Lorenz gauge? [Hint: How many independent unknowns are there?]

2–2 Consider a region of space consisting of two homogeneous media with parameters ϵ_a, μ_a, and ϵ_b, μ_b separated by a surface S as shown in the figure.

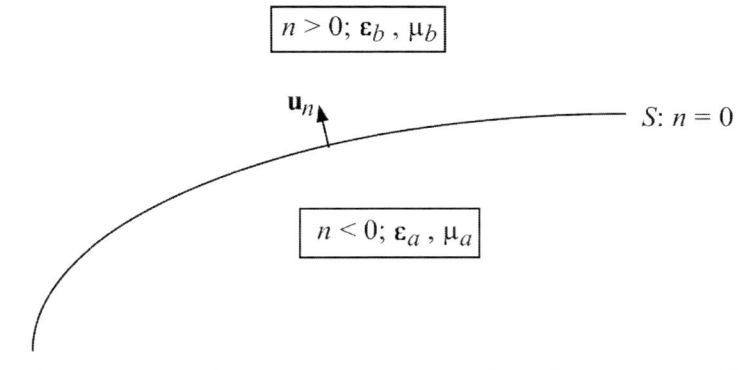

Define the material parameters to vary with position as in Problem 1-10:
$$\frac{1}{\epsilon(z)} = \frac{1}{\epsilon_a} + \left(\frac{1}{\epsilon_b} - \frac{1}{\epsilon_a}\right) \vartheta_{(\text{mat})}(z); \quad \frac{1}{\mu(z)} = \frac{1}{\mu_a} + \left(\frac{1}{\mu_b} - \frac{1}{\mu_a}\right) \vartheta_{(\text{mat})}(z)$$

With respect to the normal coordinate n shown in the figure, let the following be true in the (strong equality) sense of generalized functions:
$$\begin{aligned}
\mu\mathbf{H} &= \nabla \times \mathbf{A} \\
\mathbf{E} &= -j\omega\mathbf{A} - \nabla\Phi \\
\nabla \times \mathbf{H} &= j\omega\epsilon\mathbf{E} + \mathbf{J} \\
\nabla \times \mathbf{E} &= -j\omega\mu\mathbf{H}
\end{aligned}$$

Assume that the current density $\mathbf{J}$ and its corresponding charge density ρ contain no delta functions $\delta_S(n)$ or derivatives thereof (i.e., there are no surface charge densities ρ_S or current densities $\mathbf{J}_S$ concentrated on S). You may also assume all of the results of Problem 1-10 to be true.

(a)   Show that Φ contains no terms $\vartheta_S(n)$, $\delta_S(n)$ or derivatives thereof. Show that $\mathbf{A}$ contains no term $\delta_S(n)$ or any of its derivatives, and that the only possible step-function behavior is

$$\mathbf{A} = \mathbf{u}_n A_n^\vartheta \vartheta_S(n) + \text{smoother terms}$$

Therefore obtain the following boundary conditions for the potentials: Φ, $\mathbf{u}_n \times \mathbf{A}$, $(\nabla \times \mathbf{A})_{\text{tan}}/\mu = \mathbf{u}_n \times (\partial \mathbf{A}/\partial n - \nabla A_n)/\mu$, and $\epsilon(-j\omega A_n - \partial \Phi/\partial n)$ are all continuous at S.

(b)   If a gauge condition is not imposed on the potentials, not only are the differential equations for Φ and $\mathbf{A}$ coupled, but as seen in part (a) above, so are the boundary conditions. Unfortunately, if ϵ and μ are not constant, there is in general no gauge condition capable of decoupling both differential equations and boundary conditions (that is the same as saying that there is no choice of gauge which will fully decouple the equations for Φ and $\mathbf{A}$ in the sense of generalized functions).

Choose the following gauge condition to relate Φ to $\mathbf{A}$ in the sense of generalized functions:

$$-j\omega\Phi = \frac{1}{\epsilon}\nabla \cdot \left(\frac{1}{\mu}\mathbf{A}\right)$$

Find the boundary conditions on A_n and $\partial \Phi/\partial n$ at S that are implied by this gauge condition.[5]

2–3   Verify in detail Equation (2.29).

2–4   In a source-free region of space exterior to a general cylindrical surface S (i.e., one to which $\mathbf{u}_z$ is always tangential as shown above), where either $\mathbf{E}_{tan}$ or $\mathbf{H}_{tan}$ is given at S, show that the Whittaker potentials U and V are by themselves capable of representing the field everywhere outside S.

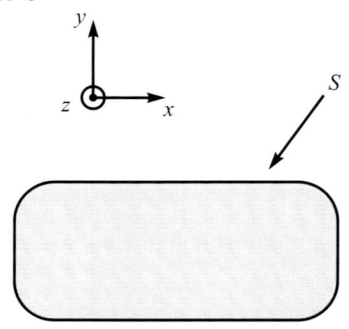

2–5   Show that the Debye potentials defined by (2.51) satisfy (2.52).

2–6   Derive (2.49) from (2.48).

---

[5]S. V. Mushenko, *Izv. VUZ Elektromekhanika*, no. 5, pp. 487-499, 1977.

3 Fundamental Properties of the Electromagnetic Field

3.1 CAUSALITY; DOMAIN OF DEPENDENCE

Solutions of the time-dependent Maxwell equations are observed (macroscopically) to obey the principle of causality. This means that if ρ_{ext} and $\mathbf{J}_{\text{ext}}$ are zero for $t < 0$, and if $\mathbf{E}$ and $\mathbf{H}$ are identically zero for $t < t_0$, where t_0 is some value of time less than zero, then $\mathbf{E}$ and $\mathbf{H}$ are in fact zero for all $t < 0$. In other words, there is no effect (field) before the cause (source). When solving initial/boundary-value problems for fields, causality must often be used to select the correct, unique solution of Maxwell's equations out of several possible ones. In this section, we study the implications of causality and their connection with the so-called radiation condition.

3.1.1 DOMAIN OF DEPENDENCE

Consider the differential form of the Poynting identity (1.14). Instead of integrating over a time-independent volume (as is customary), let us carry out the integration over a volume $V(t)$ that changes with time. Denote the closed surface bounding this volume by $S(t)$, and let it be defined by Equation (1.139) as in Chapter 1. The Poynting identity is

$$-\int_{V(t)} \mathbf{J}_{\text{tot}} \cdot \mathbf{E} \, dV = \oint_{S(t)} \left(\mathbf{E} \times \frac{\mathbf{B}}{\mu_0} \right) \cdot \mathbf{u}_n \, dS + \int_{V(t)} \frac{\partial}{\partial t} \left[\frac{B^2}{2\mu_0} + \frac{\epsilon_0 E^2}{2} \right] dV \tag{3.1}$$

Observe that we cannot simply bring the partial derivative with respect to time outside the last volume integral, since the region being integrated over varies with time. The appropriate way to extract the time derivative from the volume integral is to use identity (D.67), which introduces another term to the surface integral:

$$-\int_{V(t)} \mathbf{J}_{\text{tot}} \cdot \mathbf{E} \, dV = \oint_{S(t)} \left[\mathbf{E} \times \frac{\mathbf{B}}{\mu_0} - \mathbf{v} \left(\frac{B^2}{2\mu_0} + \frac{\epsilon_0 E^2}{2} \right) \right] \cdot \mathbf{u}_n \, dS$$
$$+ \frac{d}{dt} \int_{V(t)} \left[\frac{B^2}{2\mu_0} + \frac{\epsilon_0 E^2}{2} \right] dV \tag{3.2}$$

where $\mathbf{v}$ is the velocity of a point on the moving surface $S(t)$.

To utilize this result, let the currents $\mathbf{J}_{\text{tot}}$ be confined to a finite region of space V_J (i.e., be zero outside this region) at a given time t, and suppose that $\mathbf{J}_{\text{tot}}$ is zero for $t < 0$. Let $\mathbf{r}_0$ be a point outside V_J, and let R_0 be the shortest

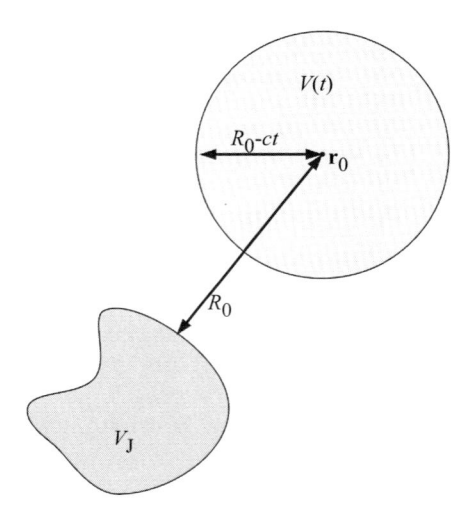

Figure 3.1 The regions V_J and $V(t)$ in the domain of dependence theorem.

distance from any point in V_J to $\mathbf{r}_0$. Choose $V(t)$ to be a sphere centered at $\mathbf{r}_0$ with radius $R(t) = R_0 - ct$, where $c = 1/\sqrt{\mu_0\epsilon_0}$ is the speed of light in vacuum. This $V(t)$ contains $\mathbf{r}_0$ for times $0 < t < R_0/c$, but contains no points of V_J (Figure 3.1). The outward unit vector $\mathbf{u}_n$ normal to $S(t)$ is in the radial direction, so for this shrinking surface the normal component of surface velocity is $\mathbf{v} \cdot \mathbf{u}_n = -c$. Therefore, for $0 < t < R_0/c$, (3.2) becomes

$$0 = \oint_{S(t)} \left[\mathbf{E} \times \frac{\mathbf{B}}{\mu_0} \cdot \mathbf{u}_n + c \left(\frac{B^2}{2\mu_0} + \frac{\epsilon_0 E^2}{2} \right) \right] dS + \frac{d}{dt} \int_{V(t)} \left[\frac{B^2}{2\mu_0} + \frac{\epsilon_0 E^2}{2} \right] dV$$

$$(3.3)$$

Now integrate from $t = 0^-$ (at which time $\mathbf{E}$ and $\mathbf{B}$ are zero) to a time t_1 between 0 and R_0/c:

$$0 = \int_{0^-}^{t_1} \oint_{S(t)} \left[\mathbf{E} \times \frac{\mathbf{B}}{\mu_0} \cdot \mathbf{u}_n + c \left(\frac{B^2}{2\mu_0} + \frac{\epsilon_0 E^2}{2} \right) \right] dS \, dt$$

$$+ \int_{V(t_1)} \left[\frac{B^2}{2\mu_0} + \frac{\epsilon_0 E^2}{2} \right] dV \qquad (3.4)$$

But

$$\oint_{S(t)} \left[\mathbf{E} \times \frac{\mathbf{B}}{\mu_0} \cdot \mathbf{u}_n + c \left(\frac{B^2}{2\mu_0} + \frac{\epsilon_0 E^2}{2} \right) \right] dS$$

$$\geq \oint_{S(t)} \left[-E\frac{B}{\mu_0} + c \left(\frac{B^2}{2\mu_0} + \frac{\epsilon_0 E^2}{2} \right) \right] dS$$

$$= \frac{1}{2\zeta_0} \oint_{S(t)} (E - cB)^2 \, dS \geq 0 \qquad (3.5)$$

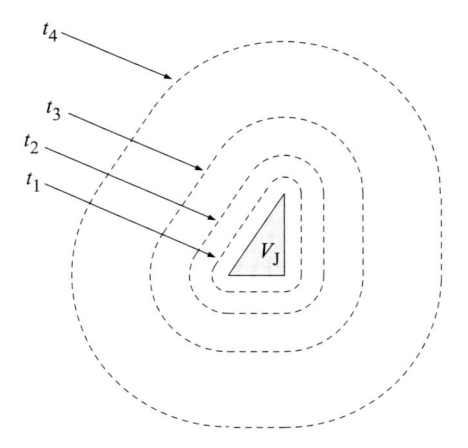

Figure 3.2 Variation of the domain of dependence at successive times $t = 0$ (when it is equal to V_J), $t_1 < t_2 < t_3 < t_4$.

where $\zeta_0 = \sqrt{\mu_0/\epsilon_0}$ is the wave impedance of free space.

Thus both terms on the right side of (3.4) are nonnegative, but the left side is zero. We conclude that both terms on the right side are zero, and since their integrands are nonnegative, these integrands must be zero throughout the domains of integration. In particular, $\mathbf{E}(\mathbf{r}_0, t) \equiv 0$ and $\mathbf{B}(\mathbf{r}_0, t) \equiv 0$ for $0 < t < R_0/c$. Physically, this means that no field can travel from the currents $\mathbf{J}_{\text{tot}}$ to an observation point faster than the speed of light c. Put another way, at a given time t, the set of points whose distance from at least one point in V_J is $\leq ct$ forms what is called the *domain of dependence* of the field on the sources.[1] Any point outside this domain cannot (yet) be influenced by the sources that were switched on at $t = 0$. An example of how the domain of dependence changes as time evolves is shown in Figure 3.2. Let R_J be the radius of the smallest sphere that contains V_J, and set the origin at the center of this sphere. When $ct \gg R_J$, the domain of dependence approaches a sphere of radius ct centered at the origin, regardless of the shape of V_J.

This argument is sufficient for the case when $\mathbf{J}_{\text{tot}}$ are exclusively impressed sources confined to a fixed region V_J. However, if some of the currents are induced by the response of the medium to the fields, we have to modify our

---

[1]See

A. Rubinowicz, *Phys. Zeits.*, vol. 27, pp. 707-710, 1926.

— , *Acta Phys. Polon.*, vol. 14, pp. 209-224, 1955.

For the wave equation, a nice treatment in English is given in

P. R. Garabedian, *Partial Differential Equations.* New York: Chelsea, 1986, Section 6.1.

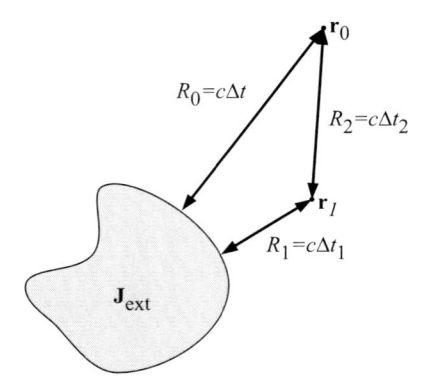

Figure 3.3 Domain of dependence theorem for a causal medium.

discussion a bit.[2] Consider a typical point $\mathbf{r}_1$ located in a causal medium as shown in Figure 3.3. There can be no induced current at $\mathbf{r}_1$ until a field has arrived at this point, a time delay of $\Delta t_1 = R_1/c$, say. For a causal medium, the induced current at $\mathbf{r}_1$ will not begin to act until time $t = \Delta t_1$, and cannot produce a field at an observation point $\mathbf{r}_0$ until a further time delay $\Delta t_2 = R_2/c$ has elapsed. Since $R_1 + R_2$ cannot be less than the direct distance R_0 from the impressed sources to $\mathbf{r}_0$, the presence of induced sources at $\mathbf{r}_1$ in a material cannot result in fields that arrive earlier than they would in free space. By superposition of the effects at all such points of the medium, we thus conclude that no signal in any causal medium may propagate faster than the speed of light in vacuum. The domain of dependence for a given source distribution in a causal medium is therefore the same as that for the same source distribution in free space.

3.1.2 MOTION OF WAVEFRONTS

In the previous subsection, we established that the fields produced by a source of finite extent vanish outside the domain of dependence at a given instant of time. However, it does not necessarily follow that the fields will be nonzero everywhere within the domain of dependence. The boundary of the smallest region outside of which the fields vanish is called a *wavefront*, and in this subsection we will study how wavefronts evolve with time.

---

[2]Mathematically more rigorous treatments of the domain of dependence for materials can be found in:

P. W. Karlsson, *Zeits. Angew. Math. Phys.*, vol. 21, pp. 246-252, 1970.

——, *J. Inst. Math. Appl.*, vol. 8, pp. 1-7, 1971.

M. Fabrizio and A. Morro, *Electromagnetism of Continuous Media.* Oxford, UK: Oxford University Press, 2003, Section 7.11.

We consider a region of free space in which no sources (surface-concentrated or not) are present. Suppose that $\mathbf{E}$ and $\mathbf{H}$ are zero on one side of a surface $S(t)$ such as the one described by (1.139), and possess nonzero step-function discontinuities $\mathbf{E}^\vartheta$ and $\mathbf{H}^\vartheta$ across $S(t)$ given by (1.142)-(1.143) of Section 1.5.3. We seek the constraints that are necessary for the function $f(\mathbf{r}, t)$ to represent an allowable (moving) discontinuity of the field (i.e., for the moving surface to be a wavefront). Setting the source terms in (1.144) and (1.147)-(1.148) equal to zero, we obtain

$$\begin{aligned}
\mathbf{u}_n \cdot \mathbf{H}^\vartheta &= 0 \\
\mathbf{u}_n \cdot \mathbf{E}^\vartheta &= 0 \\
\mathbf{u}_n \times \mathbf{H}^\vartheta &= -\epsilon_0 v_n \mathbf{E}^\vartheta \\
\mathbf{u}_n \times \mathbf{E}^\vartheta &= \mu_0 v_n \mathbf{H}^\vartheta
\end{aligned} \tag{3.6}$$

where $v_n = \mathbf{v} \cdot \mathbf{u}_n$ is the normal component of the velocity of the surface. Taking the cross product of the third of (3.6) with $\mathbf{u}_n$ and using the first and fourth equations, we get

$$\mathbf{H}^\vartheta = \mu_0 \epsilon_0 v_n^2 \mathbf{H}^\vartheta \tag{3.7}$$

Unless $\mathbf{H}^\vartheta$ (and thus also $\mathbf{E}^\vartheta$) is zero, we require the constraint

$$v_n^2 = c^2 \tag{3.8}$$

or $v_n = \pm c$, where $c = 1/\sqrt{\mu_0 \epsilon_0}$ is the velocity of light in vacuum. We see that the wavefront must travel with the speed of light.

From these results, we observe a number of properties of the field near a wavefront. From (3.6) and (3.8), we have

$$\mathbf{u}_n \times \mathbf{H}^\vartheta = \mp \frac{\mathbf{E}^\vartheta}{\zeta_0} \tag{3.9}$$

where $\zeta_0 = \sqrt{\mu_0/\epsilon_0}$ is the wave impedance of free space. Moreover, from (3.6) we find that

$$\mathbf{u}_n \cdot \mathbf{E}^\vartheta = \mathbf{u}_n \cdot \mathbf{H}^\vartheta = \mathbf{E}^\vartheta \cdot \mathbf{H}^\vartheta = 0 \tag{3.10}$$

so that the discontinuities in $\mathbf{E}$ and $\mathbf{H}$ are perpendicular to each other and to the normal to the wavefront.[3] This behavior is locally the same as that of a plane wave, but now the wavefront normal direction $\pm\mathbf{u}_n$ is not necessarily fixed.

---

[3]This result is due to:

A. E. H. Love, *Proc. London Math. Soc.*, ser. 2, vol. 1, pp. 37-62, 1904.

and the use of the generalized function technique to obtain these and further results can be found in:

D. E. Betounes, *J. Math. Phys.*, vol. 23, pp. 2304-2311, 1982.

If instead of free space the field exists in a lossless nondispersive stationary inhomogeneous medium with isotropic medium parameters $\epsilon(\mathbf{r})$ and $\mu(\mathbf{r})$, then similar analysis shows that $\mathbf{E}^\vartheta$, $\mathbf{H}^\vartheta$, and $\mathbf{u}_n$ are all perpendicular, and that

$$1 = \mu(\mathbf{r})\epsilon(\mathbf{r})v_n^2 \qquad (3.11)$$

or

$$v_n = \pm\frac{1}{\sqrt{\mu(\mathbf{r})\epsilon(\mathbf{r})}} \qquad (3.12)$$

Thus, a wavefront (or by analogous considerations, any transition between regions of zero field and nonzero field) in an inhomogeneous, isotropic source-free region must travel locally as a plane wave and with the local velocity of light in the medium. If a source is "turned on" at time $t = 0$, the field at a point in space not occupied by sources will not become nonzero until the disturbance has had time to travel at this velocity from the source region to the point of observation of the field (this was shown in the previous subsection). This is a stronger statement than mere causality, and is the basis for the derivation of the radiation condition.

3.1.3 THE RAY EQUATION AND THE EIKONAL

A slightly different analysis of the wavefront problem leads to another form of this result, more useful for some purposes. Instead of characterizing the wavefront surface $S(t)$ by the function f in (1.139), which is directly related to the surface normal, we express the equation of the surface as an explicit expression (at least locally) for the time variable:

$$t = \psi(\mathbf{r}) \qquad (3.13)$$

where the function ψ is called the *eikonal*. Our previous derivation can be repeated if in (1.142) and (1.143) we replace f by $f_0 = t - \psi(\mathbf{r})$. Now we no longer have $\nabla f = \mathbf{u}_n$ as before, but instead $\nabla f_0 = -\nabla\psi$, while $\partial f_0/\partial t = 1$. The result for the step discontinuities in the field for the case of variable μ and ϵ is:

$$\begin{aligned}
\nabla\psi \cdot \mathbf{H}^\vartheta &= 0 \\
\nabla\psi \cdot \mathbf{E}^\vartheta &= 0 \\
\nabla\psi \times \mathbf{H}^\vartheta &= -\epsilon\mathbf{E}^\vartheta \\
\nabla\psi \times \mathbf{E}^\vartheta &= \mu\mathbf{H}^\vartheta
\end{aligned} \qquad (3.14)$$

Eliminating either $\mathbf{E}^\vartheta$ or $\mathbf{H}^\vartheta$ from the third and fourth of these equations, and assuming nonzero values for the field jumps, we obtain the so-called *eikonal equation* for ψ:

$$\nabla\psi \cdot \nabla\psi = (\nabla\psi)^2 = \mu(\mathbf{r})\epsilon(\mathbf{r}) \qquad (3.15)$$

This equation is essentially equivalent to (3.11), when the time-explicit description (3.13) for the wavefront is used.

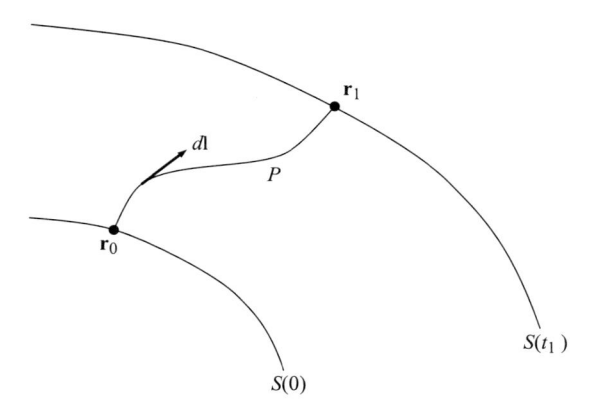

Figure 3.4 Points $\mathbf{r}_0$ and $\mathbf{r}_1$ on two wavefronts, and a path P between them.

The physical meaning of the eikonal function $\psi(\mathbf{r})$ is that it is the arrival time of a wavefront at the point $\mathbf{r}$ in space. Its gradient, we see, is directed parallel to $\mathbf{u}_n$, the unit vector normal to a wavefront at each point in space. Indeed,

$$\nabla\psi = \pm\mathbf{u}_n\sqrt{\mu\epsilon} \tag{3.16}$$

by (3.15). Consider a point $\mathbf{r}_0$ on the wavefront $S(0)$ that exists at $t = 0$, and another point $\mathbf{r}_1$ on the wavefront $S(t_1)$ that exists at $t = t_1$, as shown in Figure 3.4. The *propagation time* t_{prop} from $\mathbf{r}_0$ to $\mathbf{r}_1$ is the value of the eikonal at $\mathbf{r}_1$:

$$t_{\mathrm{prop}} = \psi(\mathbf{r}_1) = \int_{\mathbf{r}_0}^{\mathbf{r}_1} \nabla\psi \cdot d\mathbf{l} \le \int_{\mathbf{r}_0}^{\mathbf{r}_1} |\nabla\psi|\, dl = \int_{\mathbf{r}_0}^{\mathbf{r}_1} \sqrt{\mu\epsilon}\, dl \tag{3.17}$$

where the last equality follows from (3.15). Equality in (3.17) occurs only if $\nabla\psi$ is always in the same direction as $d\mathbf{l}$ along the entire path of integration. We can thus turn this result around and state *Fermat's Principle*:

The value of the *optical path length* of a path P between the fixed points $\mathbf{r}_0$ and $\mathbf{r}_1$:

$$\Theta(\mathbf{r}_0, \mathbf{r}_1) \equiv \int_P \sqrt{\mu\epsilon}\, dl \tag{3.18}$$

is a minimum only if P has $d\mathbf{l} = \mathbf{u}_n dl$ at all points along the path, where $\mathbf{u}_n$ is the unit vector perpendicular to the wavefront at each point of P. Such a path that is always directed perpendicular to wavefronts is known as a *ray*.

We will see examples of Fermat's principle when we examine the problem of reflection and refraction of the wave emitted by a point source above a material half-space in Section 5.1.

Define the refractive index as $n(\mathbf{r}) = c\sqrt{\mu(\mathbf{r})\epsilon(\mathbf{r})}$. Let $\mathbf{r}(l)$ denote the position vector $\mathbf{r}$ of a point on a ray, as a function of the arc length l measured along the ray from some conveniently chosen reference point on the ray. The arc length increases in the direction of the unit vector $\mathbf{u}_l$ tangent to the ray at each point. Then

$$\frac{d\mathbf{r}}{dl} = \mathbf{u}_l$$

by construction. But by (3.16) (renaming $\mathbf{u}_n$ as $\mathbf{u}_l$ in this context), we have

$$n(\mathbf{r})\mathbf{u}_l = c\nabla\psi \tag{3.19}$$

Now we differentiate both sides of this equation with respect to l and use (3.19):

$$\frac{d}{dl}\left[n(\mathbf{r})\mathbf{u}_l\right] = c\frac{d}{dl}\nabla\psi = c\left(\mathbf{u}_l \cdot \nabla\right)\nabla\psi = \frac{c^2}{n(\mathbf{r})}\left(\nabla\psi \cdot \nabla\right)\nabla\psi$$

From vector identities (D.6) and (D.8) we have

$$\nabla\left[(\nabla\psi)^2\right] = 2\left(\nabla\psi \cdot \nabla\right)\nabla\psi$$

so using (3.15) we can therefore write

$$\frac{d}{dl}\left[n(\mathbf{r})\mathbf{u}_l\right] = \frac{c^2}{2n(\mathbf{r})}\nabla\left[(\nabla\psi)^2\right] = \frac{c^2}{2n(\mathbf{r})}\nabla\left(\frac{n^2(\mathbf{r})}{c^2}\right) = \nabla n$$

We thus obtain the differential equation of a ray[4]

$$\frac{d}{dl}\left\{n[\mathbf{r}(l)]\frac{d\mathbf{r}(l)}{dl}\right\} = \nabla n[\mathbf{r}(l)] \tag{3.20}$$

or, more compactly,

$$\frac{d}{dl}(n\mathbf{u}_l) = \nabla n \tag{3.21}$$

The physical meaning of this result is that the direction $\mathbf{u}_l$ of a ray bends (or *refracts*) towards the direction of ∇n, i.e., towards the direction of increasing n (the law of refraction). As a simple example, suppose that $n(\mathbf{r}) = $ constant, so that $\nabla n = 0$. Then by (3.21), $d\mathbf{u}_l/dl = 0$, and thus $\mathbf{u}_l$ has a constant direction. We conclude that a ray in a homogeneous medium is always a straight line. This follows also from Fermat's principle.

---

[4]See

M. Kline and I. W. Kay, *Electromagnetic Theory and Geometrical Optics*. New York: Krieger, 1979.

M. Born and E. Wolf, *Principles of Optics*, 7th ed. Cambridge, UK: Cambridge University Press, 2002, Chapter 3.

A further example is provided by propagation in a medium in which the refractive index is a function only of one Cartesian coordinate, let us say $n = n(z)$. Then $\nabla n = \mathbf{u}_z n'(z)$, and we can evaluate the derivative of the quantity $\mathbf{u}_z \times n\mathbf{u}_l$ as

$$\frac{d}{dl}(\mathbf{u}_z \times n\mathbf{u}_l) = \mathbf{u}_z \times \frac{d}{dl}(n\mathbf{u}_l) = \mathbf{u}_z \times \nabla n = 0 \tag{3.22}$$

or in other words,

$$n\mathbf{u}_z \times \mathbf{u}_l = \text{constant} \tag{3.23}$$

If it is assumed for simplicity that the wavefront has no variation in the y-direction, then we write $\mathbf{u}_l = \mathbf{u}_x \sin\theta(z) + \mathbf{u}_z \cos\theta(z)$, where $\theta(z)$ is the angle between the ray and the z-axis. Then (3.23) gives

$$n(z)\sin\theta(z) = \text{constant} \tag{3.24}$$

which is none other than Snell's law, generalized to a continuously layered medium.

We should mention in concluding this section that the time-domain behavior of a field can in general be represented by the inverse Fourier transform of the frequency domain field. When a field has a temporal discontinuity such as a wavefront, its behavior is necessarily dominated by the high frequency constituents of the Fourier spectrum. Thus, our results for rays derived above are often also used to describe high-frequency fields, whose wavelength is small compared to other physical dimensions of the environment. The reader is referred to the previously given references for details.

3.2 PASSIVITY AND UNIQUENESS

3.2.1 TIME-DOMAIN THEOREMS

An important property of the electromagnetic field established in a passive medium is that it is a passive response in a sense similar to that of (1.69) for the Joule losses of the medium. To make this precise, consider a set of impressed sources $\mathbf{J}_{\text{ext}}$ in some bounded region of space that are switched on at $t = 0$. Now we apply the differential Poynting theorem (1.14), and integrate over all space (not just a finite volume). When we convert the integral of the divergence term to a surface integral (on a surface that tends to infinity) the domain of dependence theorem gives that the field on this surface is zero for any finite time t, so the surface integral vanishes. Separating the total current into impressed and induced parts, and integrating from $t = 0^-$ to some $t_1 > 0$, we get an expression for the work $W(t_1)$ done over this time interval by the impressed sources in establishing the field:

$$W(t_1) \equiv -\int_{0-}^{t_1}\int \mathbf{E} \cdot \mathbf{J}_{\text{ext}} dV\, dt = \int_{0-}^{t_1}\int \mathbf{E} \cdot \mathbf{J}_{\text{ind}} dV\, dt + \int \left[\frac{B^2}{2\mu_0} + \frac{\epsilon_0 E^2}{2}\right] dV\bigg|_{t=t_1} \tag{3.25}$$

The term

$$\int_{0^-}^{t_1} \int \mathbf{E} \cdot \mathbf{J}_{\text{ind}} \, dV \, dt$$

is nonnegative if the material is passive. The second is nonnegative by inspection (the integrand is nonnegative everywhere). Thus the work done by the impressed sources is always nonnegative, and we say the "load" presented to the impressed sources by the electromagnetic environment is passive. We emphasize that this does not mean that the instantaneous power

$$-\int \mathbf{E} \cdot \mathbf{J}_{\text{ext}} \, dV$$

being delivered by the impressed sources can never be negative. The environment can sometimes give back some of the energy that was delivered to it by the impressed sources. This returned energy cannot, however, be larger than the amount supplied by the sources.

A slight variation on the foregoing technique allows us to obtain a uniqueness theorem for the electromagnetic field. This theorem tells us exactly how much information about the field needs to be specified in order for only one field to fit all the conditions. Suppose that two different fields both satisfied Maxwell's equations (1.1)-(1.4) in the same environment (medium and boundaries) and with the same impressed sources $\mathbf{J}_{ext}$. Assume as usual that $\mathbf{J}_{ext} = 0$ for $t \leq 0$. Call these two fields $\mathbf{E}_1, \mathbf{B}_1$ and $\mathbf{E}_2, \mathbf{B}_2$, respectively. The principle of causality asserts that these fields are all identically zero for $t \leq 0$. Their difference,

$$\mathbf{E}_d = \mathbf{E}_1 - \mathbf{E}_2$$
$$\mathbf{B}_d = \mathbf{B}_1 - \mathbf{B}_2$$

satisfies Maxwell's equations with $\mathbf{J}_{ext} = 0$. Then (3.25) applies with $W(t) = 0$ for all t. Since the two terms on the right side are nonnegative, they must be zero, and therefore their integrands (which are nonnegative) must be zero throughout space. Thus the two fields must in fact be identical throughout space for all $t > 0$. Were this not the case, the right side of (3.25) would have to be positive.

Suppose that we integrated the Poynting identity not over all space, but only over some volume V bounded by a closed surface S as shown in Figure 3.5. Instead of (3.25), we will now have an additional surface integral term containing the Poynting vector:

$$0 = \int_{0^-}^{t_1} \int \mathbf{E}_d \cdot \mathbf{J}_{d,\text{ind}} \, dV \, dt + \int \left[\frac{B^2}{2\mu_0} + \frac{\epsilon_0 E_d^2}{2} \right] dV \bigg|_{t=t_1}$$
$$+ \int_{0^-}^{t_1} \oint_S \left(\mathbf{E}_d \times \frac{\mathbf{B}_d}{\mu_0} \right) \cdot \mathbf{u}_n \, dS \, dt \tag{3.26}$$

If either $\mathbf{u}_n \times \mathbf{E}_1 = \mathbf{u}_n \times \mathbf{E}_2$ or $\mathbf{u}_n \times \mathbf{B}_1 = \mathbf{u}_n \times \mathbf{B}_2$ on S, then the surface integral on the right side of (3.26) will be zero. The remaining terms are

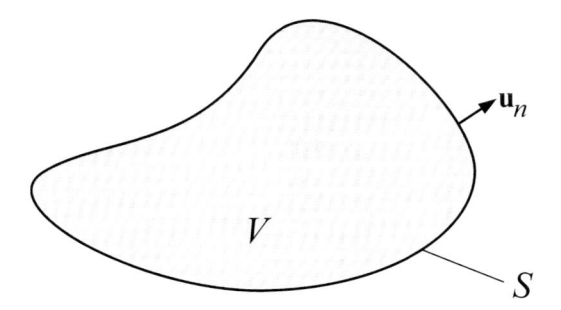

Figure 3.5 Geometry for the uniqueness theorem.

nonnegative and prove that the two fields must be the same, as before. Thus, one way of insuring a uniquely determined field everywhere is to require that either tangential $\mathbf{E}_1$ be the same as tangential $\mathbf{E}_2$ or that tangential $\mathbf{B}_1$ be the same as tangential $\mathbf{B}_2$ on S. This would be true if both fields obeyed the boundary condition for a PEC at S, for example.

3.2.2 RADIATION CONDITIONS

While causality in the time domain is a relatively straightforward concept, its implications in the frequency domain are more subtle. Let us begin by finding far-field expressions for $\mathbf{E}$ and $\mathbf{H}$ using somewhat heuristic arguments.[5]

We have seen that for $r \gg R_J$, the wavefront of the field of a current distribution confined to within a distance of R_J from the origin is approximately a sphere. If we consider for a moment only the "partial" field produced by the current at a single instant of time and at a single point in V_J, we see that it must follow an expanding spherical wavefront, and from (3.8)-(3.10) must obey:

$$\mathbf{E} \simeq -\zeta \mathbf{u}_r \times \mathbf{H}; \qquad \mathbf{H} \simeq \frac{1}{\zeta} \mathbf{u}_r \times \mathbf{E} \qquad (3.27)$$

in the limit as $r \to \infty$, where $\zeta = \sqrt{\mu/\epsilon}$ is the wave impedance of the medium. Moreover, since the spherical wavefront is expanding (moving outward), these partial fields must depend on time through the combination of variables $t-r/c$. Finally, these partial fields are produced by the expenditure of a finite amount of energy from the source, so the energy at a finite instant of time t radiated through the spherical surface S_r of radius r centered at the origin must be finite for any value of r:

$$\lim_{r \to \infty} \int \oint_{S_r} \mathbf{E} \times \mathbf{H} \cdot \mathbf{u}_r \, dS \, dt < \infty \qquad (3.28)$$

---

[5]In Chapter 4 we will show more rigorously how to obtain the far-field potentials of a spatially bounded current source in the time domain; see expressions (4.83) and (4.85). The far-fields themselves are easily found from these potentials.

Since the surface element on the sphere in spherical coordinates is $dS = r^2 \sin\theta\, d\theta d\phi$, this means that $\mathbf{E}$ and $\mathbf{H}$ must decay at least as fast as $1/r$ when $r \to \infty$. By superposition, these properties are also true of the total field produced by an arbitrary bounded current source. Collecting these results, we see that the far-field of such a source must have the form:

$$\mathbf{E}(\mathbf{r}, t) \sim \frac{\mathbf{F}_E(\mathbf{u}_r, t - r/c)}{r}; \quad \mathbf{H}(\mathbf{r}, t) \sim \frac{\mathbf{F}_H(\mathbf{u}_r, t - r/c)}{r}; \quad (r \to \infty) \quad (3.29)$$

If we examine how the principle of causality entered into this discussion, we see that its effect was to eliminate the possibility that fields with incoming wavefronts could have been produced by any part of the sources. Had such contributions to the field been present, they would have possessed a dependence on t through the combination of variables $t + r/c$, and hence introduced terms of the form[6]

$$\mathbf{E}_{\text{adv}}(\mathbf{r}, t) \sim \frac{\mathbf{F}_{E,\text{adv}}(\mathbf{u}_r, t + r/c)}{r}; \quad \mathbf{H}_{\text{adv}}(\mathbf{r}, t) \sim \frac{\mathbf{F}_{H,\text{adv}}(\mathbf{u}_r, t + r/c)}{r}; \quad (r \to \infty) \tag{3.30}$$

into the expressions for the far-fields $\mathbf{E}$ and $\mathbf{H}$. These so-called *advanced* solutions of Maxwell's equations (in contrast to the *retarded* solutions that obey (3.29)) are necessarily zero for all time, if forced to obey the principle of causality.

Now, if (3.29) and (3.30) are subjected to a Fourier transform (1.70), and standard properties of the transform are invoked, we have in the frequency domain that

$$\hat{\mathbf{E}}(\mathbf{r}, \omega) \sim \hat{\mathbf{F}}_E(\mathbf{u}_r, \omega) \frac{e^{-j\omega\sqrt{\mu\epsilon}r}}{r}[1 + O(r^{-1})]$$

$$\hat{\mathbf{H}}(\mathbf{r}, \omega) \sim \hat{\mathbf{F}}_H(\mathbf{u}_r, \omega) \frac{e^{-j\omega\sqrt{\mu\epsilon}r}}{r}[1 + O(r^{-1})] \tag{3.31}$$

as $r \to \infty$, while

$$\hat{\mathbf{E}}_{\text{adv}}(\mathbf{r}, \omega) \sim \hat{\mathbf{F}}_{E,\text{adv}}(\mathbf{u}_r, \omega) \frac{e^{+j\omega\sqrt{\mu\epsilon}r}}{r}[1 + O(r^{-1})]$$

$$\hat{\mathbf{H}}_{\text{adv}}(\mathbf{r}, \omega) \sim \hat{\mathbf{F}}_{H,\text{adv}}(\mathbf{u}_r, \omega) \frac{e^{+j\omega\sqrt{\mu\epsilon}r}}{r}[1 + O(r^{-1})] \tag{3.32}$$

---

[6]Other treatments of the form of the far field limit of a causal solution of Maxwell's equations are found in

H. E. Moses, R. J. Nagem, and G. v. H. Sandri, *J. Math. Phys.*, vol. 33, pp. 86-101, 1992.

A. D. Yaghjian and T. B. Hansen, *J. Appl. Phys.*, vol. 79, pp. 2822-2830, 1996.

T. B. Hansen and A. D. Yaghjian, *Plane-Wave Theory of Time-Domain Fields*. New York: IEEE Press, 1999.

in the same limit. The absence of advanced solutions in the time domain necessarily implies that

$$\hat{\mathbf{F}}_{E,\text{adv}}(\mathbf{u}_r, \omega) \equiv 0; \qquad \hat{\mathbf{F}}_{H,\text{adv}}(\mathbf{u}_r, \omega) \equiv 0$$

in the frequency domain. If we are working on a time-harmonic problem in the frequency domain from the start, it would be nice to have a way to eliminate solutions of the form (3.32) directly, without having intermediate recourse to calculations in the time domain.

This is often done using what are called *radiation conditions*. To obtain these conditions, we substitute (3.31) or (3.32) into Maxwell's equations (1.75)-(1.78) written in spherical coordinates. Examining the limits as $r \to \infty$ and grouping like powers of r^{-1}, we find that

$$\begin{aligned} \mathbf{u}_r \times \hat{\mathbf{F}}_H &= -\frac{1}{\zeta}\hat{\mathbf{F}}_E \\ \mathbf{u}_r \times \hat{\mathbf{F}}_E &= +\zeta\hat{\mathbf{F}}_H \end{aligned} \tag{3.33}$$

[which of course also follow from (3.27)], while

$$\begin{aligned} \mathbf{u}_r \times \hat{\mathbf{F}}_{H,\text{adv}} &= +\frac{1}{\zeta}\hat{\mathbf{F}}_{E,\text{adv}} \\ \mathbf{u}_r \times \hat{\mathbf{F}}_{E,\text{adv}} &= -\zeta\hat{\mathbf{F}}_{H,\text{adv}} \end{aligned} \tag{3.34}$$

Note that these equations also imply that

$$\mathbf{u}_r \cdot \hat{\mathbf{F}}_E = \mathbf{u}_r \cdot \hat{\mathbf{F}}_H = 0$$

and similarly for $\hat{\mathbf{F}}_{E,\text{adv}}$ and $\hat{\mathbf{F}}_{H,\text{adv}}$. Our radiation condition must therefore ensure that (3.33) rather than (3.34) holds, and additionally allow for the limiting nature of (3.31) as $r \to \infty$.

This is achieved by enforcing the Silver-Müller radiation conditions:[7]

$$\begin{aligned} \lim_{r\to\infty} r\left[\mathbf{u}_r \times \hat{\mathbf{H}} + \frac{1}{\zeta}\hat{\mathbf{E}}\right] &= 0 \\ \lim_{r\to\infty} r\left[\mathbf{u}_r \times \hat{\mathbf{E}} - \zeta\hat{\mathbf{H}}\right] &= 0 \end{aligned} \tag{3.35}$$

We often say that solutions of Maxwell's equations obeying (3.35) (or (3.31)) are *outgoing* waves, while those behaving like (3.32) are *incoming* waves. It is important to recognize that, although the solution selected by the radiation condition does indeed have an outgoing phase as $r \to \infty$ in the present case (where μ and ϵ are independent of $\mathbf{r}$ and ω), it is causality that must

---

[7]The Silver-Müller conditions have been shown to imply also that $\mathbf{E}$ and $\mathbf{H}$ are $O(r^{-1})$ as $r \to \infty$ by C. H. Wilcox, *Comm. Pure Appl. Math.*, vol. 9, pp. 115-134, 1956.

finally determine the proper solution of any physical problem. In cases where dispersive, absorbing or even amplifying media are present, or where boundaries between different media extend to infinity, very different forms of the radiation condition may apply.[8]

A scalar form of the radiation condition can be obtained for fields or potentials that are solutions of Helmholtz's equation based as above on causality arguments. For the electric scalar potential Φ obeying (2.13), for example, it takes the form

$$\lim_{r \to \infty} r \left(\frac{\partial \Phi}{\partial r} + jk\Phi \right) = 0 \tag{3.36}$$

A similar form holds for other potentials, even vectors such as $\mathbf{A}$, $\mathbf{F}$, $\mathbf{\Pi}_e$, and $\mathbf{\Pi}_m$.

3.2.3 TIME-HARMONIC THEOREMS

In this section we will give two uniqueness theorems for solutions of the time-harmonic version of Maxwell's equations. Such theorems tell us how much we need to say about a field before we can guarantee that it, and no other field satisfying all the same conditions, is the solution of the problem. These theorems have among their uses application to the further discussion of equivalence principles that we will discuss in the next section. We will assume throughout this subsection that $\omega \neq 0$, so that the Maxwell divergence equations (1.75)-(1.76) are implied by the curl equations (1.77)-(1.78) and need not be considered further.

Before tackling the uniqueness theorems, it is useful to present the time-harmonic form of the passivity theorem analogous to the one for time-domain fields obtained in Section 3.2.1. Before proceeding we will need to discuss first the *principle of limiting amplitude*. Suppose that no magnetic sources are present, and the impressed electric current density is time-harmonic and begins to act at $t = 0$ on a system containing only linear materials:

$$\mathbf{J}_{\text{ext}} = \vartheta(t)\text{Re}\left[\hat{\mathbf{J}}_{\text{ext}}(\mathbf{r})e^{j\omega t}\right] \tag{3.37}$$

The principle of limiting amplitude holds if, after a long time has passed, the fields produced by this source settle down to a time-harmonic response of frequency ω:

$$\mathbf{E} \to \text{Re}\left[\hat{\mathbf{E}}(\mathbf{r})e^{j\omega t}\right] \qquad \text{as } t \to \infty \tag{3.38}$$

etc. This is what we implicitly assume when dealing with time-harmonic situations. After all, when we turn on a generator, we expect the response of

---

[8]See, e.g.,

G. N. Vinokurov and V. I. Zhulin, *Sov. J. Quantum Electron.*, vol. 12, pp. 329-333, 1982.

B. Zhang, *Proc. Roy. Soc. Edinburgh*, vol. 128A, pp. 173-192, 1998.

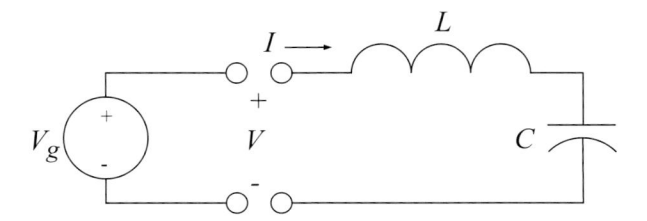

Figure 3.6 Resonant LC circuit violates the principle of limiting amplitude and uniqueness if $\omega = 1/\sqrt{LC}$.

a linear system to be time-harmonic as well, at least after the generator has had time to "warm up" and all transients have died away.

However, it is easy to construct examples where the principle of limiting amplitude is not true. Consider the lossless resonant circuit shown in Figure 3.6, with the resonant frequency $\omega = 1/\sqrt{LC}$. If a time-harmonic voltage generator $V_g = V_0 \sin \omega t$ is connected to this circuit and switched on at $t = 0$, and if the initial values of inductor current and capacitor voltage are zero at $t = 0$, then it is a standard result of elementary circuit theory that the inductor current grows without bound as it oscillates:

$$I(t) = \frac{V_0 t}{2L} \sin \omega t$$

In this example, where a system is driven at one of its resonant frequencies, the principle of limiting amplitude does not hold. It is thus an important problem in the electromagnetic case to ascertain exactly when such resonant response is absent and the principle can be accepted as valid.[9] In all practical cases when at least a small loss is present (including "losses" due to energy radiated towards infinity), the limiting amplitude principle can be assumed to hold, and we will do so from here on.

Now let the time t_1 be large enough that the field response to (3.37) can be assumed to have reached the time-harmonic steady state. Then if $T = 2\pi/\omega$ is the period, the energy delivered by the impressed sources at a time equal to an integer number N periods later than t_1 can be written

$$W(t_1 + NT) \simeq W(t_1) - NT\frac{1}{2}\mathrm{Re}\int \hat{\mathbf{E}} \cdot \hat{\mathbf{J}}_{\mathrm{ext}}^* \, dV \qquad (3.39)$$

[9]See

C. H. Wilcox, *Arch. Rat. Mech. Anal.*, vol. 3, pp. 133-148, 1959.

N. D. Kazarinoff and R. K. Ritt, *ibid.*, vol. 5, pp. 177-186, 1960.

C. S. Morawetz, *Comm. Pure Appl. Math.*, vol. 15, pp. 349-361, 1962.

E. C. Zachmanoglou, *Arch. Rat. Mech. Anal.*, vol. 14, pp. 312-325, 1963.

C. S. Morawetz, *Comm. Pure Appl. Math.*, vol. 18, pp. 183-189, 1965.

C. O. Bloom, *J. Diff. Eq.*, vol. 19, pp. 296-329, 1975.

by appealing to (3.25) and the usual time-averaging procedure for time-harmonic fields. But $W(t_1 + NT)$ must be nonnegative for any N, and in particular if we take $N > 0$ large enough, we find that we must have

$$-\frac{1}{2}\mathrm{Re} \int \hat{\mathbf{E}} \cdot \hat{\mathbf{J}}^*_{\mathrm{ext}} \, dV \geq 0 \qquad (3.40)$$

This is the passivity property of the response to a time-harmonic source: in free space there can be no time average power delivered to an impressed source by the field. We can extend this result to an impressed source in a passive medium in a similar way to the time-domain case in Section 3.2.1.

There are two versions of uniqueness theorems for the time-harmonic Maxwell equations that will be stated here and proved in the remainder of this section. The first uniqueness theorem applies to fields in a bounded region:

> Uniqueness Theorem I: *Let V be a bounded region of space with boundary surface S. Suppose that $\mathbf{E}_1$, $\mathbf{H}_1$ and $\mathbf{E}_2$, $\mathbf{H}_2$ are both solutions of Maxwell's curl equations (1.77)-(1.78) in V, and that both have identical values of tangential $\mathbf{E}$ and $\mathbf{H}$ on S. Both solutions are assumed to have been produced by the same source current densities $\mathbf{J}$ and $\mathbf{M}$. $\mathbf{E}_1$, $\mathbf{H}_1$ and $\mathbf{E}_2$, $\mathbf{H}_2$ are identical.*

On the other hand, the second uniqueness theorem deals with fields in an unbounded region of space:

> Uniqueness Theorem II: *Two fields obeying Maxwell's equations for $\omega \neq 0$ in the same infinite lossless region V (with the same constitutive parameters and the same sources), the same boundary conditions on a bounded scatterer surface S_0, and the radiation conditions (3.35), must be identical in all of V.*

The proofs of both uniqueness theorems will require a lemma that we will prove using Green's identities in Chapter 7 for the case when ϵ and μ are independent of position. The proof in the general case is rather tedious and technical, and the reader is referred to the literature for details.[10] It is as follows:

> Null-Field Lemma: *Let V be a bounded region of space with boundary surface S. If Maxwell's curl equations (1.77)-(1.78) with $\mathbf{J} = \mathbf{M} = 0$ and the constitutive equations (1.79)-(1.80) hold throughout V, and both tangential $\mathbf{E}$ and tangential $\mathbf{H}$ vanish on S, then $\mathbf{E}$ and $\mathbf{H}$ vanish identically in all of V.*

---

[10]C. Müller, *Foundations of the Mathematical Theory of Electromagnetic Waves*, Berlin, Springer-Verlag, 1969, pp. 267-281.

From this lemma we can easily get our Uniqueness Theorem I. The proof is straightforward: the difference fields

$$\mathbf{E}_d = \mathbf{E}_1 - \mathbf{E}_2$$

and

$$\mathbf{H}_d = \mathbf{H}_1 - \mathbf{H}_2$$

then obey Maxwell's equations with $\mathbf{J}$ and $\mathbf{M}$ equal to zero, and our lemma gives us that $\mathbf{E}_d$ and $\mathbf{H}_d$ must be zero, from which the uniqueness theorem follows immediately.

Notice that if *only* tangential $\mathbf{E}$ *or* tangential $\mathbf{H}$ were identical, the uniqueness would not be guaranteed, since, for example, if S is a perfectly conducting surface and V contains only lossless material, then at certain real frequencies ω there can exist nonzero, source-free resonant cavity mode fields in V, and any multiple of such a field could be added to a given solution of Maxwell's equations, still leaving a valid solution. The circuit example shown in Figure 3.6 illustrates this point nicely. If we constrain $V = 0$ at the terminals of the LC section on the right by connecting a short circuit at its terminals, it is still possible to have a nonzero current flowing through the network at its resonant frequency $\omega = 1/\sqrt{LC}$. Only by additionally forcing $I = 0$ do we guarantee zero voltages and currents throughout the network.

Next, we prove the second uniqueness theorem for solutions of Maxwell's equations in loss-free infinite space (or in the exterior of a compact obstacle).[11] Suppose again that $\mathbf{E}_1$, $\mathbf{H}_1$ and $\mathbf{E}_2$, $\mathbf{H}_2$ are both solutions of Maxwell's equations (1.77)-(1.78) in the same region of space, namely the volume V bounded by some bounded surface S_0 and a spherical surface S_R of (large) radius R as shown in Figure 3.7. We assume that for r greater than some R_0, ϵ and μ are real and independent of position, and that S_0 lies entirely inside S_{R_0}. Both solutions are assumed to have been produced by the same source current densities $\mathbf{J}$ and $\mathbf{M}$, and we wish to determine under what conditions we are assured that these two fields must be the same. The difference fields

$$\mathbf{E}_d = \mathbf{E}_1 - \mathbf{E}_2$$

and

$$\mathbf{H}_d = \mathbf{H}_1 - \mathbf{H}_2$$

again obey Maxwell's equations with $\mathbf{J}$ and $\mathbf{M}$ equal to zero.

As in Section 3.2, we start with Poynting's theorem (but now its complex version (1.85)) for the difference fields:

$$\nabla \cdot (\mathbf{E}_d \times \mathbf{H}_d^*) = j\omega(\epsilon|\mathbf{E}_d|^2 - \mu|\mathbf{H}_d|^2) \tag{3.41}$$

---

[11] For the case of media with losses (i.e., $\sigma > 0$), the proof does not require the use of the radiation condition—see Problem 3–4.

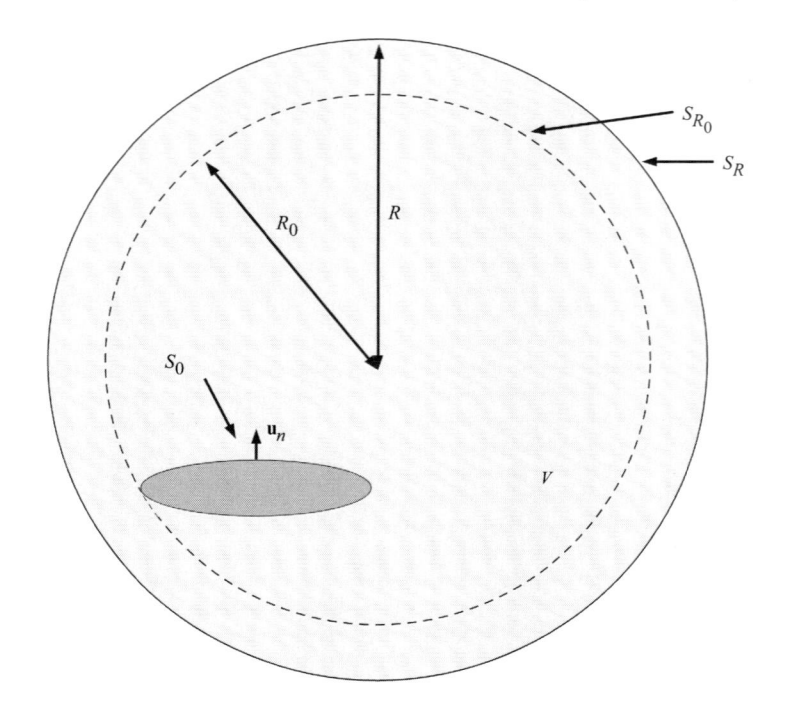

Figure 3.7 Volume V associated with the proof of the uniqueness theorem.

again assuming the validity of (1.79)-(1.80) with ϵ and μ real. Integrating (3.41) over V and using the divergence theorem gives

$$\oint_{S_R} \mathbf{E}_d \times \mathbf{H}_d^* \cdot \mathbf{u}_r \, dS - \oint_{S_0} \mathbf{E}_d \times \mathbf{H}_d^* \cdot \mathbf{u}_n \, dS = j\omega \int_V (\epsilon |\mathbf{E}_d|^2 - \mu |\mathbf{H}_d|^2) \, dV \quad (3.42)$$

where $\mathbf{u}_n$ is the *outward* unit normal from S_0, pointing *into* V. Now,

$$\mathrm{Re} \oint_{S_R} \mathbf{E}_d \times \mathbf{H}_d^* \cdot \mathbf{u}_r \, dS$$

is independent of R as can be seen from (3.42) at once. But

$$
\begin{aligned}
4\sqrt{\frac{\mu}{\epsilon}}\mathrm{Re}(\mathbf{E}_d \times \mathbf{H}_d^* \cdot \mathbf{u}_r) \;=\;& |\mathbf{E}_d|^2 + |\mathbf{u}_r \times \mathbf{E}_d|^2 \\
+\;& \frac{\mu}{\epsilon}\left[|\mathbf{H}_d|^2 + |\mathbf{u}_r \times \mathbf{H}_d|^2 \right] \\
-\;& \frac{\mu}{\epsilon}|\mathbf{u}_r \times \mathbf{H}_d + \sqrt{\frac{\epsilon}{\mu}}\mathbf{E}_d|^2 \\
-\;& |\mathbf{u}_r \times \mathbf{E}_d - \sqrt{\frac{\mu}{\epsilon}}\mathbf{H}_d|^2 \qquad (3.43)
\end{aligned}
$$

and if both $\mathbf{E}_1$, $\mathbf{H}_1$ and $\mathbf{E}_2$, $\mathbf{H}_2$ obey the radiation conditions, then as $R \to \infty$, the last two terms contribute a vanishingly small amount to the surface integral by (3.35), so we have

$$\lim_{R\to\infty} \text{Re} \oint_{S_R} \mathbf{E}_d \times \mathbf{H}_d^* \cdot \mathbf{u}_r \, dS$$

$$= \lim_{R\to\infty} \frac{1}{4}\sqrt{\frac{\epsilon}{\mu}} \oint_{S_R} \left\{ |\mathbf{E}_d|^2 + |\mathbf{u}_r \times \mathbf{E}_d|^2 + \frac{\mu}{\epsilon}[|\mathbf{H}_d|^2 + |\mathbf{u}_r \times \mathbf{H}_d|^2] \right\} dS$$

$$= \lim_{R\to\infty} \frac{1}{2} \oint_{S_R} \left[\sqrt{\frac{\epsilon}{\mu}}|\mathbf{E}_{d,\tan}|^2 + \sqrt{\frac{\mu}{\epsilon}}|\mathbf{H}_{d,\tan}|^2 \right] dS \qquad (3.44)$$

since rE_r and rH_r go to zero as $r \to \infty$.

Thus, we know that

$$\text{Re} \oint_{S_R} \mathbf{E}_d \times \mathbf{H}_d^* \cdot \mathbf{u}_r \, dS = \lim_{R\to\infty} \text{Re} \oint_{S_R} \mathbf{E}_d \times \mathbf{H}_d^* \cdot \mathbf{u}_r \, dS \geq 0 \qquad (3.45)$$

If we also knew that

$$\text{Re} \oint_{S_0} \mathbf{E}_d \times \mathbf{H}_d^* \cdot \mathbf{u}_n \, dS \leq 0 \qquad (3.46)$$

we could then conclude that in fact both (3.45) and (3.46) are exactly zero, since the real part of the left side of (3.42) must be zero. This can happen in a number of circumstances:

i) S_0 is a perfect conductor; $\mathbf{u}_n \times \mathbf{E}|_{S_0} = 0$.
ii) S_0 is an ideal magnetic wall; $\mathbf{u}_n \times \mathbf{H}|_{S_0} = 0$.
iii) S_0 is an impedance surface; $\mathbf{u}_n \times \mathbf{E}|_{S_0} = Z_S \mathbf{u}_n \times (\mathbf{u}_n \times \mathbf{H})|_{S_0}$ with $\text{Re}(Z_S) \geq 0$.
iv) S_0 is composed of some sections of perfect conductor, some sections of ideal magnetic wall, and the rest impedance surface as in iii).

Under these conditions we have from (3.44) that

$$\lim_{R\to\infty} \frac{1}{2} \oint_{S_R} \left[\sqrt{\frac{\epsilon}{\mu}}|\mathbf{E}_{d,\tan}|^2 + \sqrt{\frac{\mu}{\epsilon}}|\mathbf{H}_{d,\tan}|^2 \right] dS = 0 \qquad (3.47)$$

But Rellich's theorem (Appendix C) states that (3.47) implies that $\mathbf{E}_d$ and $\mathbf{H}_d$ must vanish identically for all $r > R_0$. Thus in particular $\mathbf{E}_{d,\tan}$ and $\mathbf{H}_{d,\tan}$ are zero at S_{R_0}. The lemma from the beginning of this section then shows that $\mathbf{E}_d$ and $\mathbf{H}_d$ must in fact vanish in all of V. We then have Uniqueness Theorem II.

An analogous uniqueness theorem for solutions of Helmholtz's equation can be stated, its proof being very similar to that of the vector version outlined before. For a finite volume V interior to a closed surface S we find that a solution Φ of Helmholtz's equation (2.13) is unique if both Φ and $\partial\Phi/\partial n$ are specified on S (say, as zero); knowledge of only one of these boundary values

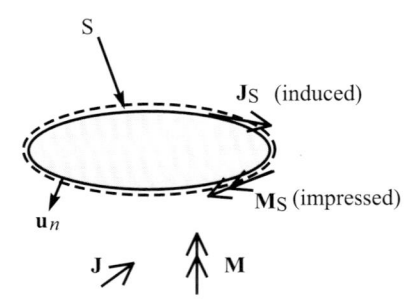

Figure 3.8 Equivalence principle based on tangential surface sources: alternate equivalent problem.

may permit an arbitrary amount of a resonant cavity mode to be present at certain frequencies, as discussed above. On the other hand, if the volume V is exterior to a bounded closed surface on which Φ *or* $\partial\Phi/\partial n$ is specified, and if Φ satisfies the radiation condition (3.36), then such a solution to Helmholtz's equation in V is unique.

It might be noted that these proofs, as well as that for the time-dependent case in Section 3.2, depend upon the finiteness of the energy integrals (for example, on the right side of (3.41)) for any finite volume V. This restriction will lead us to another condition that may need to be imposed to assure the uniqueness of a field solution—the so-called "edge condition" to be discussed in Chapter 4.

3.2.4 EQUIVALENCE PRINCIPLES AND IMAGE THEORY

We have already seen how it is possible for two different sets of sources to produce the same electromagnetic field in a certain region of space. Because of the uniqueness theorems, we can often find other ways of replacing one formulation of an electromagnetic problem in a given region of space by another one such that both formulations result in the same fields and sources in the region of interest, although they may be quite different elsewhere. In this section, we further examine equivalence principles based upon the use of equivalent surface source distributions.

Consider once more the equivalent problems of Figure 1.5. Since the fields and sources inside S in Figure 1.5(b) are identically zero, we could (for example) place a perfectly conducting surface just inside S and remove the surface electric currents $\mathbf{J}_S$ as shown in Figure 3.8. Demanding once again that $\mathbf{E}$ and $\mathbf{H}$ exterior to S be the same as in Figure 1.5(a), and imposing the radiation condition as well, the uniqueness theorem shows that these must be the fields produced by the given impressed sources. There will still be a surface electric current $\mathbf{J}_S$ at the conductor surface, but it is now induced rather than impressed, and is just behind S rather than on S proper.

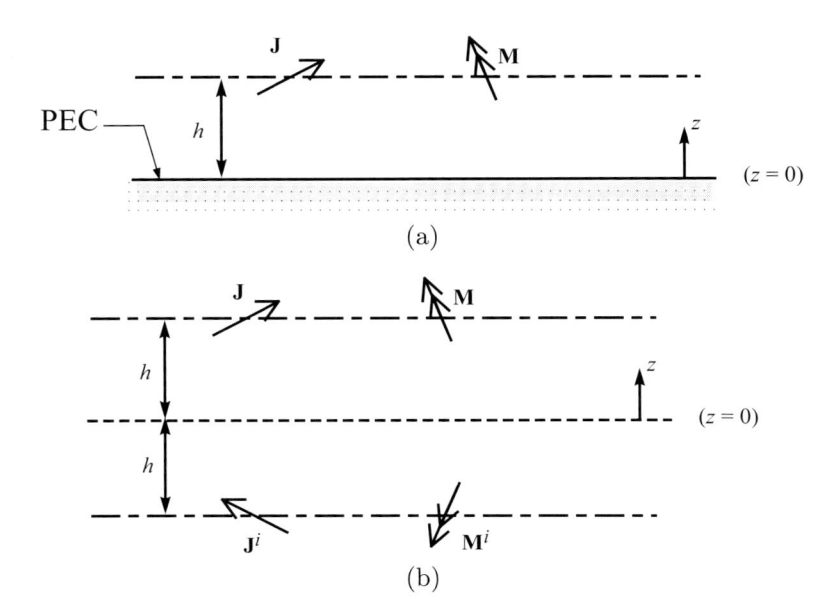

Figure 3.9 (a) Sources above a perfectly conducting plane; (b) Image equivalent for $z > 0$.

Another example of the application of equivalence theorems (actually two separate applications) is the theory of images above a perfectly conducting plane surface. Consider the electric and magnetic currents located above the plane $z = 0$, which is taken to be a PEC, as shown in Figure 3.9(a). We seek an auxiliary problem in which the ground plane is absent and additional sources are present in $z < 0$ such that the resulting fields in $z > 0$ are identical with those of case (a). These additional sources must be such that, together with the original sources in $z > 0$, they produce a zero value of $\mathbf{u}_z \times \mathbf{E}$ at $z = 0$ in the image problem (Figure 3.9).

For example, if $\mathbf{M} = 0$ and $\mathbf{J} = \mathbf{u}_z J_z(x, y, z)$ in the original problem, and if we denote the fields this source distribution would produce in *infinite free space* (in the absence of the ground plane) as $\mathbf{E}_0(x, y, z)$ and $\mathbf{H}_0(x, y, z)$, then the image sources

$$\mathbf{J}^i(x, y, z) = \mathbf{u}_z J_z(x, y, -z)$$
$$\mathbf{M}^i(x, y, z) = 0 \tag{3.48}$$

will produce the fields (also in the absence of the ground plane)

$$\mathbf{E}^i(x, y, z) = \mathbf{u}_z E_{0z}(x, y, -z) - \mathbf{E}_{0t}(x, y, -z)$$
$$\mathbf{H}^i(x, y, z) = -\mathbf{u}_z H_{0z}(x, y, -z) + \mathbf{H}_{0t}(x, y, -z) \tag{3.49}$$

Here "$_t$" denotes components of a vector transverse (perpendicular) to z in the xy-plane. By superposing these with the original fields to get the total field in case (b):

$$\begin{aligned} \mathbf{E} &= \mathbf{E}_0 + \mathbf{E}^i \\ \mathbf{H} &= \mathbf{H}_0 + \mathbf{H}^i \end{aligned} \tag{3.50}$$

we find that $\mathbf{u}_z \times \mathbf{E}|_{z=0}$ is indeed zero, so that (3.50) does give the proper total field in the region $z > 0$.

In similar fashion, we can show that the image currents for horizontal electric currents $\mathbf{J}_t$ are:

$$\mathbf{J}^i = -\mathbf{J}_t(x, y, -z) \tag{3.51}$$

while for vertical magnetic currents M_z we have

$$\mathbf{M}^i = -\mathbf{u}_z M_z(x, y, -z) \tag{3.52}$$

and for horizontal magnetic currents $\mathbf{M}_t$,

$$\mathbf{M}^i = \mathbf{M}_t(x, y, -z) \tag{3.53}$$

These results are summarized schematically in Figure 3.9.

One can also image a given set of sources if the plane $z = 0$ is a PMC. Considerations similar to those for the case of a PEC show that the proper image sources for a PMC are the negative of those for a PEC.

3.3   LORENTZ RECIPROCITY

A result of considerable interest in the formulation and solution of field problems is the Lorentz reciprocity theorem. Suppose we have a region of space bounded by one or more perfectly conducting objects[12] (whose collective surface we denote by S_0) and a large sphere of radius R centered at the origin, which we denote by S_R (Figure 3.10). Consider two possible time-harmonic electromagnetic "states" of this structure. One, which we denote by "a", is produced by the given sources $\mathbf{J}^a$, $\mathbf{M}^a$ distributed in some fashion throughout space, and results in the fields $\mathbf{E}^a$, $\mathbf{H}^a$. If these sources are removed and replaced by new ones $\mathbf{J}^b$, $\mathbf{M}^b$—in general unrelated to $\mathbf{J}^a$, $\mathbf{M}^a$—then we have the fields $\mathbf{E}^b$, $\mathbf{H}^b$ corresponding to the "b" state. Obviously, from Maxwell's equations we have

$$\begin{aligned} \nabla \times \mathbf{E}^{a,b} &= -j\omega\mu\mathbf{H}^{a,b} - \mathbf{M}^{a,b} \\ \nabla \times \mathbf{H}^{a,b} &= j\omega\epsilon\mathbf{E}^{a,b} + \mathbf{J}^{a,b} \end{aligned} \tag{3.54}$$

In addition, $\mathbf{E}^{a,b}$ and $\mathbf{H}^{a,b}$ must satisfy the radiation condition and appropriate boundary conditions.

---

[12]We could also let them be perfect magnetic walls, impedance surfaces, or certain other types of boundary, as will be clear from the derivation that follows.

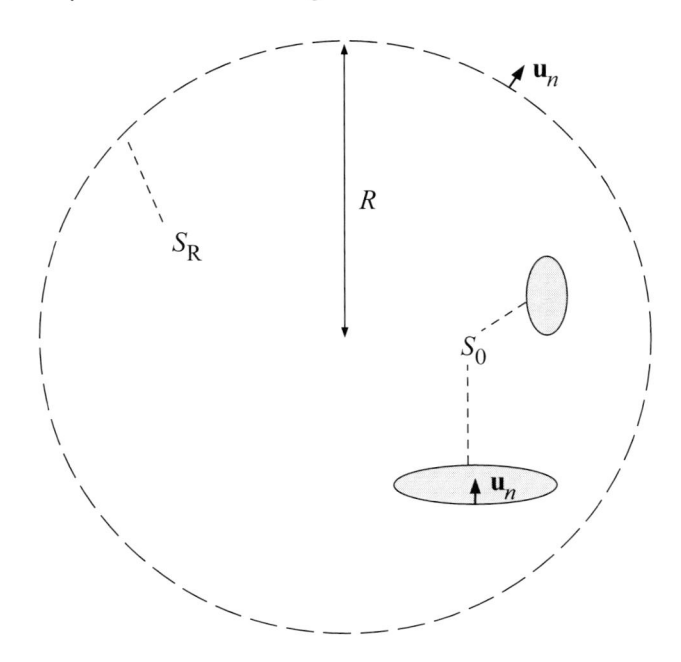

Figure 3.10 Geometry for the reciprocity theorem.

Even though $\mathbf{E}^a$, $\mathbf{H}^a$ and $\mathbf{E}^b$, $\mathbf{H}^b$ do not exist simultaneously in the space being considered, let us take the quantity

$$
\begin{aligned}
\nabla \cdot (\mathbf{E}^a \times \mathbf{H}^b - \mathbf{E}^b \times \mathbf{H}^a) &= \mathbf{H}^b \cdot \nabla \times \mathbf{E}^a - \mathbf{E}^a \cdot \nabla \times \mathbf{H}^b \\
&\quad - \mathbf{H}^a \cdot \nabla \times \mathbf{E}^b + \mathbf{E}^b \cdot \nabla \times \mathbf{H}^a \\
&= -\mathbf{H}^b \cdot \mathbf{M}^a + \mathbf{E}^b \cdot \mathbf{J}^a \\
&\quad + \mathbf{H}^a \cdot \mathbf{M}^b - \mathbf{E}^a \cdot \mathbf{J}^b
\end{aligned} \tag{3.55}
$$

which follows from (3.54) and a vector identity. Integrating (3.55) over any volume V bounded by a surface S with unit normal vector $\mathbf{u}_n$ pointing outward from V gives the integral form of the Lorentz reciprocity theorem:

$$
\begin{aligned}
\oint_S (\mathbf{E}^a \times \mathbf{H}^b - \mathbf{E}^b \times \mathbf{H}^a) \cdot \mathbf{u}_n \, dS &= \int_V (\mathbf{E}^b \cdot \mathbf{J}^a - \mathbf{H}^b \cdot \mathbf{M}^a) \, dV \\
&\quad - \int_V (\mathbf{E}^a \cdot \mathbf{J}^b - \mathbf{H}^a \cdot \mathbf{M}^b) \, dV
\end{aligned} \tag{3.56}
$$

If $S = S_0 \cup S_R$ as described above, then the portion of the surface integral in (3.56) over S_0 vanishes since $\mathbf{u}_n \times \mathbf{E}|_{S_0} = 0$. Likewise from the radiation condition

$$
\lim_{R \to \infty} \oint_{S_R} (\mathbf{E}^a \times \mathbf{H}^b - \mathbf{E}^b \times \mathbf{H}^a) \cdot \mathbf{u}_n \, dS = 0 \tag{3.57}
$$

and so (3.56) becomes in this case

$$\int (\mathbf{E}^a \cdot \mathbf{J}^b - \mathbf{H}^a \cdot \mathbf{M}^b) \, dV = \int (\mathbf{E}^b \cdot \mathbf{J}^a - \mathbf{H}^b \cdot \mathbf{M}^a) \, dV \tag{3.58}$$

where the integrals are now carried out over all space. In the absence of magnetic sources, this reduces to

$$\int \mathbf{E}^a \cdot \mathbf{J}^b \, dV = \int \mathbf{E}^b \cdot \mathbf{J}^a \, dV \tag{3.59}$$

As an example of how the reciprocity theorem can be applied, suppose that $\mathbf{J}^a$ is chosen to be a tangential surface source on one of the conducting bodies that make up S_0. Then,

$$\int \mathbf{E}^b \cdot \mathbf{J}^a \, dV \to \int_{S_0} \mathbf{E}^b_{tan} \cdot \mathbf{J}^a_S \, dS = 0$$

and we have by (3.59) that

$$\int \mathbf{E}^a \cdot \mathbf{J}^b \, dV = 0$$

no matter what the source $\mathbf{J}^b$ is chosen to be. This can only be true if $\mathbf{E}^a \equiv 0$ everywhere. That is, an electric surface current impressed at the surface of a perfect conductor produces no electric field (and for that matter no magnetic field either) exterior to that surface. In the language of circuit theory, the current source is short-circuited when placed at the conductor. This result is also implied by the example discussed in connection with the equivalence principle in Section 3.2.4.

3.4 SCATTERING PROBLEMS

In this section, we will provide statements of the most commonly encountered scattering problems in electromagnetic theory. Other, more specialized boundary problems are presented in Appendix E.

3.4.1 APERTURE RADIATION PROBLEMS

An aperture radiation problem is one where a tangential electric field $\mathbf{E}_{0t}$ is specified over a portion (the aperture) A of a given closed surface, the remainder of which (S) is a perfectly conducting body. By the equivalence theorem, we may regard this as a scattering problem (as discussed in the following subsection) in which an equivalent impressed surface magnetic current density on the conductor is the source of the incident field. But a more direct formulation is as follows. The field $\mathbf{E}$, $\mathbf{H}$ radiating from the aperture must satisfy Maxwell's equations exterior to the conductor, and obey

$$\mathbf{E}_{\tan}|_A = \mathbf{E}_{0t}$$

at the aperture, and

$$\mathbf{E}_{\tan}\big|_S = 0$$

elsewhere on the conductor. The uniqueness theorem assures us that only one solution will exist to this problem.

3.4.2 CLASSICAL SCATTERING PROBLEMS

Often, the excitation of a structure will come from outside, rather than from inside. We call this situation a scattering problem. A given set of sources is present in space, which produces an "incident" field $\mathbf{E}^i$, $\mathbf{H}^i$ in the absence of any scattering object. If these same sources are placed in the presence of a scatterer, such as a perfectly conducting surface S, the incident field may satisfy Maxwell's equations, but will not obey the boundary condition that tangential $\mathbf{E}$ must be zero at the surface of the conductor. Thus, we consider that the total field must be the sum of the incident field and a "scattered" field:

$$\mathbf{E}^{\text{tot}} = \mathbf{E}^i + \mathbf{E}^s \qquad \mathbf{H}^{\text{tot}} = \mathbf{H}^i + \mathbf{H}^s \qquad (3.60)$$

As the total tangential $\mathbf{E}$ field must vanish at S, we have that $\mathbf{E}^s$ and $\mathbf{H}^s$ must satisfy Maxwell's equations outside the conductor, while

$$\mathbf{E}^s_{\tan}\big|_S = -\,\mathbf{E}^i_{\tan}\big|_S \qquad (3.61)$$

at the conductor. Our uniqueness theorems tell us that this suffices to determine the scattered field everywhere. Note that the scattered field also exists inside the conductor, and must be exactly equal to the negative of the incident field there in order to produce a total field in the conductor that is zero. This is automatically guaranteed by the boundary condition above at all but a certain set of special frequencies (see Chapter 7). In any case the exterior field is always unique.

If, on the other hand, the scatterer is a material body with permittivity ϵ and permeability μ that differ from the background values (those of the environment in which the incident field is assumed to be produced), then the scattered field must be assumed to be unknown inside the scatterer, and additional conditions are required for its full determination. Calling the scattered field outside the scatterer $\mathbf{E}^s_{\text{out}}$, $\mathbf{H}^s_{\text{out}}$ and that inside the scatterer $\mathbf{E}^s_{\text{in}}$, $\mathbf{H}^s_{\text{in}}$, we need to impose continuity of both tangential field components at the surface S of the scatterer to determine the solution:

$$\begin{aligned}
\mathbf{E}^s_{\text{out},\tan}\big|_S + \mathbf{E}^i_{\tan}\big|_S &= \mathbf{E}^s_{\text{in},\tan}\big|_S \\
\mathbf{H}^s_{\text{out},\tan}\big|_S + \mathbf{H}^i_{\tan}\big|_S &= \mathbf{H}^s_{\text{in},\tan}\big|_S
\end{aligned} \qquad (3.62)$$

The scattered field in any event can be viewed as being produced by the currents induced on the scatterer. Consider the special case of a scatterer of bounded spatial extent located in an infinite homogeneous space. If the incident wave is a plane wave of unit amplitude propagating in the direction of a unit vector $\mathbf{u}_i$,

$$\mathbf{E}^i = \mathbf{u}_e e^{-jk\mathbf{R}\cdot\mathbf{u}_i} \tag{3.63}$$

where the unit vector $\mathbf{u}_e$ indicates the direction of polarization of the incident wave, then by (3.31) we know that in the far field,

$$\mathbf{E}^s \sim \mathbf{F}_E^s(\mathbf{u}_r, \mathbf{u}_i)\frac{e^{-jkr}}{r} \tag{3.64}$$

as $r \to \infty$. In this context, the function $\mathbf{F}_E^s(\mathbf{u}_r, \mathbf{u}_i)$ is called the *scattering amplitude*. The *bistatic differential scattering cross section* is then defined as

$$\sigma(\mathbf{u}_r, \mathbf{u}_i) = \lim_{r\to\infty} 4\pi r^2 \frac{|\mathbf{S}^s|}{|\mathbf{S}^i|} = 4\pi |\mathbf{F}_E^s(\mathbf{u}_r, \mathbf{u}_i)|^2 \tag{3.65}$$

where

$$\mathbf{S}^i = \frac{1}{2}\mathbf{E}^i \times \mathbf{H}^{i*}; \qquad \mathbf{S}^s = \frac{1}{2}\mathbf{E}^s \times \mathbf{H}^{s*} \tag{3.66}$$

are the complex Poynting vectors of the incident and scattered field, respectively.

If we further restrict ourselves to a lossless scatterer in a lossless medium, the physical meaning of the bistatic differential scattering cross section is the area that, when multiplied by the power density of the incident wave, would give the total power radiated by the scatterer if the scattered power density were made equal to that in the direction $\mathbf{u}_r$ for all directions. Other special terminologies are also often introduced; there is the *backscattering cross section*

$$\sigma_b(\mathbf{u}_i) = \sigma(-\mathbf{u}_i, \mathbf{u}_i) \tag{3.67}$$

the *forward scattering cross section*

$$\sigma_f(\mathbf{u}_i) = \sigma(\mathbf{u}_i, \mathbf{u}_i) \tag{3.68}$$

and the *(total) scattering cross section*

$$\sigma_s(\mathbf{u}_i) = \frac{1}{4\pi} \int_{\theta=0}^{\pi} \int_{\phi=0}^{2\pi} \sigma(\mathbf{u}_r, \mathbf{u}_i) \sin\theta \, d\theta d\phi = \lim_{r\to\infty} \frac{1}{|\mathbf{S}^i|} \oint_{S_r} |\mathbf{S}^s(\mathbf{u}_r, \mathbf{u}_i)| \, dS \tag{3.69}$$

where the angular coordinates (θ, ϕ) are those of the scattering direction $\mathbf{u}_r$, and S_r is the surface of a sphere of radius r centered at the origin. An important relation exists between the total scattering cross section and the forward scattering amplitude:

$$\sigma_s(\mathbf{u}_i) = -\frac{4\pi}{k}\text{Im}\left[\mathbf{u}_e \cdot \mathbf{F}_E^s(\mathbf{u}_i, \mathbf{u}_i)\right] \tag{3.70}$$

This result is known as the *optical theorem* or *forward scattering theorem*, and its proof is left as an exercise.

3.4.3 APERTURE SCATTERING PROBLEMS

The aperture radiation problem as formulated in Section 3.4.1 does not naturally arise in practical applications. This is because an aperture field cannot in general be determined ahead of time, before Maxwell's equations have been solved—the sources or an incident field are what is known. Sometimes one makes a reasonable approximation to the aperture field, such as the field of the incident wave (as in Kirchhoff's approximation), but this is strictly only an approximation. The correct way to formulate an aperture problem is as one of scattering.

Let there be a thin, perfectly conducting shell S_0 in a region of space. Let $\mathbf{u}_n$ be the unit normal vector pointing outwards from S_0. If given sources are placed near S_0 (either inside it, outside it, or both), a certain field $\mathbf{E}_0$ and $\mathbf{H}_0$ will be produced. These are the solution of two scattering problems as described in the previous subsection: one for the interior of S_0 and the other for the exterior. We will presume these fields to be known.

In the aperture scattering problem, it is not S_0 that is thought of as causing the scattering of the field, but the action of cutting an aperture A into S_0, removing a portion of the conductor so that the inside and outside regions can interact with each other. Thus $\mathbf{E}_0$ and $\mathbf{H}_0$ are the incident field, and the scattered field $\mathbf{E}^s$ and $\mathbf{H}^s$ is what must be added to the incident field in order to get a total field that satisfies the boundary conditions on both A and S. Since the incident and total field must satisfy Maxwell's equations, so also must the scattered field. From the boundary conditions on the total and incident field, we have that $\mathbf{E}^s$ is continuous across all of S_0, while

$$\mathbf{u}_n \times \mathbf{E}^s\big|_S = 0$$
$$\mathbf{u}_n \times \mathbf{H}^s\big|_{n=0^-}^{n=0^+} = -\mathbf{J}_{S0} \qquad \text{in } A \qquad (3.71)$$

where $\mathbf{J}_{S0}$ is the known total surface current density on the shell S_0 associated with the incident field:

$$\mathbf{J}_{S0} = \mathbf{u}_n \times \mathbf{H}_0\big|_{n=0^-}^{n=0^+}$$

3.5 PLANAR SCATTERERS AND BABINET'S PRINCIPLE

For the case of a PEC scatterer that is a disk of zero thickness lying in the plane $z = 0$, some special properties of the fields follow. Let the disk occupy the region S in the plane $z = 0$, and we denote the remaining portion of the xy-plane as A. A set of sources $\mathbf{J}$, $\mathbf{M}$ radiates in the presence of this disk, and produces the incident field $\mathbf{E}^i$, $\mathbf{H}^i$ if the disk is absent. Now decompose the sources into

$$\mathbf{J} = \frac{1}{2}\left(\mathbf{J}_{ei} + \mathbf{J}_{mi}\right) \qquad (3.72)$$
$$\mathbf{M} = \frac{1}{2}\left(\mathbf{M}_{ei} + \mathbf{M}_{mi}\right)$$

where

$$\mathbf{J}_{ei} = \mathbf{J} + \mathbf{J}^i \qquad (3.73)$$

and

$$\mathbf{M}_{ei} = \mathbf{M} + \mathbf{M}^i \qquad (3.74)$$

are the original sources together with the images of them that would be produced by a PEC occupying the entire plane $z = 0$, while

$$\mathbf{J}_{mi} = \mathbf{J} - \mathbf{J}^i \qquad (3.75)$$

and

$$\mathbf{M}_{mi} = \mathbf{M} - \mathbf{M}^i \qquad (3.76)$$

are the original sources together with the images of them that would be produced by a PMC occupying the entire plane $z = 0$. These image sources are given by (3.48) and (3.51)-(3.53).

Now, the field produced by $\mathbf{J}_{ei}$ and $\mathbf{M}_{ei}$ already obeys $\mathbf{u}_n \times \mathbf{E} = 0$ at S (over all of $z = 0$, in fact), so the insertion of the disk into this part of the incident field will produce no scattered field. On the other hand, symmetry arguments show that the field produced by $\mathbf{J}_{mi}$ and $\mathbf{M}_{mi}$ is such that $\mathbf{H}_t$ is an odd function of z. The insertion of the disk into this part of the incident field does not disturb this anti-symmetry of the $\mathbf{H}_t$ field, but will additionally induce surface electric currents on S, producing the scattered field that is needed to satisfy the condition that total tangential electric field on S must vanish. Since $\mathbf{H}_t$ must be continuous across A, as an odd function of z it must therefore be equal to zero there (the same is not necessarily true on S, across which $\mathbf{H}_t$ may be discontinuous). Thus the total field $\mathbf{E}_1$, $\mathbf{H}_1$ produced by $\mathbf{J}$ and $\mathbf{M}$ in the presence of a PEC disk at S obeys

$$\mathbf{H}_{1t} = \mathbf{H}^i \qquad \text{on } A \qquad (3.77)$$

$$\mathbf{E}_{1t} = 0 \qquad \text{on } S \qquad (3.78)$$

Suppose now we remove the PEC from S, and place a PMC instead at A. We call this the complementary magnetic screen to the original PEC disk. Similar considerations to those presented above lead us to conclude that the total field $\mathbf{E}_2$, $\mathbf{H}_2$ produced by the same sources in the presence of this complementary screen will obey

$$\mathbf{E}_{2t} = \mathbf{E}^i \qquad \text{on } S \qquad (3.79)$$

$$\mathbf{H}_{2t} = 0 \qquad \text{on } A \qquad (3.80)$$

The sum of the fields $\mathbf{E}_1 + \mathbf{E}_2$, $\mathbf{H}_1 + \mathbf{H}_2$ from these two boundary problems thus obeys

$$\mathbf{E}_{1t} + \mathbf{E}_{2t} = \mathbf{E}^i \qquad \text{on } S \qquad (3.81)$$

$$\mathbf{H}_{1t} + \mathbf{H}_{2t} = \mathbf{H}^i \qquad \text{on } A \qquad (3.82)$$

But by the uniqueness theorem of Section 3.2.3, the sum of the two fields must be equal to the incident field everywhere:

$$\begin{aligned} \mathbf{E}_1 + \mathbf{E}_2 &= \mathbf{E}^i \\ \mathbf{H}_1 + \mathbf{H}_2 &= \mathbf{H}^i \end{aligned} \tag{3.83}$$

or in other words

$$\begin{aligned} \mathbf{E}_2 &= \mathbf{E}^i - \mathbf{E}_1 \\ \mathbf{H}_2 &= \mathbf{H}^i - \mathbf{H}_1 \end{aligned} \tag{3.84}$$

This result is known as *Babinet's Principle*[13] relating the fields of these complementary scattering problems.

If we further invoke duality on the complementary magnetic screen, we can obtain the field $\mathbf{E}_3$, $\mathbf{H}_3$ produced by the dual sources (1.88) and (1.89), in the presence of a PEC screen located at A. In other words, the original PEC disk at S, and a complementary PEC screen at A, excited by mutually dual sources, will produce fields related to the original incident field as

$$\mathbf{E}_1 + \mathbf{H}_3 = \mathbf{E}^i$$

$$\mathbf{H}_1 - \mathbf{E}_3 = \mathbf{H}^i$$

There is thus no need to formulate and solve the aperture-scattering problem from scratch—its solution is easily found from that of the complementary PEC disk.

---

[13]See

E. T. Copson, *Proc. Roy. Soc. London*, ser. A, vol. 186, pp. 100-118, 1946.

J.-P. Vasseur, *C. R. Acad. Sci. (Paris)*, vol. 229, pp. 586-587, 1949.

R. F. Harrington, *Time-Harmonic Electromagnetic Fields*. New York: McGraw-Hill, 1961, pp. 365-367.

D. S. Jones, *The Theory of Electromagnetism*. Oxford: Pergamon Press, 1964, Section 9.3.

Extension to the case when the medium is inhomogeneous has been made in:

B. D. Popović and A. Nesić, *IEE Proc. Part H*, vol. 132, pp. 131-137, 1985.

3.6 PROBLEMS

3–1 Suppose the path of a certain ray is a circle of radius r_0:

$$\mathbf{r}(l) = r_0 \left(\mathbf{u}_x \cos \frac{l}{r_0} + \mathbf{u}_y \sin \frac{l}{r_0} \right)$$

where in cylindrical coordinates the arc length is $l = r_0 \phi$.

(a) From the differential equation for a ray, find the general condition on the refractive index function $n(\mathbf{r})$ necessary for this circular ray path to exist.

(b) If n is a spherically symmetric function of radial distance only $n = n(r)$, show that the general condition obtained in part (a) reduces to $r_0 n'(r_0) + n(r_0) = 0$. What does this condition mean physically?

(c) For a square-law medium with $n(r) = n_0(1 - r^2/a^2)$, where n_0 and a are positive constants, what must the radius of the circular ray path be?

3–2 Derive Equations (3.33) in detail using the spherical-coordinate form of Maxwell's equations.

3–3 Show that Equations (3.49) are solutions of Maxwell's equations when the sources are given by (3.48).

3–4 If $\sigma \neq 0$ throughout the region of space under consideration, prove Uniqueness Theorem II of Section 3.2.3 *without* using Rellich's theorem. This result can be used to obtain a unique solution for lossless media by taking the limit of the unique solution determined by this criterion as $\sigma \to 0^+$. This method is called the principle of limiting absorption, and is examined more fully in Section 6.1.

3–5 Prove (3.36).

3–6 In the context of the uniqueness theorems of Section 3.2.3, consider the condition (3.46). Suppose that a *Generalized Impedance Boundary Condition* (GIBC)[14] of the form

$$\mathbf{u}_n \times \mathbf{E}\big|_{S_0} = -\left\{ Z_S \mathbf{H}_{\text{tan}} + \mathbf{u}_n \times \nabla \left(h E_n \right) \right\}_{S_0}$$

is imposed on the fields at the surface S_0, where Z_S and h are complex quantities that may be functions of position. What restrictions must be imposed on Z_S and h in order to guarantee that (3.46) is true for all possible fields that are solutions of Maxwell's equations, and therefore that the GIBC along with the radiation condition guarantees a unique solution to Maxwell's equations?

---

[14]T. B. A. Senior and J. L. Volakis, *Approximate Boundary Conditions in Electromagnetics.* London: Institution of Electrical Engineers, 1995.

3–7 The Poynting vector and energy density are sometimes used to define a local, time-dependent energy velocity for a given electromagnetic field:

$$\mathbf{v}_E = \frac{\mathcal{P}}{U}$$

where

$$\mathcal{P} = \mathbf{E} \times \mathbf{H}$$

is the Poynting vector, and

$$U = \frac{1}{2}\epsilon_0 E^2 + \frac{1}{2}\mu_0 H^2$$

is the energy density of a field in free space. Show that $|\mathbf{v}_E| \leq c$, where c is the velocity of light in free space. Under what conditions does equality hold?

3–8 (a) If $\mathbf{E}^a$, $\mathbf{H}^a$ and $\mathbf{E}^b$, $\mathbf{H}^b$ are solutions of Maxwell's equations satisfying the radiation condition (3.35), show that (3.57) holds.
(b) In fact, if the interior of S_R contains all points at which $\mathbf{J}^a$, $\mathbf{M}^a$, $\mathbf{J}^b$, or $\mathbf{M}^b$ are not equal to zero, show that

$$\oint_{S_R} (\mathbf{E}^a \times \mathbf{H}^b - \mathbf{E}^b \times \mathbf{H}^a) \cdot \mathbf{u}_n \, dS = 0$$

without the need to take the limit as $R \to \infty$.

3–9 Prove the optical theorem (3.70). [*Hint: Start with the fact that for the total field*

$$\mathrm{Re} \int_S \mathbf{E} \times \mathbf{H}^* \cdot \mathbf{u}_n \, dS = 0$$

for a lossless scatterer whose surface is S.]

4 Radiation by Simple Sources and Structures

In this chapter, we will examine the formulation and solution of electromagnetic *radiation* problems, in which a given source exists in infinite space and we want to find the resultant field. The results of this chapter will be used in the next two chapters to formulate and solve the problem of scattering from various objects.

4.1 POINT AND LINE SOURCES IN UNBOUNDED SPACE

4.1.1 STATIC POINT CHARGE

Consider the simple example of a static point charge of q coulombs in an unbounded homogeneous medium (Figure 4.1). The charge density of such a point charge can be written as

$$\rho(\mathbf{r}) = q\delta(x - x_0)\delta(y - y_0)\delta(z - z_0) \equiv q\delta(\mathbf{r} - \mathbf{r}_0) \tag{4.1}$$

where $\mathbf{r}_0 = x_0\mathbf{u}_x + y_0\mathbf{u}_y + z_0\mathbf{u}_z$ is its location. From (2.13) we have

$$\nabla^2\Phi = -\frac{q}{\epsilon}\delta(\mathbf{r} - \mathbf{r}_0) \tag{4.2}$$

Now it is permitted to take $\mathbf{r}_0 = 0$ without loss of generality because no physical change in the problem results from doing so: Only our "labeling of the map" for the Cartesian coordinate system is changed and not the structure itself. The delta-function $\delta(\mathbf{r})$ is spherically symmetric, and symmetry of the problem demands that the potential Φ be a function of r only in a spherical coordinate system (r, θ, ϕ). Then (4.2) becomes

$$\frac{1}{r^2}\frac{d}{dr}\left(r^2\frac{d\Phi}{dr}\right) = -\frac{q}{\epsilon}\delta(\mathbf{r}) \tag{4.3}$$

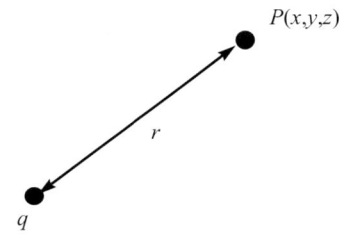

Figure 4.1 Point charge in unbounded medium.

In the region $r > 0$, the right side of (4.3) is zero, and we may readily integrate it to obtain the homogeneous solution

$$\Phi(r) = \frac{C_1}{r} + C_2 \qquad (4.4)$$

where C_1 and C_2 are (as yet) undetermined constants. We may set $C_2 = 0$ since its value does not affect the field calculated from the potential Φ. We obtain the value of C_1 by integrating both sides of (4.3) over a spherical volume enclosing the point charge:

$$4\pi r^2 \frac{d\Phi}{dr} = -\frac{q}{\epsilon}$$

or, $C_1 = q/4\pi\epsilon$. Thus,

$$\Phi = \frac{q}{4\pi\epsilon r} \qquad (4.5)$$

which is the well-known potential for this point charge. This result is obtained in a different way in Example 21 of Appendix A.

4.1.2 POTENTIAL OF A PULSED DIPOLE IN FREE SPACE

We next examine the case of an elementary (or Hertz) electric dipole suddenly created at time $t = 0$ and located at the origin:

$$\mathbf{p}(t) = \mathbf{p}_0 \vartheta(t) \qquad (4.6)$$

so that the current density is:

$$\mathbf{J} = \frac{\partial \mathbf{p}}{\partial t} \delta(\mathbf{r}) = \mathbf{p}_0 \delta(\mathbf{r}) \delta(t) \qquad (4.7)$$

as in the first term of (1.20). The vector potential produced by this source is $\mathbf{A} = \mu_0 \mathbf{p}_0 G$, where G satisfies the inhomogeneous wave equation:

$$\nabla^2 G - \frac{1}{c^2} \frac{\partial^2 G}{\partial t^2} = -\delta(\mathbf{r})\delta(t) \qquad (4.8)$$

where $c = 1/\sqrt{\mu_0 \epsilon_0}$ is the speed of light in vacuum. After G is found, the electric and magnetic fields are obtained from (2.1) and (2.3):

$$\mathbf{B} = \nabla \times \mathbf{A}; \qquad \mathbf{E} = -\nabla\Phi - \frac{\partial \mathbf{A}}{\partial t} \qquad (4.9)$$

where Φ is determined from the Lorenz gauge condition (2.11). Although the solution of general partial differential equations is best done with transform and series expansion techniques, we can use ideas similar to those employed for the static point charge to solve (4.8) more directly.

First, we observe that G is spherically symmetric; that is, $G = G(r, t)$ only. Now introduce the new unknown function $g = rG$, which obeys

$$\frac{1}{r}\left[\frac{\partial^2 g}{\partial r^2} - \frac{1}{c^2}\frac{\partial^2 g}{\partial t^2}\right] = -\delta(\mathbf{r})\delta(t) \tag{4.10}$$

Next, we notice that for $t \neq 0$ or $r \neq 0$, the right side of (4.10) is zero, and in that variable range we can obtain the homogeneous solution by making the changes of variable $u = t - r/c$ and $v = t + r/c$. The resulting equation is

$$\frac{\partial^2 g}{\partial u \partial v} = 0 \tag{4.11}$$

whose most general solution is

$$g = g_1(u) + g_2(v) = g_1(t - r/c) + g_2(t + r/c) \tag{4.12}$$

where g_1 and g_2 are, so far, arbitrary functions of their arguments. The full solution of (4.8) should thus have the form:

$$G(r, t) = \frac{1}{r}\left[g_1(t - r/c) + g_2(t + r/c)\right] \tag{4.13}$$

Substituting (4.13) into (4.8) and eliminating terms that cancel out, we have

$$-\delta(\mathbf{r})\delta(t) = \nabla^2 G - \frac{1}{c^2}\frac{\partial^2 G}{\partial t^2} = -4\pi\delta(\mathbf{r})\left[g_1(t - r/c) + g_2(t + r/c)\right] \tag{4.14}$$

where we have used the result (A.92) from Example 21 of Appendix A, or equivalently, the result of Section 4.1.1.

The coefficients of $\delta(\mathbf{r})$ in (4.14) must be equal at $r = 0$ by (A.21). Thus,

$$\delta(t) = 4\pi\left[g_1(t) + g_2(t)\right]$$

so that

$$G(r, t) = \frac{\delta(t - r/c)}{4\pi r} + \frac{1}{r}\left[g_2(t + r/c) - g_2(t - r/c)\right]$$

But we now impose the requirement of causality: Since there should be zero response observed before the source is "turned on" at $t = 0$, we must have $G(r, t) \equiv 0$ for all r and any $t < 0$. This means

$$g_2(t + r/c) = g_2(t - r/c)$$

for $t < 0$ and $r \geq 0$, meaning that $g_2(t)$ must be a constant and thus,

$$G(r, t) = \frac{\delta(t - r/c)}{4\pi r} \tag{4.15}$$

This is observed to be an expanding spherical wave emanating from the origin.

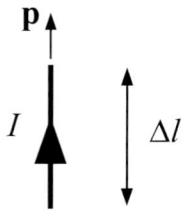

Figure 4.2 Approximate model of a Hertz dipole.

If the location of the dipole had been moved to a point $\mathbf{r}' = x'\mathbf{u}_x + y'\mathbf{u}_y + z'\mathbf{u}_z$, and turned on at the instant t', this solution would have been displaced in space and time to give the result

$$G(|\mathbf{r} - \mathbf{r}'|, t - t') = \frac{\delta(t - t' - |\mathbf{r} - \mathbf{r}'|/c)}{4\pi|\mathbf{r} - \mathbf{r}'|} \tag{4.16}$$

which obeys the differential equation

$$\nabla^2 G(|\mathbf{r} - \mathbf{r}'|, t - t') - \frac{1}{c^2}\frac{\partial^2 G(|\mathbf{r} - \mathbf{r}'|, t - t')}{\partial t^2} = -\delta(\mathbf{r} - \mathbf{r}')\delta(t - t') \tag{4.17}$$

The electric and magnetic fields are then reconstructed from (4.9) as indicated above. We observe from these results that before the current is turned on, these fields are zero as demanded by causality. After initiation of the current, the response obeys the domain of dependence relation of Section 3.1.1—it remains zero at any point whose distance from the current is larger than ct. The spherical boundary $r = ct$ between zero and nonzero response is the wavefront, which, once initiated, expands in the manner described in Section 3.1.2.

4.1.3 TIME-HARMONIC DIPOLE

Next, consider a time-harmonic dipole current element. An electric Hertz dipole is now described by the electric current density

$$\mathbf{J} = j\omega\mathbf{p}\delta(\mathbf{r} - \mathbf{r}_0) \tag{4.18}$$

where $\mathbf{p}$ is the vector dipole moment of this source. This dipole can be considered as the limiting case of a linear current I flowing in the same direction as $\mathbf{p}$ over a length Δl such that

$$j\omega p = j\omega|\mathbf{p}| = \lim_{\Delta l \to 0} I\Delta l$$

approaches a finite, nonzero constant (Figure 4.2). If $\mathbf{p} = p\mathbf{u}_z$, say, then we can take the stream potentials $\mathbf{T}_m$ and $\mathbf{T}_e$ to be zero and express its fields in terms of Whittaker potentials U and V, and furthermore we can put $U = 0$.

For, consider Equations (2.35). Since $\mathbf{M} = 0$, we put $\boldsymbol{\Pi}_m = 0$, and since $\mathbf{J} = j\omega p \mathbf{u}_z \delta(\mathbf{r} - \mathbf{r}_0)$, we can put $\boldsymbol{\Pi}_e = \mathbf{u}_z V$, and get

$$(\nabla^2 + k^2)V = -\frac{p}{\epsilon}\delta(\mathbf{r} - \mathbf{r}_0) \tag{4.19}$$

and we have a completely scalar problem.[1]

Our scalar problem is spherically symmetric, so as before we choose $\mathbf{r}_0 = 0$ with no loss of generality, and put $V = V(r)$:

$$\left(\frac{1}{r^2}\frac{d}{dr}r^2\frac{d}{dr} + k^2\right)V = -\frac{p}{\epsilon}\delta(\mathbf{r}) \tag{4.20}$$

The substitution $f = rV$ reduces (4.20) to an easily solved form. In $r > 0$, we have

$$V = \frac{1}{r}\left[C_1 e^{-jkr} + C_2 e^{jkr}\right] \tag{4.21}$$

We set $C_2 = 0$ by the radiation condition, and determine C_1 by integration of (4.20) over a small sphere centered about $r = 0$. Calculation is easiest if we let the radius of this sphere approach zero:

$$-\frac{p}{\epsilon} = \lim_{R\to 0} 4\pi \int_0^R r^2\, dr \left[\frac{1}{r^2}\frac{d}{dr}r^2\frac{d}{dr} + k^2\right]\frac{C_1}{r}e^{-jkr} = -4\pi C_1$$

Thus,

$$V = V_0 \frac{e^{-jkr}}{r} \tag{4.22}$$

where

$$V_0 = \frac{p}{4\pi\epsilon} \tag{4.23}$$

and substitution into (2.33)-(2.34) gives the well-known fields of an elementary dipole. If $\mathbf{r}_0 \neq 0$, then (4.22) must be replaced by

$$V = V_0 \frac{e^{-jk|\mathbf{r}-\mathbf{r}_0|}}{|\mathbf{r} - \mathbf{r}_0|} \tag{4.24}$$

We note that this result also follows if we take the Fourier transform of the result (4.15) with respect to t.

---

[1] It may not be obvious that these five components of the Hertz vectors *must* vanish. In fact, our exposition here is based on a certain amount of hindsight, and its validity depends on verifying (when all is said and done) that the final expression indeed constitutes a solution of all the conditions of our problem. The appropriate uniqueness theorem then assures us that this valid solution is in fact the *only* one. The philosophy here is similar to that of the method of separation of variables: Our assumptions may fail, and we would then have to abandon them, but if they are not contradicted by any of the conditions of the problem, they must have been justified.

4.1.4 LINE SOURCES IN UNBOUNDED SPACE

The fields of an electric line current source can be found in a similar way. We put

$$\mathbf{J} = I_0 \mathbf{u}_z \delta(x)\delta(y) \equiv I_0 \mathbf{u}_z \delta(\boldsymbol{\rho}) \tag{4.25}$$

so that the current I_0 flows along the z-axis. We have again $U = 0$, and V satisfies

$$(\nabla^2 + k^2)V = -\frac{I_0}{j\omega\epsilon}\delta(\boldsymbol{\rho}) \tag{4.26}$$

where $\boldsymbol{\rho} = x\mathbf{u}_x + y\mathbf{u}_y$. The symmetry of this problem dictates that $V = V(\rho)$, so

$$\left(\frac{1}{\rho}\frac{d}{d\rho}\rho\frac{d}{d\rho} + k^2\right)V = -\frac{I_0}{j\omega\epsilon}\delta(\boldsymbol{\rho}) \tag{4.27}$$

The general solution for $\rho > 0$ is

$$V = C_1 H_0^{(1)}(k\rho) + C_2 H_0^{(2)}(k\rho) \tag{4.28}$$

where $H_0^{(1),(2)}$ are the Hankel functions of first and second kinds, respectively.

For $k\rho \to \infty$, the asymptotic behaviors of $H_\nu^{(1),(2)}$ are obtained from (B.14). We see that we must set $C_1 = 0$ in (4.28) in order to satisfy the (two-dimensional version of the) radiation condition. The source condition is enforced by integrating (4.27) over a small circle about the origin. To carry out this integral, we make use of the small-argument behavior of $H_0^{(2)}(z)$ given in (B.18). Thus

$$H_0^{(2)\prime}(z) \sim -\frac{2j}{\pi z} \qquad (z \to 0) \tag{4.29}$$

and (4.27) yields, upon integration:

$$-\frac{I_0}{j\omega\epsilon} = 2\pi C_2 \left(-\frac{2j}{\pi}\right)$$

so that

$$V = \frac{I_0}{j\omega\epsilon}\left(\frac{1}{4j}\right)H_0^{(2)}(k\rho) \tag{4.30}$$

If the source location is not at the origin ($\boldsymbol{\rho}_0 \neq 0$), then we have

$$V = \frac{I_0}{j\omega\epsilon}\left(\frac{1}{4j}\right)H_0^{(2)}(k|\boldsymbol{\rho} - \boldsymbol{\rho}_0|) \tag{4.31}$$

Since $\partial/\partial\phi = \partial/\partial z = 0$ in this problem, the fields are relatively easy to get from (2.33), (2.34), and (2.38):

$$\mathbf{E} = k^2\mathbf{u}_z V = -j\omega\mu I_0\mathbf{u}_z\left(\frac{1}{4j}\right)H_0^{(2)}(k\rho) \tag{4.32}$$

$$\mathbf{H} = j\omega\epsilon\nabla \times (\mathbf{u}_z V) = \frac{I_0}{4j}(-\mathbf{u}_\phi)kH_0^{(2)\prime}(k\rho) \tag{4.33}$$

4.2 ALTERNATE REPRESENTATIONS FOR POINT AND LINE SOURCE POTENTIALS

The solutions obtained in the previous section made essential use of the symmetry of the problem to reduce the scalar Helmholtz equation to an ordinary differential equation that could be readily solved. In other situations we may not have this symmetry and must rely on other techniques to solve the equation. We illustrate these techniques here on examples from the previous section.

4.2.1 TIME-HARMONIC LINE SOURCE

Consider the line source problem (4.26), generalized so that the source is at an arbitrary location $\rho_0 = x_0\mathbf{u}_x + y_0\mathbf{u}_y$:

$$(\nabla^2 + k^2)V = -\frac{I_0}{j\omega\epsilon}\delta(x - x_0)\delta(y - y_0) \tag{4.34}$$

Introduce the Fourier transform pair

$$\tilde{V}(\alpha, y) = \frac{1}{2\pi}\int_{-\infty}^{\infty} e^{j\alpha x}V(x, y)\,dx$$

$$V(x, y) = \int_{-\infty}^{\infty} e^{-j\alpha x}\tilde{V}(\alpha, y)\,d\alpha \tag{4.35}$$

as in (A.60). The transform of (4.34) is

$$\left(\frac{\partial^2}{\partial y^2} + k^2 - \alpha^2\right)\tilde{V} = -\frac{I_0}{j\omega\epsilon}\frac{e^{j\alpha x_0}}{2\pi}\delta(y - y_0) \tag{4.36}$$

The problem for $\tilde{V}$ is now an ordinary differential equation corresponding to a sheet source at $y = y_0$. We solve this as in the previous section. For $y > y_0$, we have

$$\tilde{V} = C_1 e^{j\xi y} + C_2 e^{-j\xi y} \tag{4.37}$$

and for $y < y_0$ likewise,

$$\tilde{V} = C_3 e^{j\xi y} + C_4 e^{-j\xi y} \tag{4.38}$$

where $\xi = \sqrt{k^2 - \alpha^2}$. For definiteness, we must decide upon which sign (in other words, which *branch*) of the square root to choose. We take

$$\mathrm{Re}(\xi) > 0 \quad \text{for } \alpha^2 < k^2$$

$$\text{and} \tag{4.39}$$

$$\mathrm{Im}(\xi) < 0 \quad \text{for } \alpha^2 > k^2$$

Other choices are possible (though sometimes less convenient); they would simply alter the values of C_1 through C_4 that are obtained below.[2]

We must have a solution that satisfies the radiation condition. Now as $y \to \infty$, the term $C_1 e^{j\xi y}$ represents an incoming[3] (noncausal) wave, and our superposition of such waves in the inverse Fourier transform integral should not include these. Therefore, $C_1 = 0$. Likewise we set $C_4 = 0$ to have only outgoing waves as $y \to -\infty$. Our solution is now

$$\begin{aligned}
\tilde{V} &= C_2 e^{-j\xi y} \quad (y > y_0) \\
&= C_3 e^{j\xi y} \quad (y < y_0)
\end{aligned} \tag{4.40}$$

Since V is proportional to E_z, which is a tangential field component with respect to the plane $y = y_0$, and since there are no magnetic surface currents present, V (and hence also $\tilde{V}$) must be continuous at $y = y_0$. We can thus write

$$\tilde{V} = C e^{-j\xi|y - y_0|} \tag{4.41}$$

where C must yet be determined.

Next, we enforce the source condition—here, a "jump" condition across the Fourier transformed sheet of electric surface current at $y = y_0$. Integrate (4.36) across the source sheet from $y = y_0^-$ to $y = y_0^+$. Since $\tilde{V}$ is itself continuous, we get

$$\left.\frac{d\tilde{V}}{dy}\right|_{y_0^-}^{y_0^+} = -\frac{I_0}{j\omega\epsilon} \frac{e^{j\alpha x_0}}{2\pi} \tag{4.42}$$

(note that this also follows from the general result (2.32)), or from (4.41)

$$-2j\xi C = -\frac{I_0}{j\omega\epsilon} \frac{e^{j\alpha x_0}}{2\pi}$$

and thus

$$\tilde{V} = \frac{I_o}{j\omega\epsilon} \frac{1}{4\pi j} \frac{e^{-j\xi|y - y_0| + j\alpha x_0}}{\xi} \tag{4.43}$$

From the inversion formula in (4.35),

$$V(x, y) = \frac{I_o}{j\omega\epsilon} \frac{1}{4\pi j} \int_{-\infty}^{\infty} \frac{e^{-j\xi|y - y_0| - j\alpha(x - x_0)}}{\xi} \, d\alpha \tag{4.44}$$

---

[2]Keep in mind that while C_1 through C_4 are constants as far as y is concerned, they are allowed to be (and in fact must be) functions of the spectral variable α. This should be remembered whenever we refer to such numbers as "constants."

[3]If $\alpha^2 < k^2$; when $\alpha^2 > k^2$, the situation is even worse—we have a field that grows exponentially with distance from the source. This argument is somewhat heuristic: see Section 4.3.3 for a more rigorous one.

This is a so-called *plane-wave spectrum* or *Sommerfeld integral* representation of V. Since it must be identically equal to the function expressed by formula (4.31), we have obtained the identity:

$$H_0^{(2)}(k|\boldsymbol{\rho} - \boldsymbol{\rho}_0|) = \frac{1}{\pi} \int_{-\infty}^{\infty} \frac{e^{-j\xi|y-y_0|-j\alpha(x-x_0)}}{\xi} \, d\alpha \qquad (4.45)$$

which plays an important role in subsequent applications.

Another representation of the line source potential can be obtained in the following way. For $\rho_0 \neq 0$,

$$\delta(\boldsymbol{\rho} - \boldsymbol{\rho}_0) = \frac{\delta(\rho - \rho_0)\delta_{2\pi}(\phi - \phi_0)}{\rho}$$

and thus the Helmholtz equation can be written as

$$\left(\frac{1}{\rho}\frac{\partial}{\partial\rho}\rho\frac{\partial}{\partial\rho} + \frac{1}{\rho^2}\frac{\partial^2}{\partial\phi^2} + k^2\right) V = -\frac{I_o}{j\omega\epsilon}\frac{\delta(\rho - \rho_0)\delta_{2\pi}(\phi - \phi_0)}{\rho} \qquad (4.46)$$

where $\delta_{2\pi}$ is a periodic replication of the delta function as defined in (A.73).

We now use, instead of a Fourier transform, a Fourier series representation in the angular coordinate ϕ. The relevant transform pair is:

$$V_m(\rho) = \frac{1}{2\pi}\int_0^{2\pi} V(\rho, \phi)e^{-jm\phi} \, d\phi$$

$$V(\rho, \phi) = \sum_{m=-\infty}^{\infty} V_m(\rho)e^{jm\phi} \qquad (4.47)$$

We can understand this representation in the context of generalized functions if necessary (see Appendix A).

Applying the transform (4.47) to (4.46), we get

$$\left(\frac{1}{\rho}\frac{\partial}{\partial\rho}\rho\frac{\partial}{\partial\rho} - \frac{m^2}{\rho^2} + k^2\right) V_m = -\frac{I_o}{j\omega\epsilon}\frac{e^{-jm\phi_0}\delta(\rho - \rho_0)}{2\pi\rho} \qquad (4.48)$$

where use has been made of (A.74). The homogeneous form of this equation is the Bessel equation of order m analogous to that of order zero encountered before. For $\rho < \rho_0$ it is appropriate to write

$$V_m = C_1 J_m(k\rho) + C_2 Y_m(k\rho) \qquad (4.49)$$

while for $\rho > \rho_0$ we write

$$V_m = C_3 H_m^{(1)}(k\rho) + C_4 H_m^{(2)}(k\rho) \qquad (4.50)$$

To ensure that the field will satisfy the radiation condition, we set $C_3 = 0$. Moreover, Maxwell's equations demand that the field (and thus in this case the potential) remain finite at points of space not containing concentrated sources. Thus we demand that $V_m(0)$ be finite, and we must set $C_2 = 0$, because of Equations (B.15)-(B.17) in the Appendix. Moreover, $V_m(\rho)$ must be continuous at $\rho = \rho_0$, and thus we can write, compactly,

$$V_m(\rho) = \tilde{C}_m J_m(k\rho_<) H_m^{(2)}(k\rho_>) \tag{4.51}$$

where

$$\rho_> = \max(\rho, \rho_0), \quad \rho_< = \min(\rho, \rho_0) \tag{4.52}$$

and $\tilde{C}_m$ is a constant to be determined.

We determine $\tilde{C}_m$ from the jump condition, integrating (4.48), multiplied by ρ, from ρ_0^- to ρ_0^+:

$$\rho_0 \left. \frac{dV_m}{d\rho} \right|_{\rho_0^-}^{\rho_0^+} = -\frac{I_0}{j\omega\epsilon} \frac{e^{-jm\phi_0}}{2\pi} \tag{4.53}$$

$$= k\rho_0 \tilde{C}_m \left\{ H_m^{(2)\prime}(k\rho_0) J_m(k\rho_0) - J_m'(k\rho_0) H_m^{(2)}(k\rho_0) \right\}$$

But from the Wronskian relation (B.21) for this combination of Bessel and Hankel functions, (4.53) becomes

$$\tilde{C}_m = \frac{I_0}{j\omega\epsilon} \frac{e^{-jm\phi_0}}{4j} \tag{4.54}$$

so that

$$V(\rho, \phi) = \frac{I_0}{j\omega\epsilon} \frac{1}{4j} \sum_{m=-\infty}^{\infty} e^{jm(\phi-\phi_0)} J_m(k\rho_<) H_m^{(2)}(k\rho_>) \tag{4.55}$$

Again, this must coincide with our previously obtained form of the same solution in (4.30). We have thus proved the identity

$$H_0^{(2)}(k|\boldsymbol{\rho} - \boldsymbol{\rho}_0|) = \sum_{m=-\infty}^{\infty} e^{jm(\phi-\phi_0)} J_m(k\rho_<) H_m^{(2)}(k\rho_>) \tag{4.56}$$

and $\rho_<$, $\rho_>$ are defined by (4.52). This is one version of a so-called *addition theorem* for Bessel functions. It shows that a solution for the potential of a line source located at an arbitrary point ($\boldsymbol{\rho} = \boldsymbol{\rho}_0$) can be found not only in the form of a single cylindrical wave centered about the location of the source, but also as a sum of cylindrical wave functions centered about some other point (here, $\boldsymbol{\rho} = 0$).

A limiting case of (4.56) yields the expansion of a plane wave as a series of Bessel functions. By (B.14), the cylindrical wave $H_0^{(2)}(k|\boldsymbol{\rho} - \boldsymbol{\rho}_0|)$ behaves locally like a plane wave if the source location $\rho_0 \to \infty$. More precisely, referring

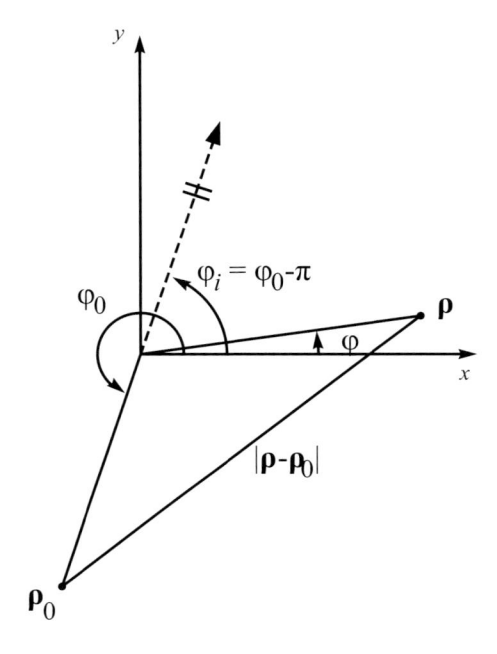

Figure 4.3 Geometry for addition theorem and plane wave expansion.

to Figure 4.3, since

$$\begin{aligned}
|\boldsymbol{\rho} - \boldsymbol{\rho}_0| &= \sqrt{\rho^2 - 2\boldsymbol{\rho} \cdot \boldsymbol{\rho}_0 + \rho_0^2} \\
&= \rho_0 \sqrt{1 - \frac{2\boldsymbol{\rho} \cdot \boldsymbol{\rho}_0}{\rho_0^2} + \frac{\rho^2}{\rho_0^2}} \\
&\simeq \rho_0 \left[1 - \frac{\boldsymbol{\rho} \cdot \boldsymbol{\rho}_0}{\rho_0^2} \right] \\
&= \rho_0 - \rho \cos(\phi - \phi_0)
\end{aligned} \tag{4.57}$$

for $\rho_0 \gg \rho$, we have from (B.14) that

$$H_0^{(2)}(k|\boldsymbol{\rho} - \boldsymbol{\rho}_0|) \simeq \sqrt{\frac{2}{\pi k \rho_0}} e^{-j(k\rho_0 - \pi/4)} e^{-jk\rho \cos(\phi - \phi_i)} \tag{4.58}$$

where $\phi_i = \phi_0 - \pi$ is the angle of incidence indicated in Figure 4.3. We know this because

$$e^{-jk\rho \cos(\phi - \phi_i)} = e^{-jk(x \cos \phi_i + y \sin \phi_i)}$$

is a plane wave traveling at the angle ϕ_i to the x-axis, so that regarding the multiplying factor

$$\sqrt{\frac{2}{\pi k \rho_0}} e^{-j(k\rho_0 - \pi/4)}$$

which is independent of ρ and ϕ, as a constant amplitude, (4.58) indeed represents a simple plane wave.

Now, the Hankel functions on the right side of (4.56) can also be replaced by their asymptotic expansions for $k\rho_> = k\rho_0 \gg 1$ (Equation (B.14)). Equating the result to (4.58) and letting $\rho_0 \to \infty$, we get

$$e^{-jk\rho\cos(\phi-\phi_i)} = \sum_{m=-\infty}^{\infty} e^{jm(\phi-\phi_i-\pi/2)} J_m(k\rho) \tag{4.59}$$

which is the desired expansion of a plane wave into Bessel functions, and will prove useful in problems involving scattering from cylinders.

On the other hand, we might notice that (4.59) (as indeed was (4.56)) is a Fourier series expansion like (4.47), and we could also attempt to obtain it using the formal theory of these series. So, for instance, by (4.47), we can write

$$e^{-jk\rho\cos\phi} = \sum_{m=-\infty}^{\infty} a_m e^{jm\phi} \tag{4.60}$$

where the coefficients a_m are given by

$$a_m = \frac{1}{2\pi} \int_0^{2\pi} e^{-jk\rho\cos\phi} e^{-jm\phi} \, d\phi \tag{4.61}$$

On comparing with (4.59), we obtain the integral representation (B.22) for J_m. A similar though more complicated identity could also be obtained based on (4.56), but is less useful in practice.

4.2.2 TIME-HARMONIC POINT SOURCE

Alternative representations for the spherical wave function that is the potential of a Hertz dipole source can be found in very similar ways. For example, consider Equation (4.19) in cylindrical coordinates:

$$\left(\frac{1}{\rho}\frac{\partial}{\partial\rho}\rho\frac{\partial}{\partial\rho} + \frac{1}{\rho^2}\frac{\partial^2}{\partial\phi^2} + \frac{\partial^2}{\partial z^2} + k^2\right) V = -\frac{p}{\epsilon}\delta(z-z_0)\delta(\boldsymbol{\rho}-\boldsymbol{\rho}_0) \tag{4.62}$$

Once again, we use a Fourier transform pair:

$$\tilde{V}(\boldsymbol{\rho},\beta) = \frac{1}{2\pi}\int_{-\infty}^{\infty} e^{j\beta z} V(\boldsymbol{\rho},z)\, dz$$

$$V(\boldsymbol{\rho},z) = \int_{-\infty}^{\infty} e^{-j\beta z}\tilde{V}(\boldsymbol{\rho},\beta)\, d\beta \tag{4.63}$$

The transform of (4.62) is

$$\left(\frac{1}{\rho}\frac{\partial}{\partial\rho}\rho\frac{\partial}{\partial\rho} + \frac{1}{\rho^2}\frac{\partial^2}{\partial\phi^2} + \eta^2\right)\tilde{V} = -\frac{p}{\epsilon}\frac{e^{j\beta z_0}}{2\pi}\delta(\boldsymbol{\rho}-\boldsymbol{\rho}_0) \tag{4.64}$$

where

$$\begin{aligned}
\eta &= \sqrt{k^2 - \beta^2}, \quad \text{Re}(\eta) > 0 \quad (\text{for } \beta^2 < k^2) \\
&= -j\sqrt{\beta^2 - k^2}, \quad \text{Im}(\eta) < 0 \quad (\text{for } \beta^2 > k^2)
\end{aligned} \quad (4.65)$$

analogous to (4.39). But this transformed equation is identical with that of the line source problem solved in (4.30) if we make the replacements $k \to \eta$ and $I_0 \to \frac{p}{2\pi} e^{j\beta z_0}$. Thus, the solution to (4.64) must be

$$\tilde{V} = \frac{p}{\epsilon} \frac{e^{j\beta z_0}}{2\pi} \frac{1}{4j} H_0^{(2)}(\eta|\boldsymbol{\rho} - \boldsymbol{\rho}_0|) \quad (4.66)$$

and using the inverse transform in (4.63):

$$V(\boldsymbol{\rho}, z) = \frac{p}{\epsilon} \frac{1}{8\pi j} \int_{-\infty}^{\infty} e^{-j\beta(z-z_0)} H_0^{(2)}(\eta|\boldsymbol{\rho} - \boldsymbol{\rho}_0|) \, d\beta \quad (4.67)$$

But this must be the same as the solution we obtained in the form (4.22), so we have the representation

$$\frac{e^{-jkR}}{R} = \frac{1}{2j} \int_{-\infty}^{\infty} e^{-j\beta(z-z_0)} H_0^{(2)}(\eta|\boldsymbol{\rho} - \boldsymbol{\rho}_0|) \, d\beta \quad (4.68)$$

of a spherical wave as a spectrum of cylindrical (or line source) waves. Here,

$$R = \sqrt{(z - z_0)^2 + |\boldsymbol{\rho} - \boldsymbol{\rho}_0|^2} \quad (4.69)$$

A further variant of this representation is obtained if we use the representation (4.45) for $H_0^{(2)}$ in (4.68), but replacing k by η. We have

$$\frac{e^{-jkR}}{R} = \frac{1}{2\pi j} \int_{-\infty}^{\infty} \int_{-\infty}^{\infty} \frac{e^{-j[\alpha(x-x_0)+\beta(z-z_0)+\gamma|y-y_0|]}}{\gamma} \, d\alpha d\beta \quad (4.70)$$

where $\gamma = \sqrt{k^2 - \alpha^2 - \beta^2}$, and $\text{Re}(\gamma) > 0$ when $\alpha^2 + \beta^2 < k^2$, and $\text{Im}(\gamma) < 0$ when $\alpha^2 + \beta^2 > k^2$. This again is a plane-wave spectrum representation, or double Sommerfeld integral. Observe, moreover, that since

$$R = \sqrt{(x - x_0)^2 + (y - y_0)^2 + (z - z_0)^2}$$

there is the freedom to interchange any of the quantities $x - x_0$, $y - y_0$ and $z - z_0$ with each other without changing the value of (4.70). For example,

$$\frac{e^{-jkR}}{R} = \frac{1}{2\pi j} \int_{-\infty}^{\infty} \int_{-\infty}^{\infty} \frac{e^{-j[\alpha(x-x_0)+\beta(y-y_0)+\gamma|z-z_0|]}}{\gamma} \, d\alpha d\beta \quad (4.71)$$

Equation (4.71) can be cast into an alternate form involving a single integral by making a change of integration variables into "spectral polar coordinates":

$$\alpha = \lambda \cos \chi, \quad \beta = \lambda \sin \chi \quad (4.72)$$

Then (4.71) becomes

$$\frac{e^{-jkR}}{R} = \frac{1}{2\pi j} \int_0^\infty \lambda\, d\lambda \int_0^{2\pi} d\chi \frac{e^{-j\gamma|z-z_0|}e^{-j\lambda|\boldsymbol{\rho}-\boldsymbol{\rho}_0|\cos(\chi-\phi_1)}}{\gamma}$$

where

$$\gamma = \sqrt{k^2 - \lambda^2} \qquad\qquad (4.73)$$

and

$$
\begin{aligned}
x - x_0 &= |\boldsymbol{\rho} - \boldsymbol{\rho}_0|\cos\phi_1 \\
y - y_0 &= |\boldsymbol{\rho} - \boldsymbol{\rho}_0|\sin\phi_1
\end{aligned}
\qquad\qquad (4.74)
$$

But the χ integral can be recognized (after a change of integration variable) as the integral representation of the Bessel function given in (B.22). Thus we arrive at the Sommerfeld integral

$$\frac{e^{-jkR}}{R} = \int_0^\infty e^{-j\gamma|z-z_0|} J_0(\lambda|\boldsymbol{\rho} - \boldsymbol{\rho}_0|)\frac{\lambda\, d\lambda}{j\gamma} \qquad\qquad (4.75)$$

A final variant, the need for which sometimes arises, is obtained by substituting the real part of (4.56) into (4.75), which gives

$$\frac{e^{-jkR}}{R} = \sum_{m=-\infty}^{\infty} e^{jm(\phi-\phi_0)} \int_0^\infty e^{-j\gamma|z-z_0|} J_m(\lambda\rho)J_m(\lambda\rho_0)\frac{\lambda\, d\lambda}{j\gamma} \qquad\qquad (4.76)$$

All of the representations obtained in this section are useful in formulating and solving various types of scattering problems.

4.3 RADIATION FROM SOURCES OF FINITE EXTENT; THE FRAUNHOFER FAR-FIELD APPROXIMATION

4.3.1 FAR FIELD BY SUPERPOSITION

Let us return to the consideration of a time-dependent current source $\mathbf{J}(\mathbf{r}, t)$ in infinite free space, this time fairly arbitrarily distributed in space and time. We stipulate that the source is zero for $t < 0$, and that it occupies a bounded region of space. The potential for this current source satisfies the time-domain equivalent of (2.12):

$$\nabla^2 \mathbf{A} - \frac{1}{c^2}\frac{\partial^2 \mathbf{A}}{\partial t^2} = -\mu_0 \mathbf{J} \qquad\qquad (4.77)$$

We can construct the solution of (4.77) using superposition of the solutions obtained in Section 4.1.2 for suddenly created dipoles located at various positions in space and time.

We begin by multiplying both sides of (4.17) by $\mathbf{J}(\mathbf{r}',t')$ and integrating over all space and time and using (4.16). The corresponding vector potential $\mathbf{A}$ is:

$$
\begin{aligned}
\mathbf{A}(\mathbf{r},t) &= \mu_0 \int \int_{\text{space}} G(|\mathbf{r}-\mathbf{r}'|,t-t')\mathbf{J}(\mathbf{r}',t')\,dV'dt' \\
&= \mu_0 \int_{\text{space}} \frac{\mathbf{J}(\mathbf{r}',t-|\mathbf{r}-\mathbf{r}'|/c)}{4\pi|\mathbf{r}-\mathbf{r}'|}dV'
\end{aligned}
\tag{4.78}
$$

An analogous derivation produces the scalar potential Φ from the charge distribution that corresponds to $\mathbf{J}$:

$$
\begin{aligned}
\Phi(\mathbf{r},t) &= \frac{1}{\epsilon_0} \int \int_{\text{space}} G(|\mathbf{r}-\mathbf{r}'|,t-t')\rho(\mathbf{r}',t')\,dV'dt' \\
&= \frac{1}{\epsilon_0} \int_{\text{space}} \frac{\rho(\mathbf{r}',t-|\mathbf{r}-\mathbf{r}'|/c)}{4\pi|\mathbf{r}-\mathbf{r}'|}dV'
\end{aligned}
\tag{4.79}
$$

If the origin of coordinates is located within the source distribution, and the observation distance r is much larger than the distance r' of any source point from the origin, then we can approximate

$$
|\mathbf{r}-\mathbf{r}'| \simeq r
\tag{4.80}
$$

in the denominators of the integrands of (4.78) and (4.79), but in the arguments of ρ and $\mathbf{J}$ we need to use the more accurate approximation

$$
|\mathbf{r}-\mathbf{r}'| = r\sqrt{1-2\frac{\mathbf{u}_r\cdot\mathbf{r}'}{r}+\left(\frac{r'}{r}\right)^2} \simeq r\left(1-\frac{\mathbf{u}_r\cdot\mathbf{r}'}{r}\right) = r-\mathbf{u}_r\cdot\mathbf{r}'
\tag{4.81}
$$

analogous to (4.57), where we have used the first two terms of the binomial expansion

$$
\sqrt{1+x} \simeq 1+\frac{x}{2}; \qquad (|x| \ll 1)
\tag{4.82}
$$

The so-called *far field* approximation of the potentials is now obtained:

$$
\mathbf{A}(\mathbf{r},t) \simeq \frac{\mathbf{F}_A(\mathbf{u}_r,t-r/c)}{r}; \qquad (r \to \infty)
\tag{4.83}
$$

where

$$
\mathbf{F}_A(\mathbf{u}_r,t) = \frac{\mu_0}{4\pi} \int_{\text{space}} \mathbf{J}(\mathbf{r}',t+\mathbf{u}_r\cdot\mathbf{r}'/c)dV'
\tag{4.84}
$$

and likewise

$$
\Phi(\mathbf{r},t) \simeq \frac{F_\Phi(\mathbf{u}_r,t-r/c)}{r}; \qquad (r \to \infty)
\tag{4.85}
$$

where

$$
F_\Phi(\mathbf{u}_r,t) = \frac{1}{4\pi\epsilon_0} \int_{\text{space}} \rho(\mathbf{r}',t+\mathbf{u}_r\cdot\mathbf{r}'/c)dV'
\tag{4.86}
$$

Here, $\mathbf{u}_r = \mathbf{r}/r$ is a shorthand for the dependence of a function on the angular coordinates θ and ϕ in a spherical coordinate system. An important point to note is that the "radiation patterns" $\mathbf{F}$ are not functions of r.

4.3.2   FAR FIELD VIA FOURIER TRANSFORM

Consider next an electric current line source of finite extent in the z-direction that is possibly nonuniformly distributed along its length:

$$\mathbf{J} = I(z)\mathbf{u}_z\delta(\boldsymbol{\rho}) \tag{4.87}$$

The function $I(z)$ is relatively arbitrary, except that we require that it vanishes for $|z| > l$, where $2l$ is the extent of the source in z. The Whittaker potential of this source is found by using the same approach that led to Equation (4.68). The transformed equation is

$$\left(\frac{1}{\rho}\frac{\partial}{\partial\rho}\rho\frac{\partial}{\partial\rho} + \eta^2\right)\tilde{V} = -\frac{\tilde{I}(\beta)}{j\omega\epsilon}\delta(\boldsymbol{\rho}) \tag{4.88}$$

The solution is thus

$$\tilde{V} = \frac{\tilde{I}(\beta)}{j\omega\epsilon}\left(\frac{1}{4j}\right)H_0^{(2)}(\eta\rho) \tag{4.89}$$

and so

$$V = -\frac{1}{4\omega\epsilon}\int_{-\infty}^{\infty}e^{-j\beta z}\tilde{I}(\beta)H_0^{(2)}(\eta\rho)\,d\beta \tag{4.90}$$

An alternate form of (4.90), which is probably more familiar, is obtained from the convolution theorem (A.61) for Fourier transforms, along with (4.68):

$$V = \frac{1}{j\omega\epsilon}\int_{-l}^{l}I(z')\frac{e^{-jkR}}{4\pi R}\,dz' \tag{4.91}$$

where

$$R = \sqrt{\rho^2 + (z - z')^2}$$

Equation (4.91) is convenient for extracting the approximate far-field behavior of V. Specifically, let $r = \sqrt{\rho^2 + z^2}$ be "large" in a sense to be made more precise in a moment. Then we can write

$$\begin{aligned}
R &= \sqrt{r^2 - 2rz'\cos\theta + z'^2} \\
&= r\sqrt{1 - 2\frac{z'}{r}\cos\theta + \left(\frac{z'}{r}\right)^2} \\
&\simeq r - z'\cos\theta + \frac{z'^2}{2r}\sin^2\theta + rO\left(\frac{z'^3}{r^3}\right)
\end{aligned} \tag{4.92}$$

where we have used

$$\sqrt{1 + x} \simeq 1 + \frac{x}{2} - \frac{x^2}{8} \tag{4.93}$$

which is a more accurate binomial expansion than (4.82), valid for small $|x|$. The angle θ is that of the spherical coordinate system as shown in Figure 4.4.

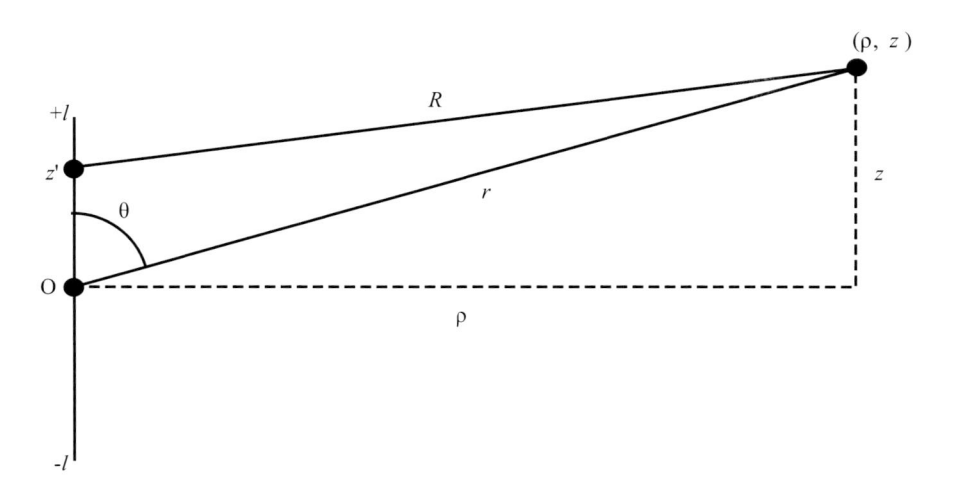

Figure 4.4 Geometry for the far-field approximation.

Since $|z'| \le l$, it seems reasonable that (4.92) should be accurate if l/r is small enough.

More precisely, if we approximate R by the first two terms of (4.92) in the exponent in the numerator of (4.91), while merely putting $R \simeq r$ in the denominator, we get the *Fraunhofer* approximation:

$$V \simeq \frac{e^{-jkr}}{j\omega\epsilon} \int_{-l}^{l} I(z') \frac{e^{jkz'\cos\theta}}{4\pi r} \, dz'$$

$$= \frac{\tilde{I}(k\cos\theta)}{j\omega\epsilon} \frac{e^{-jkr}}{2r} \qquad (4.94)$$

This expression can be expected to be accurate if the effect of the neglected terms is small. The approximation $R \simeq r$ in the denominator of (4.91) requires that

$$r \gg l \qquad (4.95)$$

while the neglect of the quadratic term in z' in the exponent of the numerator requires $r \gg kl^2$, or

$$\frac{r}{l} \gg kl \qquad (4.96)$$

Taken together, these conditions define the *Fraunhofer region*. If $kl \le 1$, then (4.95) implies (4.96), and the approximation is valid at distances from the source that are large compared to its transverse extent.

Upon comparison of (4.94) and (4.90), we see that we have actually obtained a general mathematical result about certain types of functions represented as Fourier integrals quite separate from the physical connotations of the original problem. Thus, making use only of the convolution theorem, we have:

$$\int_{-\infty}^{\infty} e^{-j\beta z} f(\beta) H_0^{(2)}(\eta\rho)\, d\beta \sim 2j \frac{e^{-jkr}}{r} f(k\cos\theta) \tag{4.97}$$

as $r \to \infty$, where $\eta = (k^2 - \beta^2)^{1/2}$ (the root taken as described earlier), and $\cos\theta = z/r$. An alternative way of expressing this formally is:

$$e^{-j\beta r\cos\theta} H_0^{(2)}(\eta r\sin\theta) \sim 2j\frac{e^{-jkr}}{r}\delta(\beta - k\cos\theta) \tag{4.98}$$

as $r \to \infty$. It is as if the only contribution to the integral on the left side of (4.97) is from values of β near $k\cos\theta$.[4]

4.3.3   THE STATIONARY PHASE PRINCIPLE

What is so special about this value of β? If we examine the integrand of (4.97), we see that when kr is large, there is a phase term that varies rapidly by comparison with the remaining part of the integrand: by (B.14), we have

$$e^{-j\beta r\cos\theta} H_0^{(2)}(\eta r\sin\theta) \sim \sqrt{\frac{2}{\pi\eta r\sin\theta}}\, e^{-j(\eta r\sin\theta + \beta r\cos\theta - \pi/4)}$$

$$= \sqrt{\frac{2}{\pi\eta r\sin\theta}}\, e^{-j(krP(\beta,\theta) - \pi/4)}$$

where

$$P(\beta,\theta) = \frac{\beta}{k}\cos\theta + \frac{\eta}{k}\sin\theta$$

If $kr \gg 1$, the integrand of (4.97) oscillates extremely rapidly as β varies, causing nearly complete cancelation so that almost nothing is contributed to the value of the integral. This behavior is shown in Figure 4.5. The only exception to this is near points where the function P becomes *stationary*; that is, when

$$\frac{\partial P}{\partial \beta} = 0$$

In the vicinity of such a value of $\beta = \beta_s$ (called a *stationary point*), there will be an important contribution to the value of the integral. This is known as the *principle of stationary phase*.[5] It is easily verified that for the function $P(\beta,\theta)$ given above, the point of stationary phase is indeed $\beta_s = k\cos\theta$.

---

[4]For a more detailed exploration of this point of view, see

R. Estrada and R. P. Kanwal, *Asymptotic Analysis: A Distributional Approach.* Boston: Birkhäuser, 1994.

[5]See:

A. Erdélyi, *Asymptotic Expansions.* New York: Dover, 1956.

F. W. J. Olver, *Asymptotics and Special Functions.* New York: Academic Press, 1974.

N. Bleistein and R. A. Handelsman, *Asymptotic Expansions of Integrals.* New York: Holt, Rinehart and Winston, 1975.

V. A. Borovikov, *Uniform Stationary Phase Method.* London: IEE, 1994.

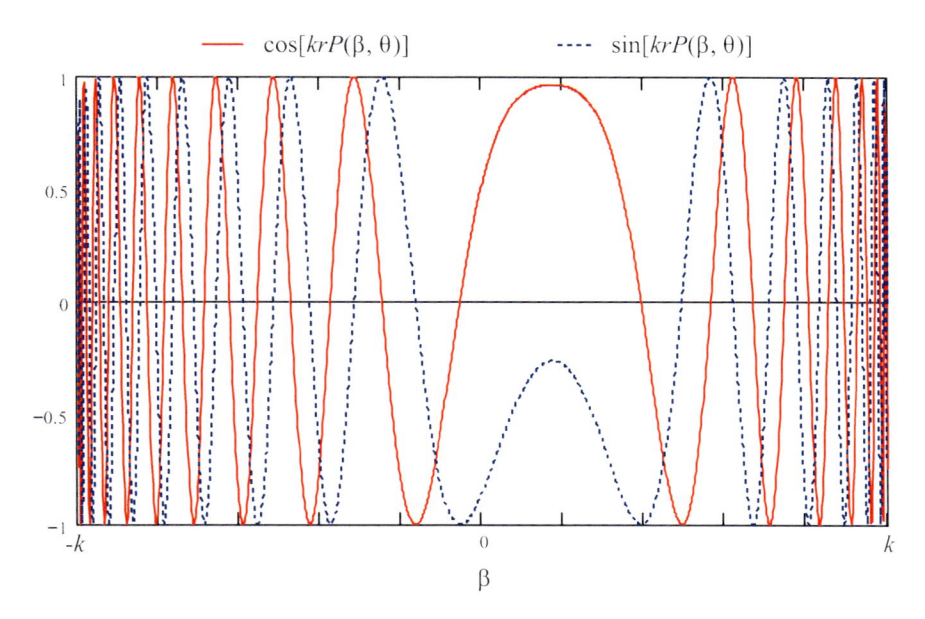

$$\begin{array}{ll} \text{——} \quad \cos[krP(\beta,\theta)] & \text{----} \quad \sin[krP(\beta,\theta)] \end{array}$$

Figure 4.5 Real part and (negative of the) imaginary part of $e^{-jkrP(\beta,\theta)}$ for $\theta = 80°$ and $kr = 50$.

A simple example is furnished by the integral:

$$I(Q) = \int_{-\infty}^{\infty} e^{-jQt^2} f(t)\, dt \tag{4.99}$$

where $f(t)$ is well-behaved near $t = 0$ and Q is large and positive. Clearly $t_s = 0$ is the only point of stationary phase here. For large values of Q, contributions to the value of the integral other than from a small neighborhood of the stationary–phase point are small, as argued above. Thus, we may approximate $f(t) \simeq f(0)$ in (4.99), and using (B.39) we obtain:

$$I(Q) \sim f(0)\sqrt{\frac{\pi}{Q}}\, e^{-j\pi/4} \tag{4.100}$$

A more complicated case is the integral

$$J(Q) = \int_{-\infty}^{\infty} f(t) e^{-jQP(t)}\, dt \tag{4.101}$$

where $P(t)$ is a real function of t possessing a single stationary point t_s at which $P'(t_s) = 0$. We now approximate $P(t) \simeq P(t_s) + (t - t_s)^2 P''(t_s)/2$ near the stationary point: this is the first three terms of the Taylor series (the second one is zero by the stationary phase condition). When t is far removed from t_s, we expect the rapid oscillation of the exponential function to give little contribution to the integral, so no large error is incurred by this approximation. We thus have, similar to the first example,

$$J(Q) \sim f(t_s) \sqrt{\frac{2\pi}{\pm QP''(t_s)}} e^{-jQP(t_s)\mp j\pi/4} \qquad (4.102)$$

where the upper signs are taken if $P''(t_s) > 0$ and the lower signs if $P''(t_s) < 0$.

Other similar results can also be obtained in this way. For example, application of the convolution theorem to integrals of the type (4.45) yields

$$\int_{-\infty}^{\infty} \frac{e^{-j\alpha x - j\xi|y|}}{\xi} f(\alpha)\, d\alpha \sim e^{-jk\rho + j\pi/4} \sqrt{\frac{2\pi}{k\rho}} f(k\cos\phi) \qquad (4.103)$$

as $k\rho \to \infty$, where $\xi = (k^2 - \alpha^2)^{1/2}$ and $\cos\phi = x/\rho$. Formally, this is equivalent to

$$\frac{e^{-j\alpha\rho\cos\phi - j\xi\rho|\sin\phi|}}{\xi} \sim e^{-jk\rho + j\pi/4} \sqrt{\frac{2\pi}{k\rho}} \delta(\alpha - k\cos\phi) \qquad (4.104)$$

This result could also have been obtained by the method of stationary phase. Both (4.97) and (4.103) required for their derivation that f was the Fourier transform of a finite function—one that vanishes outside a finite interval. Although it is possible to relax this requirement somewhat by requiring that the inverse transform of f merely decays to zero "sufficiently rapidly" as its argument approaches infinity, the use of these relations for functions that do not have the properties assumed in the derivation may lead to invalid results.

We are now in a position to reconsider the choices of sign made in (4.40) when obtaining the solution for the potential of a line source. Indeed, suppose that instead of (4.103) we investigate the far-field limit of

$$\int_{-\infty}^{\infty} \frac{e^{-j\alpha x + j\xi|y|}}{\xi} f(\alpha)\, d\alpha = \int_{-\infty}^{\infty} \frac{e^{-jk\rho P(\alpha)}}{\xi} f(\alpha)\, d\alpha \qquad (4.105)$$

as $k\rho \to \infty$, where now

$$P(\alpha) = \frac{\alpha}{k}\cos\phi - \frac{\xi}{k}|\sin\phi|$$

The stationary phase point is found from

$$0 = P'(\alpha) = \frac{1}{k}\left(\cos\phi + \frac{\alpha}{\xi}|\sin\phi|\right)$$

whose solution is found to be $\alpha_s = -k\cos\phi$. But then $P(\alpha_s) = -1$, and the far-field limit of (4.105) is

$$\int_{-\infty}^{\infty} \frac{e^{-j\alpha x + j\xi|y|}}{\xi} f(\alpha)\, d\alpha \sim e^{jk\rho - j\pi/4} \sqrt{\frac{2\pi}{k\rho}} f(-k\cos\phi) \qquad (4.106)$$

The exponential behavior in the variable ρ clearly violates the Sommerfeld radiation condition, and justifies the choice of sign in (4.40).

Double Sommerfeld integrals can also be treated by methods of this type. Integrals of the type (4.71) have a Fraunhofer far-field expansion of the form

$$\int_{-\infty}^{\infty} \int_{-\infty}^{\infty} f(\alpha, \beta) \frac{e^{-j(\alpha x + \beta y + \gamma|z|)}}{\gamma} \, d\alpha d\beta$$

$$\sim \quad 2\pi j \frac{e^{-jkr}}{r} f(k \sin \theta \cos \phi, k \sin \theta \sin \phi) \tag{4.107}$$

as $kr \to \infty$, where $\gamma = (k^2 - \alpha^2 - \beta^2)^{1/2}$. More generally, the stationary phase method can be extended to the double integral

$$J_2(Q) = \int_{-\infty}^{\infty} \int_{-\infty}^{\infty} e^{-jQP(t,u)} f(t, u) \, dt du \tag{4.108}$$

in which the phase function $P(t, u)$ has only the single stationary point (t_s, u_s) at which

$$\frac{\partial P}{\partial t} = \frac{\partial P}{\partial u} = 0$$

Then as $Q \to \infty$,

$$J_2(Q) \sim \frac{2\pi f(t_s, u_s)}{Q} \frac{e^{-jq_s \pi/2}}{\sqrt{|P_{tt} P_{uu} - P_{tu}^2|}} e^{-jQP(t_s, u_s)} \tag{4.109}$$

where

$$P_{tt} = \left. \frac{\partial^2 P}{\partial t^2} \right|_{(t_s, u_s)} \qquad P_{tu} = \left. \frac{\partial^2 P}{\partial t \partial u} \right|_{(t_s, u_s)} \qquad P_{uu} = \left. \frac{\partial^2 P}{\partial u^2} \right|_{(t_s, u_s)}$$

and q_s is either $+1$, -1 or 0, depending upon whether the eigenvalues of the matrix

$$\begin{bmatrix} P_{tt} & P_{tu} \\ P_{tu} & P_{uu} \end{bmatrix}$$

are both positive, both negative, or of opposite sign, respectively.

As we shall see, there are many useful applications of the foregoing results to radiation and diffraction problems.

4.4 RADIATION IN PLANAR REGIONS

Consider an aperture S in a conducting plane located at $z = 0$ (Figure 4.6). The aperture may be of either finite or infinite extent. Suppose there are sources of some kind behind the aperture (in $z < 0$—it may be fed by a waveguide, for example) that produce a tangential electric field $\mathbf{E}_{0t}(x, y)$ at $z = 0$ that vanishes outside the aperture. Leaving aside for now the question of how $\mathbf{E}_{0t}$ would be determined in practice—this question is addressed in

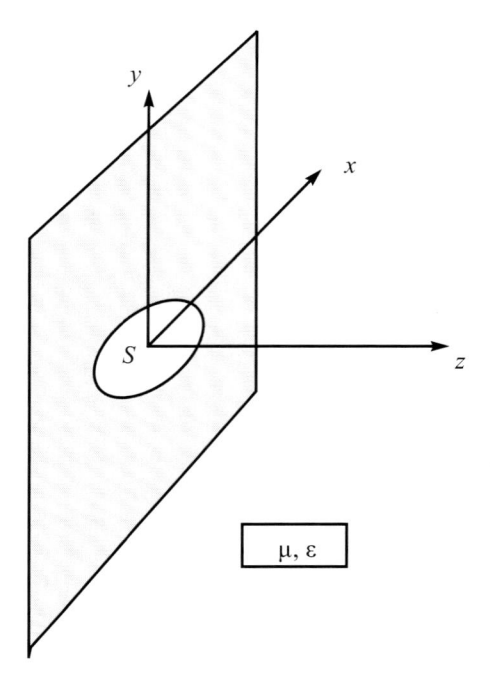

Figure 4.6   Aperture in a conducting plane.

Section 7.7—we assume that it is known, and ask how to find the field radiated by it into the forward half-space $z > 0$.

Although we could treat this problem using potentials, let us use the field directly in order to illustrate how to address a problem without using potentials. We have

$$
\begin{aligned}
(\nabla^2 + k^2)\mathbf{E} &= 0 \quad (z > 0) \\
\mathbf{E}_{\text{tan}}\big|_{z=0} &= \mathbf{E}_{0t}
\end{aligned} \tag{4.110}
$$

We represent $\mathbf{E}$ as a double Fourier transform analogous to (4.71):

$$
\mathbf{E} = \int_{-\infty}^{\infty} \int_{-\infty}^{\infty} \tilde{\mathbf{E}}(\alpha, \beta, z) e^{-j(\alpha x + \beta y)} \, d\alpha \, d\beta \tag{4.111}
$$

Then in the transform domain,

$$
\begin{aligned}
\left(\frac{d^2}{dz^2} + \gamma^2\right) \tilde{\mathbf{E}} &= 0 \quad (z > 0) \\
\tilde{\mathbf{E}}_{\text{tan}}(\alpha, \beta, 0) &= \tilde{\mathbf{E}}_{0t}(\alpha, \beta)
\end{aligned} \tag{4.112}
$$

and thus

$$
\tilde{\mathbf{E}} = \tilde{\mathbf{E}}_0(\alpha, \beta) e^{-j\gamma z} \quad (z > 0) \tag{4.113}
$$

The tangential part of $\tilde{\mathbf{E}}_0(\alpha, \beta)$ is determined from the second of Equations (4.112); the z-component must be found from other considerations. We will use the Maxwell divergence equation for $\mathbf{E}$:

$$\nabla \cdot \mathbf{E} = 0 \quad \Rightarrow \quad \frac{d\tilde{E}_z}{dz} - j(\alpha\tilde{E}_x + \beta\tilde{E}_y) = 0 \tag{4.114}$$

or, by (4.113),

$$\tilde{E}_{0z}(\alpha, \beta) = -\frac{\alpha}{\gamma}\tilde{E}_{0x}(\alpha, \beta) - \frac{\beta}{\gamma}\tilde{E}_{0y}(\alpha, \beta) \tag{4.115}$$

Thus,

$$\mathbf{E} = \int_{-\infty}^{\infty} \int_{-\infty}^{\infty} \left[\tilde{\mathbf{E}}_{0t}(\alpha, \beta) - \mathbf{u}_z \frac{\boldsymbol{\lambda} \cdot \tilde{\mathbf{E}}_{0t}}{\gamma} \right] e^{-j(\alpha x + \beta y + \gamma z)} \, d\alpha d\beta \tag{4.116}$$

where $\boldsymbol{\lambda} = \alpha\mathbf{u}_x + \beta\mathbf{u}_y$. The magnetic field can be found substituting this expression for $\mathbf{E}$ into Faraday's law: $\mathbf{H} = -\nabla \times \mathbf{E}/j\omega\mu$.

Now, by (4.107), the far-field of (4.116) is

$$\mathbf{E} \sim \frac{e^{-jkr}}{r}\mathbf{F}(\theta, \phi) \tag{4.117}$$

where

$$\begin{aligned}\mathbf{F}(\theta, \phi) &= 2\pi jk\cos\theta \left\{ \tilde{\mathbf{E}}_{0t}(k\sin\theta\cos\phi, k\sin\theta\sin\phi) \right. \\ &\left. -\mathbf{u}_z \tan\theta\mathbf{u}_\rho \cdot \tilde{\mathbf{E}}_{0t}(k\sin\theta\cos\phi, k\sin\theta\sin\phi) \right\} \end{aligned} \tag{4.118}$$

Converting the representation of $\mathbf{F}$ from cylindrical to spherical components, and noting that $\mathbf{u}_z \cdot \tilde{\mathbf{E}}_{0t} = 0$, we can (after some algebra) obtain

$$\mathbf{F}(\theta, \phi) = 2\pi jk[\cos\theta\mathbf{u}_\phi\mathbf{u}_\phi + \mathbf{u}_\theta\mathbf{u}_\rho] \cdot \tilde{\mathbf{E}}_{0t}(k\sin\theta\cos\phi, k\sin\theta\sin\phi) \tag{4.119}$$

Equation (4.117) will be accurate for r large compared to the extent of S (or, more precisely, to the extent of the portion of S on which $\mathbf{E}_{0t}$ is substantially different from zero). We can obtain the far-field expression for $\mathbf{H}$ from the radiation condition (3.35).

4.5 THE FRESNEL AND PARAXIAL APPROXIMATIONS; GAUSSIAN BEAMS

4.5.1 THE FRESNEL APPROXIMATION

The regions of validity for the "far-field" approximations (4.97)-(4.107)—and therefore in particular for expressions (4.117)-(4.119) giving the far field expressions of a radiating aperture—are the Fraunhofer region (4.95)-(4.96). For

(4.103), we must replace r by ρ in these conditions. If, however, the source is many wavelengths in extent ($kl \gg 1$), condition (4.96) may result in a Fraunhofer zone that is so far from the source as to have little practical use. To get around this difficulty, we re-examine the derivations that led to these far-field expressions, and modify them so that they are accurate much closer to the source than is the Fraunhofer approximation when $kl \gg 1$.[6]

Let us return to the problem of evaluating (4.91). Suppose we now retain the first three terms of (4.92) in the exponent in the numerator of (4.91), while still setting $R \simeq r$ in the denominator. We now have

$$\int_{-\infty}^{\infty} I(z') \frac{e^{-jk\sqrt{\rho^2 + (z-z')^2}}}{4\pi \sqrt{\rho^2 + (z-z')^2}} \, dz' \simeq \frac{e^{-jkr}}{4\pi r} \int_{-\infty}^{\infty} I(z') e^{jkz'\cos\theta - jkz'^2 \sin^2\theta/2r} \, dz'$$

(4.120)

Examination of the approximations that led to (4.120) shows that, in addition to the usual far field condition $r/l \gg 1$, we have the additional constraint $r^2 \gg kl^3$, or

$$\frac{r}{l} \gg (kl)^{1/2}$$

(4.121)

which is considerably less stringent than (4.96) if $kl \gg 1$. This new approximation is called the *Fresnel approximation*, and is often encountered in optics. It is not so simply related to the Fourier transform of the source distribution as was the Fraunhofer approximation. Similar expressions can be written for other wave functions, such as those on the left sides of (4.103) and (4.107), but their derivation will be left as an exercise for the reader.

4.5.2   THE PARAXIAL APPROXIMATION

An approximation related to (but distinct from) the Fresnel approximation is the *paraxial approximation*. Instead of (4.92), which is based upon $z' \ll r$ (i.e., $l \ll r$), we attempt a quadratic approximation of the terms in the exponent, which will be most accurate near a point of stationary phase. If $I(z')$ has no significant variations of phase, the stationary phase points of (4.91) are found from

$$\frac{\partial R}{\partial z'} = \frac{z' - z}{R} = 0$$

or at $z' = z$. Thus, let us write R as the first three terms of a Taylor series expansion about this stationary phase point (note that, as with our treatment of the integral (4.101) by the method of stationary phase, the linear term in

---

[6]A. Papoulis, *Systems and Transforms with Applications in Optics.* New York: McGraw-Hill, 1968, Chapter 9; S. Solimeno, B. Crosignani, and P. DiPorto, *Guiding, Diffraction and Confinement of Optical Radiation.* Orlando: Academic Press, 1986, Section 4.10.

this expansion vanishes):

$$
\begin{aligned}
R &= \sqrt{\rho^2 + (z - z')^2} \\
&\simeq \rho + \frac{(z - z')^2}{2\rho} + \rho O\left(\frac{(z - z')^4}{\rho^4}\right)
\end{aligned}
\tag{4.122}
$$

the validity of which now requires that $|z - z'| \ll \rho$ (or equivalently, $|z - z'| \ll r$). Like (4.92), this approximation retains quadratic terms in z', but unlike the previous expression, it poses additional constraints on the observation point: namely, z/r cannot be too large—the angle θ of observation must be close to $\pi/2$. In exchange for this restriction, the error we commit in the approximation (4.122) is smaller than that of (4.92). Thus, (4.122) is to be preferred in those cases when only observation points near a certain direction (the *paraxial region*) are of interest.

Now using (4.122) to approximate the phase term of (4.91), and $R \simeq \rho$ in the denominator of the integrand, we have

$$
\int_{-\infty}^{\infty} I(z') \frac{e^{-jk\sqrt{\rho^2 + (z - z')^2}}}{4\pi\sqrt{\rho^2 + (z - z')^2}} \, dz' \simeq \frac{e^{-jk\rho}}{4\pi\rho} \int_{-\infty}^{\infty} I(z') e^{-jk(z-z')^2/2\rho} \, dz'
\tag{4.123}
$$

and the conditions for validity of (4.123) are

$$
\begin{array}{cc}
\rho \gg |z| & \rho^3 \gg k|z|^4 \\
\rho \gg l & \rho/l \gg (kl)^{1/3}
\end{array}
\tag{4.124}
$$

of which the first two conditions are the paraxial constraint, while the second two are of the usual far-field variety. The last of these is even less restrictive than the corresponding ones from the Fraunhofer or Fresnel approximations if $kl \gg 1$.

Interestingly, the paraxial approximation (4.123) may be obtained by a completely different method as follows. From (4.90) and (4.91) the integral can be written as

$$
\int_{-\infty}^{\infty} I(z') \frac{e^{-jk\sqrt{\rho^2 + (z - z')^2}}}{4\pi\sqrt{\rho^2 + (z - z')^2}} \, dz' = \frac{1}{4j} \int_{-\infty}^{\infty} e^{-j\beta z} \tilde{I}(\beta) H_0^{(2)}(\eta\rho) \, d\beta
\tag{4.125}
$$

Now suppose that $\tilde{I}(\beta)$ differs significantly from zero only for $|\beta| \ll k$. To make this statement more precise, we suppose that

$$
\frac{\int_{-\infty}^{\infty} \beta^2 |\tilde{I}(\beta)|^2 \, d\beta}{\int_{-\infty}^{\infty} |\tilde{I}(\beta)|^2 \, d\beta} \equiv \beta_0^2 \ll k^2
$$

According to the uncertainty principle (A.62) for Fourier transforms, we have $\beta_0 z_0 \geq \frac{1}{2}$, where

$$
z_0^2 = \frac{\int_{-\infty}^{\infty} z^2 |I(z)|^2 \, dz}{\int_{-\infty}^{\infty} |I(z)|^2 \, dz}
$$

But if $I(z)$ is appreciably different from zero only in $|z| < l$, then $z_0 \leq l$, and combining this with the inequalities obtained above we get $kl \gg \frac{1}{2}$—a necessary condition for our assumption about $\tilde{I}(\beta)$ to be true. For the important range of β therefore, when $|\beta| \ll k$, we may approximate

$$\eta = \sqrt{k^2 - \beta^2} \simeq k - \frac{\beta^2}{2k} \tag{4.126}$$

Then from (B.14) if $k\rho \gg 1$, we can put

$$H_0^{(2)}(\eta\rho) \simeq \sqrt{\frac{2}{\pi k\rho}} e^{-j(k\rho - \beta^2 \rho/2k - \pi/4)}$$

so that (4.125) becomes

$$\int_{-\infty}^{\infty} I(z') \frac{e^{-jk\sqrt{\rho^2 + (z-z')^2}}}{4\pi\sqrt{\rho^2 + (z-z')^2}} \, dz' \tag{4.127}$$

$$\simeq \frac{1}{4j} \sqrt{\frac{2}{\pi k\rho}} e^{-j(k\rho - \pi/4)} \int_{-\infty}^{\infty} e^{-j\beta z + j\beta^2 \rho/2k} \tilde{I}(\beta) \, d\beta$$

By the convolution theorem and the identity

$$\int_{-\infty}^{\infty} e^{-t^2} \, dt = \sqrt{\pi} \tag{4.128}$$

this expression can be reduced to (4.123). But in deriving it this way, we employed only the restrictions

$$kl \gg 1 \quad \text{and} \quad \frac{\rho}{l} \ll (kl)^3$$

(in addition to $k\rho \gg 1$, which was necessary to invoke the large argument form of the Hankel function), and these are in essence *wide-source* and *near-field* conditions, rather than Fresnel or Fraunhofer-style far-field conditions. Hence (4.123) will often hold for all (or nearly all) ρ in the paraxial region.

Using similar techniques, Equation (4.116) can also be reduced to one of the two equivalent forms

$$\begin{aligned} \mathbf{E} &\simeq e^{-jkz} \int_{-\infty}^{\infty} \int_{-\infty}^{\infty} \tilde{\mathbf{E}}_{0t}(\alpha, \beta) e^{-j(\alpha x + \beta y) + j(\alpha^2 + \beta^2)z/2k} \, d\alpha d\beta \\ &= \frac{jk}{2\pi} \frac{e^{-jkz}}{z} \int_{S} \mathbf{E}_{0t}(x', y') e^{-jk|\boldsymbol{\rho} - \boldsymbol{\rho}'|^2/2z} \, dS' \end{aligned} \tag{4.129}$$

since $|\alpha/\gamma|$ and $|\beta/\gamma|$ are small compared to one whenever we can make the paraxial approximation $\gamma \simeq k - (\alpha^2 + \beta^2)/2k$.

4.5.3 GAUSSIAN BEAMS

An example of paraxial propagation is furnished by the *Gaussian beam*. Consider the linearly polarized aperture field

$$\mathbf{E}_{0t} = E_0 \mathbf{u}_x e^{-(x^2+y^2)/w_0^2} \tag{4.130}$$

in the plane $z = 0$. Although, strictly speaking, it is of infinite extent, this distribution decays rapidly as we move away from $\rho = 0$, reaching e^{-2} of its energy density by the time we reach $\rho = w_0$. We call w_0 the beamwidth or waist size of this distribution, and it will play the same role as the parameter l used above to characterize the width of the aperture. The radiated field in $z > 0$ that is produced by (4.130) is called a Gaussian beam.

Using (4.128) we can calculate the Fourier transform

$$
\begin{aligned}
\tilde{\tilde{\mathbf{E}}}_{0t}(\alpha, \beta) &= \mathbf{u}_x \frac{E_0}{4\pi^2} \int_{-\infty}^{\infty} \int_{-\infty}^{\infty} e^{-(x^2+y^2)/w_0^2} e^{j(\alpha x + \beta y)} \, dx \, dy \\
&= \mathbf{u}_x E_0 \frac{w_0^2}{4\pi} e^{-w_0^2(\alpha^2+\beta^2)/4}
\end{aligned}
\tag{4.131}
$$

Substitution into (4.129) gives another integral, which can be calculated with the help of (4.128). In fact,

$$
\begin{aligned}
\mathbf{E} &\simeq e^{-jkz} \int_{-\infty}^{\infty} \int_{-\infty}^{\infty} \mathbf{u}_x E_0 \frac{w_0^2}{4\pi} e^{-j(\alpha x + \beta y)} e^{-(\alpha^2+\beta^2)(w_0^2/4 - jz/2k)} \, d\alpha d\beta \\
&= \mathbf{u}_x E_0 \frac{w_0^2}{w_0^2 - j2z/k} e^{-jkz} e^{-(x^2+y^2)/(w_0^2 - j2z/k)}
\end{aligned}
\tag{4.132}
$$

Note that the magnitude of $\mathbf{E}$ is

$$|E_x| \simeq E_0 \frac{w_0}{w(z)} e^{-(x^2+y^2)/w^2(z)} \tag{4.133}$$

where $w(z)$ is the beamwidth of the broadened field in the plane $z = \text{const}$:

$$w^2(z) = w_0^2 + 4z^2/k^2 w_0^2 \tag{4.134}$$

When $z \ll k w_0^2$, there is negligible broadening of the beam. As z becomes comparable to $k w_0^2$ (called the *confocal parameter*), however, the beam begins to widen considerably, and its amplitude decreases (though not at first as rapidly as the $1/r$ decay it will eventually exhibit in the Fraunhofer region).

4.6   PROBLEMS

4–1 Consider the vector potential $\mathbf{A} = \mu_0\mathbf{p}_0 G$ of the pulsed dipole in free space, where G is given by (4.15). Show that a scalar potential Φ that obeys the Lorenz gauge condition (2.11) is given by

$$\Phi(\mathbf{r}, t) = \frac{\mathbf{p}_0 \cdot \mathbf{r}}{4\pi\epsilon_0 r^2}\left[\frac{\vartheta(t - r/c)}{r} + \frac{\delta(t - r/c)}{c}\right]$$

and is unique up to an additive constant. What is the meaning of the limiting value of this expression as $t \to \infty$?

4–2 Derive (4.103) by a convolution theorem method, analogous to that used to derive (4.97).

4–3 Repeat problem 4-2 using the method of stationary phase.

4–4 Fill in the details of the derivation of (4.106).

4–5 Show that the far-field result (4.97) can be obtained from the general stationary phase formula (4.102).

4–6 Derive (4.109).

4–7 Obtain expressions for the time-domain "far-field radiation patterns" $\mathbf{F}_E(\mathbf{u}_r, t)$ and $\mathbf{F}_H(\mathbf{u}_r, t)$ appearing in (3.29) in terms of the current density $\mathbf{J}(\mathbf{r}, t)$ and charge density $\rho(\mathbf{r}, t)$. Can you give conditions under which these far-field patterns will be zero?

4–8 A point charge Q is in motion, being located at the point $\mathbf{r}_0(t)$, where the position vector $\mathbf{r}_0$ is some given function of t. The charge density corresponding to this is thus

$$\rho(\mathbf{r}, t) = Q\delta[\mathbf{r} - \mathbf{r}_0(t)]$$

(a) Show that the current density of this moving charge is

$$\mathbf{J}(\mathbf{r}, t) = \rho(\mathbf{r}, t)\mathbf{v}_0(t)$$

i.e., show that this $\mathbf{J}$ together with the charge density above obeys conservation of charge. Here

$$\mathbf{v}_0(t) = \frac{d\mathbf{r}_0(t)}{dt}$$

is the velocity of the charge.

(b) Assume that the charge is moving more slowly than the speed of light ($|\mathbf{v}_0(t)| < c$). Find the scalar potential $\Phi(\mathbf{r}, t)$ and vector potential $\mathbf{A}(\mathbf{r}, t)$ produced by this moving charge, using (4.78) and (4.79). [These are called the Liénard-Wiechert potentials.]

4–9 Use (4.109) to prove (4.107).

4–10 If the far-field limits of a set of Whittaker potentials U and V are given as

$$U \sim \frac{e^{-jkr}}{r}F_U(\theta, \phi); \quad V \sim \frac{e^{-jkr}}{r}F_V(\theta, \phi)$$

obtain the pattern coefficient functions $\mathbf{F}_E$ and $\mathbf{F}_H$ in the far-field expansions

$$\mathbf{E} \sim \frac{e^{-jkr}}{r} \mathbf{F}_E(\theta, \phi); \quad \mathbf{H} \sim \frac{e^{-jkr}}{r} \mathbf{F}_H(\theta, \phi)$$

in terms of F_U and F_V.

4–11 (a) Derive Equation (4.119) from (4.118).

(b) Suppose that a TM_{11} rectangular waveguide mode field exists in a rectangular aperture $S: 0 \le x \le a, \ 0 \le y \le b$ in the plane $z = 0$:

$$\mathbf{E}_{0t} = E_0 \left[\mathbf{u}_x \frac{\cos \frac{\pi x}{a} \sin \frac{\pi y}{b}}{a} + \mathbf{u}_y \frac{\sin \frac{\pi x}{a} \cos \frac{\pi y}{b}}{b} \right]$$

Find the far-zone $\mathbf{E}$-field for this aperture distribution.

4–12 A circular aperture of radius a exists centered about the origin in an otherwise perfectly conducting plane $z = 0$. In this aperture, the tangential electric field is that of the TE_{01} mode of a circular metallic waveguide of radius a:

$$\mathbf{E}_{0t} = E_0 \mathbf{u}_\phi J_1 \left(j_{11} \frac{\rho}{a} \right)$$

where $j_{11} = 3.8317\ldots$ is the first root of the Bessel function J_1. Obtain an explicit far-field formula for $\mathbf{E}$ radiated by this aperture by using (4.117)-(4.119).

4–13 A time-harmonic field is produced by an aperture field

$$\mathbf{E}_{0t} = \mathbf{u}_z E_{0z} = \text{constant}$$

located on the surface of an infinite cylinder $\rho = a$. This field radiates into the infinite homogeneous space in $\rho > a$, whose parameters are ϵ and μ. Obtain expressions for the time-harmonic fields $\mathbf{E}$ and $\mathbf{H}$ produced by this aperture field. [Hint: Use a method analogous to that of Section 4.4.]

4–14 Repeat Problem 4-13, but find the field *inside* the surface of the cylinder in $\rho < a$. Under what conditions does the field become infinite? Give a physical explanation of that behavior.

4–15 Instead of (4.130), suppose that an aperture field in the plane $z = 0$ has the form of a Gaussian with a linear phase distribution:

$$\mathbf{E}_{0t} = E_0 \mathbf{u}_x e^{-(x^2+y^2)/w_0^2} e^{-j(k_x x + k_y y)}$$

where k_x and k_y are given real constants such that $k_x^2 + k_y^2 < k^2$.

(a) Compute the Fourier transform of $\mathbf{E}_{0t}$ with respect to x and y, and identify the values of α and β around which this spatial spectrum is concentrated.

(b) An approximation more appropriate to the result of part (a) for $\gamma = \sqrt{k^2 - \alpha^2 - \beta^2}$ than that used in (4.129) would be to expand in a Taylor series about $\alpha = k_x$ and $\beta = k_y$. Proceeding in this way, find a paraxial approximation to the field $\mathbf{E}$ in $z > 0$ due to this aperture field, and give a physical description of its behavior. Note that the paraxial direction will not necessarily be the z-direction.

4–16 Consider again the aperture field $\mathbf{E}_{0t}(x, y)$ given in Problem 4-15. In this problem, we will evaluate the radiated field directly in the spatial domain under a paraxial approximation, rather than going to the Fourier transform domain.

(a) Obtain an expression for $\mathbf{E}$ in $z > 0$ as an integral over the aperture field $\mathbf{E}_{0t}(x', y')$ multiplied by an appropriate function. [Hint: Use the Love equivalence principle to express the field in terms of equivalent magnetic surface currents located just in front of a perfectly conducting surface at $z = 0$.]

(b) Identify the phase term $e^{-jP(x', y')}$ of this integrand, and find the point of stationary phase that obeys

$$\frac{\partial P}{\partial x'} = \frac{\partial P}{\partial y'} = 0$$

(c) As a paraxial approximation to $\mathbf{E}$ in $z > 0$, expand the phase term as a quadratic function of x' and y' about the stationary point found above, and compute the resulting integral. Interpret your result physically.

4–17 Derive Fresnel approximations for the integrals on the left sides of (4.103) and (4.107).

4–18 Derive (4.129) from (4.116), stating the conditions for its validity.

5 Scattering by Simple Structures

In this chapter, we will examine the formulation and solution of the electromagnetic *scattering problem*, wherein a given source or incident field radiates in the presence of a material obstacle. Once again, the focus in this chapter will be on the use of differential equations to describe the fields; in Chapter 7 the same classes of problems will be tackled by integral equation methods.

5.1 DIPOLE RADIATION OVER A HALF-SPACE

Consider a vertical electric dipole (VED) located at a height z_0 above the interface between two media: a half-space $z < 0$ whose medium parameters are ($\hat{\epsilon}_2 = \epsilon_2 - j\sigma_2/\omega, \mu_2$) and a region above the interface whose parameters are ($\hat{\epsilon}_1, \mu_1$). For definiteness, we might think of medium 1 as air, and medium 2 as a lossy medium such as the earth. The problem is then that of finding the field produced by a small antenna located above the surface of the earth. If $\sigma_2 \to \infty$, the earth is a perfect conductor, and the method of image theory (Section 3.2.4) applies. Our interest is in what happens when a real earth is considered.

We use Whittaker potentials U and V. From (2.43)-(2.44), we find that the horizontal components of the field are expressed as

$$
\begin{aligned}
\mathbf{E}_t &= \nabla_t \frac{\partial V}{\partial z} - j\omega\mu \nabla \times (\mathbf{u}_z U) \\
\mathbf{H}_t &= \nabla_t \frac{\partial U}{\partial z} + j\omega\epsilon \nabla \times (\mathbf{u}_z V)
\end{aligned}
\tag{5.1}
$$

where the subscript t refers to the xy components of a vector or operator, and in the regions of space free from both actual and equivalent sources (that is, except at the source point $(0, 0, z_0)$ and at the interface), we have

$$
(\nabla^2 + k^2) \left\{ \begin{array}{c} U \\ V \end{array} \right\} = 0
\tag{5.2}
$$

We use k^2 to designate the quantity $\omega^2\mu\hat{\epsilon}$, which is equal to k_1^2 in $z > 0$, and k_2^2 in $z < 0$.

We again use the double Fourier transform:

$$
\left\{ \begin{array}{c} U \\ V \end{array} \right\} = \int_{-\infty}^{\infty} \int_{-\infty}^{\infty} \left\{ \begin{array}{c} \tilde{\tilde{U}}(\alpha, \beta, z) \\ \tilde{\tilde{V}}(\alpha, \beta, z) \end{array} \right\} e^{-j(\alpha x + \beta y)} \, d\alpha d\beta
\tag{5.3}
$$

We write the solutions to the resulting ordinary differential equations in general form as before:

$$
\begin{aligned}
\tilde{U} &= A_U(\alpha,\beta)e^{-j\gamma_1 z} \quad (z > z_0) \\
&= B_U(\alpha,\beta)e^{j\gamma_1 z} + C_U(\alpha,\beta)e^{-j\gamma_1 z} \quad (0 < z < z_0) \\
&= D_U(\alpha,\beta)e^{j\gamma_2 z} \quad (z < 0)
\end{aligned}
\tag{5.4}
$$

where

$$
\gamma_{1,2} = \sqrt{k_{1,2}^2 - \lambda^2},
$$

$\lambda^2 = \alpha^2 + \beta^2$ and square roots are defined as was done for the quantity γ used in Chapter 4. Similar expressions are used for $\tilde{V}$.

Boundary conditions are imposed on the tangential components of $\mathbf{E}$ and $\mathbf{H}$, or equivalently on $\tilde{\mathbf{E}}_t$ and $\tilde{\mathbf{H}}_t$ at the planes $z = 0$ and $z = z_0$. But

$$
\begin{aligned}
\tilde{\mathbf{E}}_t &= -j(\mathbf{u}_x\alpha + \mathbf{u}_y\beta)\frac{\partial\tilde{V}}{\partial z} - \omega\mu(-\mathbf{u}_y\alpha + \mathbf{u}_x\beta)\tilde{U} \\
\tilde{\mathbf{H}}_t &= -j(\mathbf{u}_x\alpha + \mathbf{u}_y\beta)\frac{\partial\tilde{U}}{\partial z} + \omega\hat{\epsilon}(-\mathbf{u}_y\alpha + \mathbf{u}_x\beta)\tilde{V}
\end{aligned}
\tag{5.5}
$$

For these components to be continuous at the plane $z = 0$, it can be shown that this requires[1]

$$
\frac{\partial\tilde{U}}{\partial z}, \quad \frac{\partial\tilde{V}}{\partial z}, \quad \mu\tilde{U} \quad \text{and} \quad \hat{\epsilon}\tilde{V} \quad \text{are continuous at } z = 0
\tag{5.6}
$$

Thus,

$$
\begin{aligned}
\mu_1(B_U + C_U) &= \mu_2 D_U \\
\hat{\epsilon}_1(B_V + C_V) &= \hat{\epsilon}_2 D_V \\
\gamma_1(B_U - C_U) &= \gamma_2 D_U \\
\gamma_1(B_V - C_V) &= \gamma_2 D_V
\end{aligned}
\tag{5.7}
$$

There now remains enforcement of continuity and jump conditions at $z = z_0$. Now, U has no source term associated with it, while V obeys Equation (4.62) with $\boldsymbol{\rho}_0 = 0$. Thus,

$$
\left(\frac{d^2}{dz^2} + \gamma_1^2\right)\tilde{V} = -\frac{p}{\hat{\epsilon}_1}\frac{\delta(z - z_0)}{4\pi^2} \quad (z > 0)
\tag{5.8}
$$

---

[1]Unless $\alpha = \beta = 0$. But this could only result in constant terms (in x and y) being added to the potentials, and these would have no effect on the values of the tangential fields. However, the normal field components are affected by the addition of such constants. and the enforcement of D_z and B_z continuous is consistent with the above conditions on the potential, even at $\alpha = \beta = 0$, so we will enforce them equally for all values of the spectral variables.

From (2.29) and (2.32), $\tilde{U}$ will be continuous together with its normal derivative at $z = z_0$, because of the absence of magnetic currents:

$$
\begin{aligned}
A_U e^{-j\gamma_1 z_0} &= B_U e^{j\gamma_1 z_0} + C_U e^{-j\gamma_1 z_0} \\
A_U e^{-j\gamma_1 z_0} &= -B_U e^{j\gamma_1 z_0} + C_U e^{-j\gamma_1 z_0}
\end{aligned}
\tag{5.9}
$$

On the other hand, $\tilde{V}$ is continuous, but its normal derivative has the jump

$$
\left. \frac{d\tilde{V}}{dz} \right|_{z_0^-}^{z_0^+} = -\frac{p}{\hat{\epsilon}_1} \frac{1}{4\pi^2}
\tag{5.10}
$$

so that

$$
\begin{aligned}
A_V e^{-j\gamma_1 z_0} - B_V e^{j\gamma_1 z_0} - C_V e^{-j\gamma_1 z_0} &= 0 \\
A_V e^{-j\gamma_1 z_0} + B_V e^{j\gamma_1 z_0} - C_V e^{-j\gamma_1 z_0} &= -\frac{jp}{\gamma_1 \hat{\epsilon}_1} \frac{1}{4\pi^2}
\end{aligned}
\tag{5.11}
$$

From (5.7), (5.9) and (5.11), we find that $A_U = B_U = C_U = D_U = 0$, so $U \equiv 0$. Furthermore, we have

$$
\begin{aligned}
A_V &= \frac{1}{2\pi j} \frac{V_0}{\gamma_1} [e^{j\gamma_1 z_0} + R(\alpha, \beta) e^{-j\gamma_1 z_0}] \\
B_V &= \frac{1}{2\pi j} \frac{V_0}{\gamma_1} e^{-j\gamma_1 z_0} \\
C_V &= \frac{1}{2\pi j} \frac{V_0}{\gamma_1} R(\alpha, \beta) e^{-j\gamma_1 z_0} \\
D_V &= \frac{1}{2\pi j} \frac{V_0}{\gamma_1} T(\alpha, \beta) e^{-j\gamma_1 z_0}
\end{aligned}
\tag{5.12}
$$

where

$$
V_0 = \frac{p}{4\pi \hat{\epsilon}_1}
\tag{5.13}
$$

and

$$
\begin{aligned}
R(\alpha, \beta) &= \frac{\hat{\epsilon}_2 \gamma_1 - \hat{\epsilon}_1 \gamma_2}{\hat{\epsilon}_2 \gamma_1 + \hat{\epsilon}_1 \gamma_2} \\
T(\alpha, \beta) &= \frac{2\hat{\epsilon}_1 \gamma_1}{\hat{\epsilon}_2 \gamma_1 + \hat{\epsilon}_1 \gamma_2}
\end{aligned}
\tag{5.14}
$$

From (5.3), (5.12), (5.14), and (4.71), we arrive at

$$
V = V_0 \left\{ \frac{e^{-jk_1 R_1}}{R_1} + \frac{1}{2\pi j} \int_{-\infty}^{\infty} \int_{-\infty}^{\infty} R(\alpha, \beta) e^{-j(\alpha x + \beta y) - j\gamma_1 (z + z_0)} \frac{d\alpha d\beta}{\gamma_1} \right\}
\tag{5.15}
$$

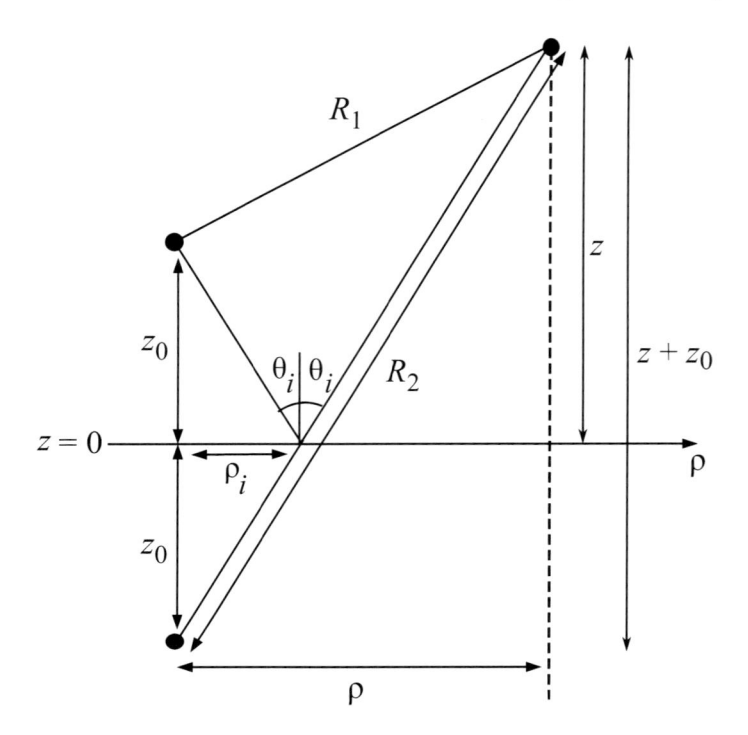

Figure 5.1 Geometry for the direct and reflected far field of a dipole above earth.

for $z > 0$, and

$$V = \frac{V_0}{2\pi j} \int_{-\infty}^{\infty} \int_{-\infty}^{\infty} T(\alpha, \beta) e^{-j(\alpha x + \beta y) - j\gamma_1 z_0 + j\gamma_2 z} \frac{d\alpha d\beta}{\gamma_1} \qquad (5.16)$$

for $z < 0$, where $R_1 = \sqrt{x^2 + y^2 + (z - z_0)^2}$ (see Figure 5.1). We see that the solution (5.15)-(5.16) consists of a spherical wave incident from the dipole, as well as reflected and transmitted waves synthesized from a spectrum of plane waves, reflected or transmitted at the interface, respectively.

5.1.1 REFLECTED WAVE IN THE FAR FIELD

We denote the reflected wave [the second term on the right side of (5.15)] by

$$V_R = \frac{V_0}{2\pi j} \int_{-\infty}^{\infty} \int_{-\infty}^{\infty} R(\alpha, \beta) e^{-j(\alpha x + \beta y) - j\gamma_1 (z + z_0)} \frac{d\alpha d\beta}{\gamma_1} \qquad (5.17)$$

It can be evaluated in the far field using the stationary phase approximation (4.107), and we have

$$V_R \sim V_0 \Gamma(\theta_i) \frac{e^{-jk_1 R_2}}{R_2} \qquad (5.18)$$

where $R_2 = \sqrt{x^2 + y^2 + (z + z_0)^2}$, $\theta_i = \tan^{-1}[\rho/(z + z_0)]$ (see Figure 5.1), and

$$\Gamma(\theta_i) = R(k_1 \sin\theta_i \cos\phi, k_1 \sin\theta_i \sin\phi) = \frac{\epsilon_2 k_1 \cos\theta_i - \epsilon_1 \sqrt{k_2^2 - k_1^2 \sin^2\theta_i}}{\epsilon_2 k_1 \cos\theta_i + \epsilon_1 \sqrt{k_2^2 - k_1^2 \sin^2\theta_i}}$$
(5.19)

Expression (5.19) will be recognized as the Fresnel reflection coefficient for a parallel-polarized plane-wave incident at an angle θ_i from the normal to the interface as shown in Figure 5.1. The reflected wave is thus equivalent to the field of a source located at the image point ($x = 0$, $y = 0$, $z = -z_0$), radiating a spherical wave whose amplitude is $\Gamma(\theta_i)$ times that of the incident wave. Alternatively, we can view the reflected wave as the result of an incident wave along a ray coming from the source at an angle θ_i, specularly reflecting at the interface $z = 0$, and propagating from there to the observation point. The fact that the spreading factor $1/R_2$ of the reflected wave in the far field is the same as that of a spherical wave is due to the law of reflection, which states that the angle of incidence is the same as the angle of reflection for a locally plane wave.

The foregoing analysis of the reflected wave tacitly assumed that there was no rapid variation in the phase of $R(\alpha, \beta)$ near the stationary phase point. In general, this assumption is not true, and this additional phase shift must be taken into account when applying the method of stationary phase. Let

$$R(\alpha, \beta) = A_R(\alpha, \beta)e^{j\chi_R(\alpha,\beta)}$$
(5.20)

where A_R and χ_R are real functions. We will modify the stationary phase approximation to (5.17) as follows. The magnitude of the reflection coefficient is approximated in the usual way by its value at the stationary phase point:

$$A_R(\alpha, \beta) \simeq A_R(\alpha_s, \beta_s)$$
(5.21)

where, as in (4.107), $\alpha_s = k_1 \sin\theta_i \cos\phi$ and $\beta_s = k_1 \sin\theta_i \sin\phi$. The phase of the reflection coefficient, on the other hand, is approximated by the first-order Taylor series approximation about the stationary phase point:

$$\chi_R(\alpha, \beta) \simeq \chi_R(\alpha_s, \beta_s) + (\alpha - \alpha_s) \left.\frac{\partial \chi_R}{\partial \alpha}\right|_{\alpha_s,\beta_s} + (\beta - \beta_s) \left.\frac{\partial \chi_R}{\partial \beta}\right|_{\alpha_s,\beta_s}$$
(5.22)

For reasons that will become apparent below, we denote the quantities

$$\Delta x = \left.\frac{\partial \chi_R}{\partial \alpha}\right|_{\alpha_s,\beta_s} ; \qquad \Delta y = \left.\frac{\partial \chi_R}{\partial \beta}\right|_{\alpha_s,\beta_s}$$
(5.23)

that have units of length. Substituting (5.22) and (5.23) into (5.17), we can pull several factors outside the integrations because they do not depend on α

or β. The result is

$$
\begin{aligned}
V_R &\simeq \frac{V_0}{2\pi j} R(\alpha_s, \beta_s) e^{-j(\alpha_s \Delta x + \beta_s \Delta y)} \\
&\quad \times \int_{-\infty}^{\infty} \int_{-\infty}^{\infty} e^{-j[\alpha(x-\Delta x)+\beta(y-\Delta y)]-j\gamma_1(z+z_0)} \frac{d\alpha d\beta}{\gamma_1} \\
&= V_0 R(\alpha_s, \beta_s) e^{-j(\alpha_s \Delta x + \beta_s \Delta y)} \frac{e^{-jk_1 R_2'}}{R_2'}
\end{aligned}
\tag{5.24}
$$

where $R_2' = \sqrt{(x - \Delta x)^2 + (y - \Delta y)^2 + (z + z_0)^2}$.

To visualize the meaning of this result, observe that the potential (5.17) is independent of the azimuthal angle ϕ in the xy-plane, so we can set $y = 0$ without loss of generality and view the wave in the xz-plane. Then $\alpha_s = k_1 \sin \theta_i$ and $\beta_s = 0$. Moreover, since $R(\alpha, \beta)$ is an even function of β, we have $\Delta y = 0$. Equation (5.24) then simplifies to:

$$
V_R \simeq V_0 \Gamma(\theta_i) e^{-jk_1 \Delta x \sin \theta_i} \frac{e^{-jk_1 R_2'}}{R_2'}
\tag{5.25}
$$

This result can be interpreted as shown in Figure 5.2. The reflected wave appears to emanate from an image source located at a point shifted horizontally

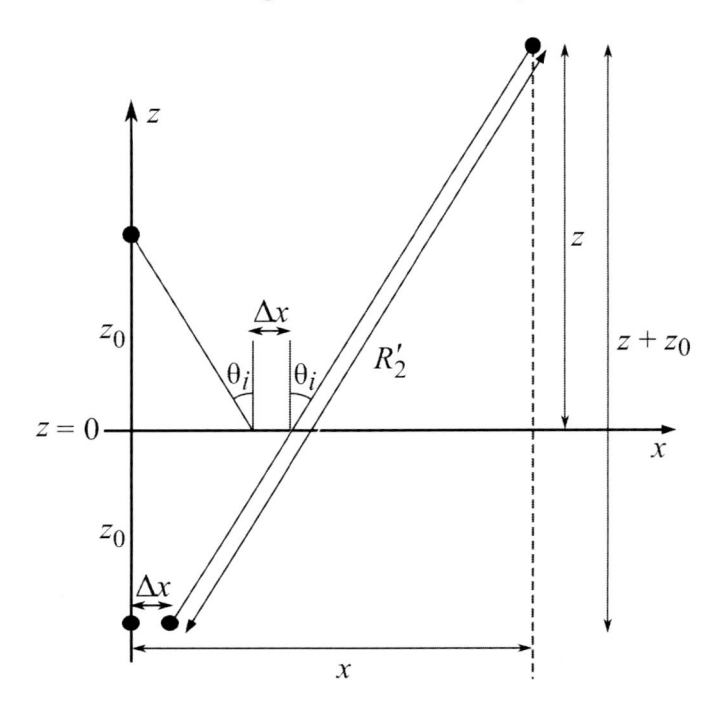

Figure 5.2   Geometry for shifted reflection from an interface.

by the distance Δx from the image location shown in Figure 5.1. Alternatively, the reflected wave can be thought of as emitted by the source dipole, traveling to the interface at $z = 0$ where it is reflected with reflection coefficient $\Gamma(\theta_i)$, and thence traveling to the observation point, having traveled along the interface a distance Δx before doing so. It is this horizontal shift that accounts for the additional factor $e^{-jk_1 \Delta x \sin \theta_i}$ in (5.25).

In the case of lossless nonmagnetic media ($\mu_1 = \mu_2 = \mu_0$ and $\epsilon_{1,2}$ are real), if $\epsilon_1 > \epsilon_2$, we can have γ_1 be real while γ_2 is imaginary, provided that $k_2 < \lambda < k_1$. For $\lambda = k_1 \sin \theta_i$, this situation produces total reflection of a plane wave. Under these conditions, $A_R \equiv 1$ in (5.20), and the phase shift of the reflection coefficient is given by

$$\chi_R(\alpha, \beta) = 2 \tan^{-1} \left(\frac{\epsilon_1}{\epsilon_2} \sqrt{\frac{\lambda^2 - k_2^2}{k_1^2 - \lambda^2}} \right) \tag{5.26}$$

For observation along the x-axis, we have from (5.23) (after some calculations) that

$$
\begin{aligned}
\Delta x &= \left. \frac{2\alpha}{\left[\alpha^2 \left(\frac{1}{k_1^2} + \frac{1}{k_2^2} \right) - 1\right] \sqrt{(\alpha^2 - k_2^2)(k_1^2 - \alpha^2)}} \right|_{\alpha = \alpha_s} \\
&= \frac{2 \tan \theta_i}{\left[\sin^2 \theta_i \left(1 + \frac{\epsilon_1}{\epsilon_2} \right) - 1\right] \sqrt{k_1^2 \sin^2 \theta_i - k_2^2}}
\end{aligned}
\tag{5.27}
$$

This distance is called the *Goos-Hänchen shift*,[2] and is observable experimentally in the reflection of Gaussian beams. Note that when θ_i approaches the critical angle $\theta_c = \sin^{-1}(k_2/k_1)$, Δx is predicted to become infinite. This is due to our assumption (5.22) becoming invalid, but more exact calculations do show a large value of the Goos-Hänchen shift in this case.

5.1.2 TRANSMITTED WAVE IN THE FAR FIELD

The far field of the transmitted potential (5.16) is found using the two-variable method of stationary phase (4.109). We will have to assume that both half-spaces are lossless (k_1 and k_2 must be real) in order to use this technique; the final result with losses present is identical, but must be derived by the more

---

[2]See

F. Goos and H. Hänchen, *Ann. Phys.*, vol. 43, pp. 383-392, 1943.

L. M. Brekhovskikh, *Waves in Layered Media*. New York: Academic Press, 1960, pp. 103-107.

K. W. Chiu and J. J. Quinn, *Amer. J. Phys.*, vol. 40, pp. 1847-1851, 1972.

J. A. Arnaud, *Beam and Fiber Optics*. New York: Academic Press, 1976, pp. 330-336.

general *method of steepest descent,* which is beyond the scope of our present treatment. In (5.16), we identify

$$QP(\alpha, \beta) = \alpha x + \beta y + \gamma_1 z_0 - \gamma_2 z \tag{5.28}$$

where Q is any convenient large parameter such as $k_1 R_1$ or $k_1 R_2$. By azimuthal symmetry, we can again set $x = \rho$ and $y = 0$ without loss of generality. The point of stationary phase is then found to be $(\alpha_s, \beta_s = 0)$, where α_s is the solution of:

$$0 = \left[\rho - z_0 \frac{\alpha}{\gamma_1} + z \frac{\alpha}{\gamma_2} \right]_{\beta=0} \tag{5.29}$$

Define two angles θ_i and θ_t from α_s as follows: $\alpha_s = k_1 \sin \theta_i = k_2 \sin \theta_t$. The stationary phase condition is expressed in terms of these angles as:

$$0 = \rho - z_0 \tan \theta_i + z \tan \theta_t \tag{5.30}$$

The physical meaning of these angles is shown in Figure 5.3. We draw a broken line between the source at $(0, 0, z_0)$ and the observation point $(\rho, 0, z)$ below the earth's surface. The position ρ_i of the break in this line is adjusted until the angles θ_i and θ_t between these lines and the z-axis obey Snell's law

$$k_1 \sin \theta_i = k_2 \sin \theta_t \tag{5.31}$$

The lengths of the two portions of this line are then:

$$L_1 = \frac{z_0}{\cos \theta_i}; \qquad L_2 = -\frac{z}{\cos \theta_t} \tag{5.32}$$

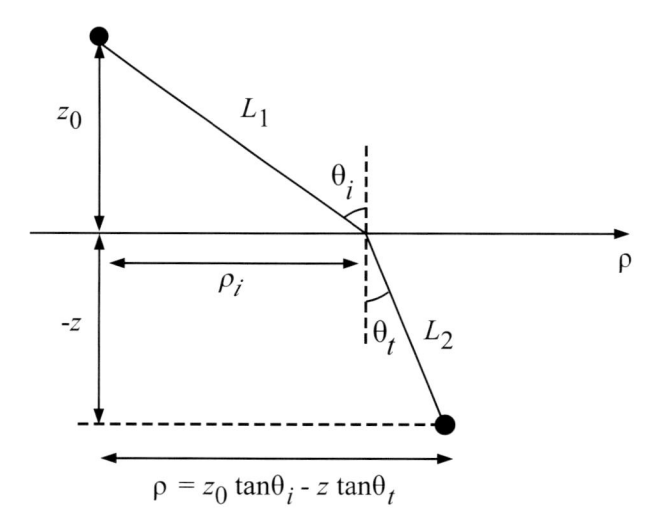

Figure 5.3   Geometry for the transmitted far field of a dipole above earth.

We now have

$$QP(\alpha_s, \beta_s) = k_1 L_1 + k_2 L_2 \tag{5.33}$$

and (leaving the remaining details to the reader),

$$V \sim \tau(\theta_i) V_0 \frac{e^{-j(k_1 L_1 + k_2 L_2)}}{\sqrt{(L_1 + k_1 L_2/k_2)[L_1 + k_1 L_2 \cos^2 \theta_i/(k_2 \cos^2 \theta_t)]}} \tag{5.34}$$

where

$$\tau(\theta_i) = T(k_1 \sin \theta_i, 0) = \frac{2\epsilon_1 k_1 \cos \theta_i}{\epsilon_2 k_1 \cos \theta_i + \epsilon_1 \sqrt{k_2^2 - k_1^2 \sin^2 \theta_i}} \tag{5.35}$$

is recognized as the Fresnel transmission coefficient for a parallel-polarized plane wave.

We see that the far field behaves *as if*: (i) the wave travelled as a local plane wave for a distance L_1 in the upper medium, and then a distance L_2 in the lower medium, as far as the phase of the wave is concerned; (ii) the amplitude of the transmitted wave is modified by the *plane-wave* transmission coefficient; (iii) in contrast with the $1/R$ amplitude dependence of a spherical wave in an infinite medium, we have here a *spreading factor*

$$\frac{1}{\sqrt{(L_1 + k_1 L_2/k_2)[L_1 + k_1 L_2 \cos^2 \theta_i/(k_2 \cos^2 \theta_t)]}} \tag{5.36}$$

which depends not only on the apparent distance traveled by the transmitted wave, but also on the relative wavenumbers of the media and the angles of incidence and transmission. This spreading factor is attributable to the variations in direction of the angle of transmission due to small variations of the incidence angle, because in fact the total field here is made up of an entire spectrum of incident plane waves, the most important of which are those near the specular directions shown in Figure 5.3. The spreading factor is more complicated than that of a plane wave because the directions of the refracted plane waves are determined by Snell's law rather than by the law of reflection. This feature is related to the Fermat principle in Problem 5-5.

5.1.3   OTHER DIPOLE SOURCES

A horizontally oriented dipole source above the earth can be treated in almost the same fashion, but one or two modifications to our technique are now necessary. The field can be expressed in terms of U and V only for $z \neq z_0$, since we are using z-directed Hertz potentials to express the fields of sources oriented in other directions. Therefore, Equation (5.8) no longer holds, and we cannot obtain the jump conditions on the potentials in the same way we did for the VED. We may appeal directly to (5.5), enforcing known discontinuities of the tangential field at a surface that contains surface currents. Details

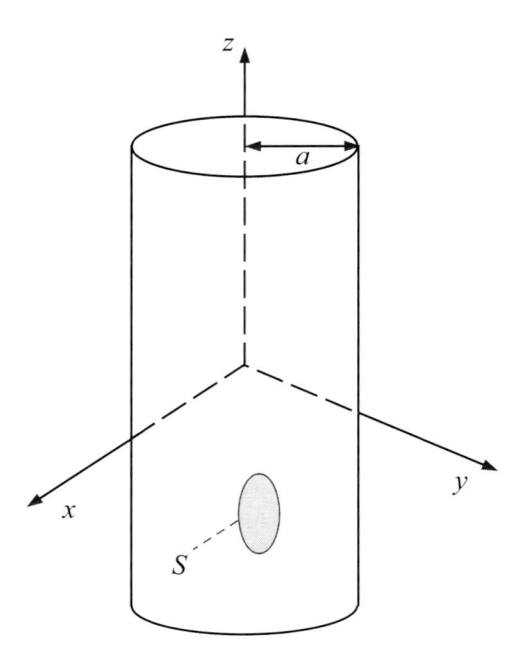

Figure 5.4   Aperture in a conducting cylinder.

are left as an exercise. We might only note that this method is tantamount to replacing the horizontally oriented current source with some equivalent vertically oriented electric and magnetic current distributions concentrated at $z = z_0$. The field thus obtained may differ from the actual field only by terms concentrated at $z = z_0$, i.e., $\delta(z - z_0)$, etc.

Alternatively, one might seek the solution in terms of Hertz potentials oriented in horizontal directions, preserving the jump condition method of the VED derivation, but requiring different expressions for the field in terms of the potentials. This method was actually the first one to be tried on the problem, and points out that while the fields themselves are uniquely determined, there is considerable flexibility in choosing potentials to represent them.

5.2   RADIATION AND SCATTERING FROM CYLINDERS

5.2.1   APERTURE RADIATION

Next, we take the case of a perfectly conducting circular cylinder that is infinitely long and has radius a. Suppose an aperture S is present in the surface, at which a known tangential electric field $\mathbf{E}_{0t}$ exists (Figure 5.4). This time, let us use the Whittaker potentials U and V to solve the problem.

By (2.33) and (2.38), we have

$$E_z \;=\; k^2 V + \frac{\partial^2 V}{\partial z^2} \tag{5.37}$$

$$E_\phi \;=\; \frac{1}{\rho}\frac{\partial^2 V}{\partial \phi \partial z} + j\omega\mu \frac{\partial U}{\partial \rho} \tag{5.38}$$

Using a Fourier transform in z and a Fourier series in ϕ, we write:

$$\left\{ \begin{array}{c} U \\ V \end{array} \right\} = \sum_{m=-\infty}^{\infty} e^{jm\phi} \int_{-\infty}^{\infty} d\beta \left\{ \begin{array}{c} \tilde{U}_m(\rho,\beta) \\ \tilde{V}_m(\rho,\beta) \end{array} \right\} e^{-j\beta z} \tag{5.39}$$

We have for $\tilde{V}_m$,

$$\left(\frac{1}{\rho}\frac{\partial}{\partial \rho}\rho\frac{\partial}{\partial \rho} + \eta^2 - \frac{m^2}{\rho^2} \right) \tilde{V}_m = 0 \quad (\rho > a) \tag{5.40}$$

and similarly for $\tilde{U}_m$. Then we get

$$\tilde{V}_m(\rho,\beta) = \frac{H_m^{(2)}(\eta\rho)}{H_m^{(2)}(\eta a)} \tilde{V}_m(a,\beta) \tag{5.41}$$

and so

$$V = \sum_{m=-\infty}^{\infty} e^{jm\phi} \int_{-\infty}^{\infty} d\beta \frac{H_m^{(2)}(\eta\rho)}{H_m^{(2)}(\eta a)} \tilde{V}_m(a,\beta) e^{-j\beta z} \tag{5.42}$$

which is a formal solution for V once $\tilde{V}_m(a,\beta)$ is determined. But this is done by transforming (5.37):

$$\eta^2 \tilde{V}_m(a,\beta) = \tilde{E}_{0z,m}\Big|_{\rho=a} \tag{5.43}$$

Likewise we construct a solution for U in the same way. However, from (5.38) we find that

$$\frac{m\beta}{a}\tilde{V}_m(a,\beta) + j\omega\mu \frac{\partial \tilde{U}_m}{\partial \rho}\Bigg|_{\rho=a} = \tilde{E}_{0\phi,m}\Big|_{\rho=a} \tag{5.44}$$

so for U we have

$$U = \sum_{m=-\infty}^{\infty} e^{jm\phi} \int_{-\infty}^{\infty} d\beta \frac{H_m^{(2)}(\eta\rho)}{\eta H_m^{(2)\prime}(\eta a)} \frac{\partial \tilde{U}_m(\rho,\beta)}{\partial \rho}\Bigg|_{\rho=a} e^{-j\beta z} \tag{5.45}$$

where the yet unknown function of β is found from (5.44).

As an example, suppose there is a uniform gap field around the circumference:

$$E_{0z}(a, \phi, z) \;=\; -\frac{V_0}{d} \quad (|z| < d/2)$$
$$=\; 0 \quad (|z| > d/2)$$
$$E_{0\phi}(a, \phi, z) \;\equiv\; 0$$

where V_0 is the slot voltage. Then $U \equiv 0$, while

$$\tilde{V}_m(a, \beta) \;=\; -\frac{V_0}{2\pi\eta^2}\frac{\sin(\beta d/2)}{\beta d/2} \quad (m = 0)$$
$$=\; 0 \quad (m \neq 0) \tag{5.46}$$

Then,

$$V = -\frac{V_0}{2\pi}\int_{-\infty}^{\infty} e^{-j\beta z}\frac{H_0^{(2)}(\eta\rho)}{H_0^{(2)}(\eta a)}\frac{\sin(\beta d/2)}{\eta^2 \beta d/2}\,d\beta \tag{5.47}$$

Appropriate differentiations of V then give the radiated field. From (4.97), we can obtain the far field as

$$V \sim -j\frac{V_0}{\pi}\frac{e^{-jkr}}{r}\frac{\sin(kd\cos\theta/2)}{k^2\sin^2\theta H_0^{(2)}(ka\sin\theta)kd\cos\theta/2} \tag{5.48}$$

as $r \to \infty$. If $kd \ll 1$, then this simplifies a little to

$$V \sim -j\frac{V_0}{\pi}\frac{e^{-jkr}}{r}\frac{1}{k^2\sin^2\theta H_0^{(2)}(ka\sin\theta)}$$

This result will be valid for any gap field that is independent of ϕ and corresponds to the same voltage V_0.

5.2.2   PLANE WAVE SCATTERING

Consider next a scattering problem: a plane wave (see (4.59))

$$E_z^{\text{inc}} = E_0 e^{-jk\rho\cos(\phi-\phi_i)} = E_0 \sum_{m=-\infty}^{\infty} e^{jm(\phi-\phi_i-\pi/2)} J_m(k\rho) \tag{5.49}$$

is incident onto a perfectly conducting circular cylinder of radius a (Figure 5.5). A surface current will be induced on the surface of the cylinder, which in turn will produce a scattered field E_z^s in space, such that the total field—the sum $E_z^{\text{inc}} + E_z^s$—is zero at $\rho = a$. We write

$$E_z^s = \sum_{m=-\infty}^{\infty} E_{zm}^s e^{jm\phi} H_m^{(2)}(k\rho) \tag{5.50}$$

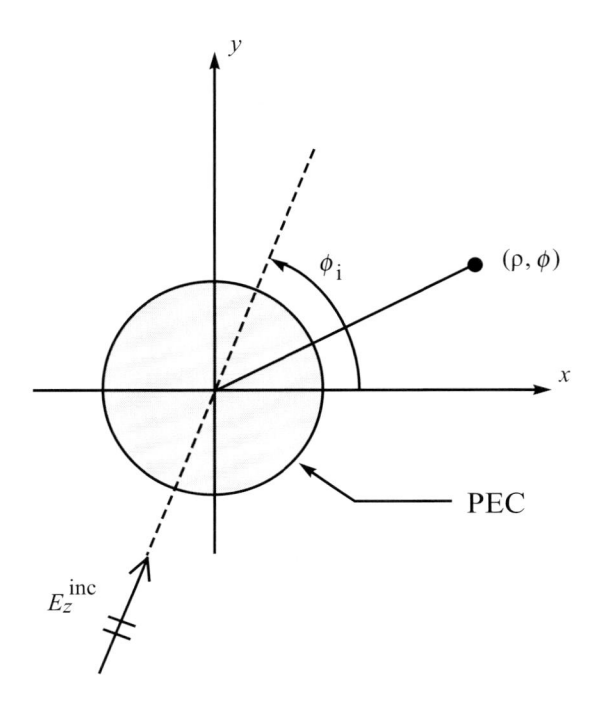

Figure 5.5 Scattering of a plane wave by a circular cylinder.

in $\rho > a$, since E_z^s satisfies the homogeneous Helmholtz equation. From (5.49), (5.50) and the boundary condition,

$$0 = \sum_{m=-\infty}^{\infty} e^{jm\phi} \left[E_{zm}^s H_m^{(2)}(ka) + E_0 J_m(ka) e^{-jm(\phi_i + \pi/2)} \right] \tag{5.51}$$

Therefore,

$$E_{zm}^s = -E_0 e^{-jm(\phi_i + \pi/2)} \frac{J_m(ka)}{H_m^{(2)}(ka)} \tag{5.52}$$

and so

$$E_z^s = -E_0 \sum_{m=-\infty}^{\infty} e^{jm(\phi - \phi_i - \pi/2)} \frac{J_m(ka)}{H_m^{(2)}(ka)} H_m^{(2)}(k\rho) \tag{5.53}$$

From the large argument expansion (B.14) for $H_m^{(2)}$, we can find the far field of the scattered wave:

$$E_z^s \sim -E_0 \sqrt{\frac{2}{\pi k\rho}} e^{-j(k\rho - \pi/4)} F(\phi - \phi_i) \tag{5.54}$$

where the radiation pattern is given by

$$F(\phi) = \sum_{m=-\infty}^{\infty} e^{jm\phi} \frac{J_m(ka)}{H_m^{(2)}(ka)} \tag{5.55}$$

If ka is large, many terms of (5.55) must be summed before practical convergence takes place, so it is inefficient for electrically large cylinders. However, if ka is small,

$$\frac{J_m(ka)}{H_m^{(2)}(ka)} = O[(ka)^{2|m|}]$$

for $m \neq 0$, so for an electrically thin cylinder, only the $m = 0$ term in the series is important. This implies that the induced current on the cylinder surface is essentially independent of the angle ϕ, and that the scattered field's radiation pattern is approximately

$$F(\phi) \simeq \frac{J_0(ka)}{H_0^{(2)}(ka)} \simeq \frac{-j\pi}{2\ln(2/ka) - 2\gamma_E - j\pi} \tag{5.56}$$

where γ_E is Euler's constant. For small ka, then, the radiation pattern also is essentially independent of ϕ.

As an example, if $ka = 0.01$ and $k\rho = 10$, then

$$\left| \frac{E_z^s}{E_0} \right| \simeq 0.08$$

which is a significant magnitude.

5.3 DIFFRACTION BY WEDGES; THE EDGE CONDITION

5.3.1 FORMULATION

We next look at diffraction of a wave by a perfectly conducting wedge of angle θ as shown in Figure 5.6.[3] As before, we will use as our source an electric line current I_0 located at the point $\boldsymbol{\rho}_0 = (\rho_0, \phi_0)$. Then,

$$\left(\frac{1}{\rho} \frac{\partial}{\partial \rho} \rho \frac{\partial}{\partial \rho} + \frac{1}{\rho^2} \frac{\partial^2}{\partial \phi^2} + k^2 \right) V = -\frac{I_0}{j\omega\epsilon} \frac{\delta(\rho - \rho_0)\delta(\phi - \phi_0)}{\rho} \tag{5.57}$$

in the region $0 < \phi < 2\pi - \theta$, but not inside the conducting wedge, nor for $\phi < 0$ or $\phi > 2\pi$. Because the Helmholtz equation for V no longer holds

---

[3]Many approaches to this problem have been taken over the years. Ours is adapted from:

W. H. Jackson, *Proc. London Math. Soc.*, ser. 2, vol. 1, pp. 393-414, 1904.

S. Blume and K. H. Wittich, *Arch. Elektrotech.*, vol. 74, pp. 321-328, 1991.

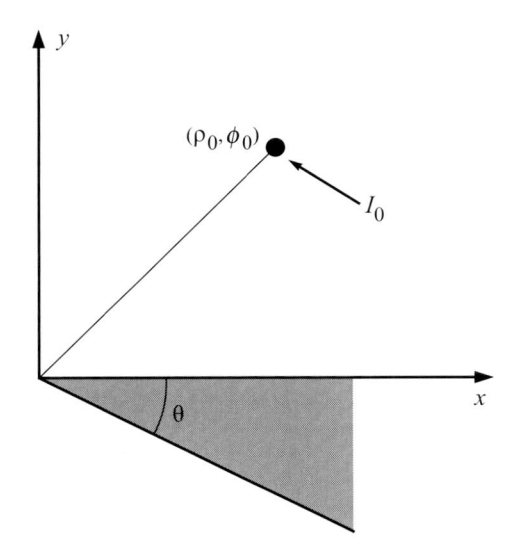

Figure 5.6 Electric line source radiating in the presence of a conducting wedge.

continuously as ϕ increases from 0 to beyond 2π, it is no longer possible to use the complex exponential Fourier series to expand the field as was done previously for problems in infinite space and exterior to the circular cylinder. However, taking account of the fact that $V = 0$ at both $\phi = 0$ and $\phi = 2\pi - \theta$, an appropriate expansion is a Fourier *sine* series as follows:

$$V = \sum_{m=1}^{\infty} V_m(\rho) \sin m\nu\phi \qquad (5.58)$$

where

$$\nu = \frac{\pi}{2\pi - \theta}$$

The transform pair for such a Fourier sine series is a simple modification of (A.72):

$$f(\phi) = \sum_{m=1}^{\infty} f_m \sin m\nu\phi$$

$$f_m = \frac{2\nu}{\pi} \int_0^{2\pi-\theta} f(\phi) \sin m\nu\phi \, d\phi \qquad (5.59)$$

so (5.57) transforms to

$$\left(\frac{1}{\rho} \frac{\partial}{\partial\rho} \rho \frac{\partial}{\partial\rho} + k^2 - \frac{m^2\nu^2}{\rho^2} \right) V_m = -\frac{I_0}{j\omega\epsilon} \frac{2\nu}{\pi} \sin m\nu\phi_0 \frac{\delta(\rho - \rho_0)}{\rho} \qquad (5.60)$$

5.3.2   THE EDGE CONDITION

The solution of (5.60) for $\rho < \rho_0$ must be treated with a little more care than in previous cases, since there is a sharp metallic edge at $\rho = 0$ that was not present before. Since charge or current buildup could conceivably occur near this edge, the exclusion of a term involving $Y_{m\nu}(k\rho)$ from our solution is no longer automatic.

Now, as noted in Section 3.2.3, we must require for uniqueness of the field solution as well as on grounds of physical reasonableness that the stored energy in any finite volume of space be finite. This means that

$$\int_V |E_u|^2 \, dV \quad , \quad \int_V |H_u|^2 \, dV$$

must all be finite, where u denotes any component of either field. But $dV = \rho \, d\rho \, d\phi \, dz$, so if a typical one of these field components varies as $O(\rho^\tau)$ when $\rho \to 0$, then the stored energy in a cylindrical volume centered about the edge behaves as

$$\int_0^{\text{const}} \rho^{2\tau} \rho \, d\rho = \left. \frac{\rho^{2\tau+2}}{2\tau + 2} \right|_0^{\text{const}}$$

which is less than infinity only if $\tau > -1$. This is a general constraint on the possible singularity of a field component independent of the physical geometry of a particular problem (e.g., this is true whether there is a wedge-like edge, a cone-like tip, or any other special feature of the scatterer). It is often referred to as the *edge condition* for the field near sharp edges or other singular features of a scatterer.

If a term $Y_{m\nu}(k\rho)$ were present in our solution for the wedge problem, the field behavior near the edge would possess the terms

$$
\begin{aligned}
E_z &\propto Y_{m\nu}(k\rho) &=& \quad O(\rho^{-m\nu}) \\
H_\phi &= -j\omega\epsilon \frac{\partial V}{\partial \rho} &=& \quad O(\rho^{-m\nu-1}) \\
H_\rho &= \frac{j\omega\epsilon}{\rho} \frac{\partial V}{\partial \phi} &=& \quad O(\rho^{-m\nu-1})
\end{aligned}
\tag{5.61}
$$

as $\rho \to 0$, since from (B.17)

$$Y_\nu(z) = O(z^{-\nu}) \qquad \text{as} \quad z \to 0$$

for $\nu \neq 0$, and in this problem, the order is equal to $m\nu$, where

$$\frac{1}{2} < \nu = \frac{\pi}{2\pi - \theta} < \infty$$

Therefore, the field behavior near the edge is of the general form described above, with $\tau = -m\nu$ or $-m\nu - 1$. But this means that $\tau = -m\nu - 1 <$

$-1 - m/2$, which is disallowed by the finite energy criterion. Thus this edge condition states that the solution for V can contain no terms in $Y_{m\nu}(k\rho)$.

Note that this is not to say that the field necessarily remains finite as $\rho \to 0$. For example, since

$$J_\nu(z) = O(z^\nu) \qquad \text{as} \quad z \to 0$$

and if the wedge has an *exterior* point—$0 < \theta < \pi$—then ν is between $1/2$ and 1, so that

$$
\begin{aligned}
E_z \propto J_\nu(k\rho) &= O(\rho^\nu) \to 0 \\
H_\phi &= O(\rho^{\nu-1}) \\
H_\rho &= O(\rho^{\nu-1})
\end{aligned}
\tag{5.62}
$$

as $\rho \to 0$. The nature of the singularity at the edge of the wedge is, however, restricted by the geometry of the edge, and for this problem we have the condition that $\tau = \nu - 1$ lies between $-1/2$ and 0.

5.3.3 FORMAL SOLUTION OF THE PROBLEM

We can thus write the formal solution of (5.60) as

$$V_m(\rho) = A_m J_{m\nu}(k\rho_<) H^{(2)}_{m\nu}(k\rho_>) \tag{5.63}$$

where $\rho_{>,<}$ have the usual meanings. We have already enforced continuity of V_m at $\rho = \rho_0$; all that remains is the jump condition:

$$
\begin{aligned}
\rho \left. \frac{dV_m}{d\rho} \right|_{\rho_0^-}^{\rho_0^+} &= -\frac{I_0}{j\omega\epsilon} \frac{2\nu}{\pi} \sin m\nu\phi_0 \\
&= k\rho_0 A_m [J_{m\nu}(k\rho_0) H^{(2)\prime}_{m\nu}(k\rho_0) - J'_{m\nu}(k\rho_0) H^{(2)}_{m\nu}(k\rho_0)] \\
&= -\frac{2j}{\pi} A_m
\end{aligned}
\tag{5.64}
$$

or,

$$A_m = \frac{I_0}{j\omega\epsilon} (-j\nu \sin m\nu\phi_0) \tag{5.65}$$

Thus,

$$V = \frac{I_0}{j\omega\epsilon} (-j\nu) \sum_{m=1}^{\infty} \sin m\nu\phi_0 \sin m\nu\phi J_{m\nu}(k\rho_<) H^{(2)}_{m\nu}(k\rho_>) \tag{5.66}$$

As in Equations (4.57) ff., let us examine the limit as the line source recedes to infinity to arrive at the case of an incident plane wave. Let I_0 vary with ρ_0 in such a way that

$$V_0 = \frac{I_0}{j\omega\epsilon} \frac{1}{4j} \sqrt{\frac{2}{\pi k\rho_0}} e^{-j(k\rho_0 - \pi/4)}$$

remains constant. Then as $\rho_0 \to \infty$, the incident wave becomes

$$V^{\mathrm{inc}} = \frac{I_0}{j\omega\epsilon}\frac{1}{4j}H_0^{(2)}(k|\boldsymbol{\rho} - \boldsymbol{\rho}_0|) \to V_0 e^{-jk\rho\cos(\phi-\phi_i)} \tag{5.67}$$

where $\phi_i = \phi_0 + \pi$ is the direction of propagation of the incident plane wave. In this case, the *total* field resulting from the plane wave incident at the wedge (by (5.66) and (B.14)) can be written as

$$\begin{aligned} V &= V_0 4\nu \sum_{m=1}^{\infty} e^{jm\nu\pi/2} J_{m\nu}(k\rho) \sin m\nu\phi_0 \sin m\nu\phi \\ &= V_0 2\nu \sum_{m=1}^{\infty} e^{jm\nu\pi/2} J_{m\nu}(k\rho)[\cos m\nu(\phi - \phi_0) - \cos m\nu(\phi + \phi_0)] \\ &\equiv V_0[W_\theta(\rho, \phi - \phi_0) - W_\theta(\rho, \phi + \phi_0)] \end{aligned} \tag{5.68}$$

where we have defined

$$W_\theta(\rho, \phi) = 2\nu \sum_{m=1}^{\infty} e^{jm\nu\pi/2} J_{m\nu}(k\rho) \cos m\nu\phi + \nu J_0(k\rho) \tag{5.69}$$

The extra term involving $J_0(k\rho)$ makes no difference to the value of V in (5.68), but is convenient during our subsequent analysis of the function W_θ.

5.3.4 THE GEOMETRICAL OPTICS FIELD

To obtain information about the far-field of the *scattered* wave from (5.68) is more difficult than in previous problems we have studied, because no portion of (5.68) is separately identifiable as the incident wave. To extract the incident wave in explicit form, we need to use the more general integral representation (B.23) for J_ν with *non-integer* order ν:

$$J_\nu(z) = \frac{e^{-j\nu\pi/2}}{2\pi}\left\{\int_{-\pi}^{\pi} e^{jz\cos\chi + j\nu\chi}\,d\chi - 2\sin\nu\pi\int_0^{\infty} e^{-\nu t - jz\cosh t}\,dt\right\} \tag{5.70}$$

By (5.70), we can split (5.69) into two parts:

$$W_\theta(\rho, \phi) = W_\theta^{(1)}(\rho, \phi) + W_\theta^{(2)}(\rho, \phi) \tag{5.71}$$

where

$$W_\theta^{(1)}(\rho, \phi) = \frac{\nu}{\pi}\left\{\sum_{m=1}^{\infty}\int_{-\pi}^{\pi} e^{jk\rho\cos\chi + jm\nu\chi}\,d\chi \cos m\nu\phi + \frac{1}{2}\int_{-\pi}^{\pi} e^{jk\rho\cos\chi}\,d\chi\right\} \tag{5.72}$$

$$W_\theta^{(2)}(\rho, \phi) = -\frac{2\nu}{\pi}\sum_{m=1}^{\infty}\cos m\nu\phi\sin m\nu\pi\int_0^{\infty} e^{-m\nu t - jk\rho\cosh t}\,dt \tag{5.73}$$

The summations over m can be brought inside the integrals if they are considered in the sense of generalized functions.

Since

$$\frac{1}{2} + \sum_{m=1}^{\infty} e^{jm\nu\chi} \cos m\nu\phi = \frac{1}{2} \sum_{m=-\infty}^{\infty} e^{jm\nu(\chi+\phi)} + j \sum_{m=1}^{\infty} e^{-jm\nu\phi} \sin m\nu\chi$$

$$= \pi\delta_{2\pi}[\nu(\chi+\phi)] + j \sum_{m=1}^{\infty} e^{-jm\nu\phi} \sin m\nu\chi \qquad (5.74)$$

where $\delta_{2\pi}$ is defined in (A.73), and since the second term in the last line of (5.74) is an odd function of χ, we can write (5.72) as:

$$W_\theta^{(1)}(\rho,\phi) = \nu \int_{-\pi}^{\pi} e^{jk\rho \cos \chi} \delta_{2\pi}[\nu(\chi+\phi)] \, d\chi \qquad (5.75)$$

This integral is readily evaluated provided we can determine which of the string of delta functions contained in $\delta_{2\pi}$ lies within the range of integration. A change of variable leads to the expression

$$W_\theta^{(1)}(\rho,\phi) = \int_{\nu(-\pi+\phi)}^{\nu(\pi+\phi)} e^{jk\rho \cos(u/\nu-\phi)} \delta_{2\pi}(u) \, du \qquad (5.76)$$

From (A.73) and the fact that

$$\int_{-\infty}^{x} f(u)\delta(u) \, du = f(0)\vartheta(x)$$

for a continuous function f, where ϑ is the Heaviside unit step function, we have

$$W_\theta^{(1)}(\rho,\phi) = \sum_{p=-\infty}^{\infty} e^{jk\rho \cos(2\pi p/\nu-\phi)} \qquad (5.77)$$

$$\{\vartheta[\nu(\pi+\phi) - 2\pi p] - \vartheta[\nu(-\pi+\phi) - 2\pi p]\}$$

The pth term of this sum will be nonzero only if $\nu(-\pi+\phi) < 2\pi p < \nu(\pi+\phi)$, or $\pi(-1 + 2p/\nu) < \phi < \pi(1 + 2p/\nu)$. Since the second argument of $W_\theta^{(1)}$ will actually be $\phi \pm \phi_0$, exactly what are present will vary depending on the values of these angles.

We will examine a particular case—that of an exterior wedge ($0 < \theta < \pi$, so that $1/2 < \nu < 1$) with a source location such that $0 < \phi_0 < \pi - \theta$. The latter condition means that only the top face of the wedge can "see" the source point; the bottom face is "shadowed" from it. Some inspection of the range of values that $\phi \pm \phi_0$ can assume reveals that only the delta function corresponding to $p = 0$ can lie in the range of integration for $W_\theta^{(1)}$, and then

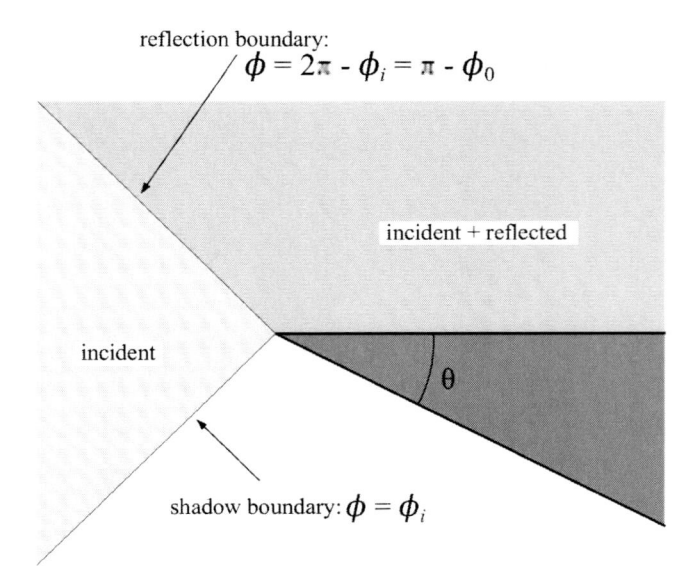

reflection boundary:
$$\phi = 2\pi - \phi_i = \pi - \phi_0$$

incident + reflected

incident

θ

shadow boundary: $\phi = \phi_i$

Figure 5.7 Coverages of the geometrical optics portion of the field scattered by a wedge.

only for some values of the argument, namely between $-\pi$ and π. Hence, we have

$$W_\theta^{(1)}(\rho, \phi) = e^{jk\rho \cos \phi} \qquad (|\phi| < \pi)$$
$$= 0 \qquad \text{(otherwise)} \qquad (5.78)$$

The contributions of $W_\theta^{(1)}$ to V, which we denote V^{GO}, are thus

$$
\begin{aligned}
V^{\text{GO}} &= V_0[W_\theta^{(1)}(\rho, \phi - \phi_0) - W_\theta^{(1)}(\rho, \phi + \phi_0)] \\
&= V_0[e^{-jk\rho \cos(\phi - \phi_i)} - e^{-jk\rho \cos(\phi + \phi_i)}] \qquad (0 < \phi < 2\pi - \phi_i) \\
&= V_0 e^{-jk\rho \cos(\phi - \phi_i)} \qquad (2\pi - \phi_i < \phi < \phi_i) \\
&= 0 \qquad (\phi_i < \phi < 2\pi - \theta)
\end{aligned}
\qquad (5.79)
$$

This is the classical result of geometrical optics (see Figure 5.7). In the first region, we have the incident plane wave plus the wave that would be reflected from the upper face of the wedge if it were a plane conductor of infinite extent. In the second region, only the incident wave is present, as the reflected wave can be produced only by the semi-infinite upper face of the wedge as shown. Finally, in the third region, not even the incident wave is present, because it is shadowed by the presence of the wedge.

5.3.5 THE DIFFRACTED FIELD

The contributions from $W_\theta^{(2)}$ constitute the *diffracted* wave from the tip of the wedge. If the summation in (5.73) is brought inside the integration, we have

$$W_\theta^{(2)}(\rho, \phi) = -\frac{\nu}{\pi} \int_0^\infty e^{-jk\rho \cosh t} \sum_{m=0}^\infty e^{-m\nu t} \left[\sin m\nu(\pi + \phi) + \sin m\nu(\pi - \phi) \right] dt$$

$$= -\frac{1}{2\pi} \int_0^\infty e^{-jk\rho \cosh t} \left\{ Q_\theta(t, \phi) + Q_\theta(t, -\phi) \right\} dt \qquad (5.80)$$

$$= -\frac{1}{4\pi} \int_{-\infty}^\infty e^{-jk\rho \cosh t} \left\{ Q_\theta(t, \phi) + Q_\theta(t, -\phi) \right\} dt$$

where

$$Q_\theta(t, \phi) = \frac{\nu \sin \nu(\pi + \phi)}{\cosh \nu t - \cos \nu(\pi + \phi)}$$

and we have used the known formula

$$\sum_{m=0}^\infty e^{-my} \sin mx = \frac{\sin x}{2[\cosh y - \cos x]} \qquad (y > 0)$$

This integral has the form of (4.101), with $P(t) = \cosh t$ and $k\rho$ as the large parameter. Thus the stationary phase approximation (4.102) can be used to evaluate it for $k\rho \gg 1$. Leaving the details to the reader, we find that the far field of the diffracted wave can be expressed as

$$W_\theta^{(2)}(\rho, \phi) \sim -\frac{1}{2} \{ Q_\theta(0, \phi) + Q_\theta(0, -\phi) \} \frac{e^{-jk\rho - j\pi/4}}{\sqrt{2\pi k\rho}}$$

$$= \frac{\nu \sin \nu\pi}{\cos \nu\pi - \cos \nu\phi} \frac{e^{-jk\rho - j\pi/4}}{\sqrt{2\pi k\rho}} \qquad (5.81)$$

The contribution of $W_\theta^{(2)}$ to V, which we denote V^{DIF}, is thus

$$V^{\mathrm{DIF}} = V_0[W_\theta^{(2)}(\rho, \phi - \phi_0) - W_\theta^{(2)}(\rho, \phi + \phi_0)]$$

$$\sim V_0 \frac{e^{-jk\rho}}{\sqrt{\rho}} D_\theta(\phi, \phi_0) \qquad (k\rho \gg 1) \qquad (5.82)$$

where D_θ is called a *diffraction coefficient*:

$$D_\theta(\phi, \phi_0) = \frac{\nu \sin \nu\pi e^{-j\pi/4}}{\sqrt{2\pi k}} \left\{ \frac{1}{\cos \nu\pi - \cos \nu(\phi - \phi_0)} \right.$$

$$\left. - \frac{1}{\cos \nu\pi - \cos \nu(\phi + \phi_0)} \right\} \qquad (5.83)$$

The diffracted field in the far zone bears a strong resemblance to a cylindrical wave (for example, the field of a line source), but is not isotropic (ϕ-independent) like the line source field. It does, however, appear to emanate from the edge, as the factor $e^{-jk\rho}/\sqrt{\rho}$ shows.

Upon results like this one, and a host of others, the so-called Geometrical Theory of Diffraction (GTD) has been built. It rests on the assumption that diffraction is an inherently local process at high enough frequency, so that waves incident on bodies of fairly general geometry can be analyzed by breaking the problem down into diffraction of waves from individual features of the object, which are found from results of canonical scattering problems such as the one we have just examined. Although approximate, this theory has been shown to be highly accurate at sufficiently short wavelengths, and away from certain portions of space where it loses validity.[4]

One such breakdown of (5.83) is that it predicts infinite fields at angles satisfying

$$\cos \nu\pi = \cos \nu(\phi \pm \phi_0) \qquad (5.84)$$

Now, for the case considered in Section 5.3.4, there are two angles at which this condition is satisfied. When $\phi - \phi_0 = \pi$ we are at the shadow boundary $\phi = \phi_i$ across which the incident wave abruptly disappears from V^{GO}, and $\phi + \phi_0 = \pi$ is the reflection boundary $\phi = 2\pi - \phi_i$ beyond which the specularly reflected wave disappears from V^{GO}. Since V itself is an overall continuous function of ϕ, whereas V^{GO} has these discontinuities, the exact value of V^{DIF} must also be discontinuous in such a way that the continuity is preserved. But the form of the cylindrical wave that V^{DIF} has in the far field could not possibly compensate the jumps in the plane waves that occur in V^{GO}. The reason for this is that the stationary-phase approximation for V^{DIF} is based on the assumption that the integrand $f(t)$ of (4.101) is smoothly varying enough so that it may be approximated by its value at $t_s = 0$. Such a function could never have a pole singularity near $t = t_s$ as does $Q_\theta(t, \pm\phi)$ when $\phi \simeq \pi$. When ϕ is not near the shadow or reflection boundary, (5.83) is accurate; in the transition zones near the ϕ which satisfy (5.84), it must be abandoned.

5.3.6   UNIFORM FAR-FIELD APPROXIMATION

To remedy this defect of the ordinary far-field expression, we must accurately evaluate the integral containing the pole in its integrand in order to obtain an expression that is not so limited in validity.[5] This will require the use of the Fresnel integral. In fact, the field diffracted from a half-plane ($\theta = 0$, $\nu = 1/2$) is expressible exactly in terms of this function. Denoting this case

---

[4]R. C. Hansen, ed., *Geometric Theory of Diffraction.* New York, IEEE Press, 1981.

[5]F. Oberhettinger, *J. Math. and Phys.*, vol. 34, pp. 245-255, 1956; *J. Res. NBS*, vol. 61, pp. 343-365, 1958.

with a subscript 0, (5.80) becomes

$$W_0^{(2)}(\rho, \phi) = -\frac{\cos\frac{\phi}{2}}{2\pi} \int_0^\infty e^{-jk\rho\cosh t} \frac{\cosh\frac{t}{2}}{\cosh^2\frac{t}{2} - \sin^2\frac{\phi}{2}} \, dt \qquad (5.85)$$

The change of variable $u = \sinh\frac{t}{2}/|\cos\frac{\phi}{2}|$ leads to

$$W_0^{(2)}(\rho, \phi) = -\frac{\mathrm{sgn}(\cos\frac{\phi}{2})}{\pi} e^{-jk\rho} \int_0^\infty e^{-2jk\rho\cos^2(\phi/2)u^2} \frac{du}{u^2 + 1} \qquad (5.86)$$

where $\mathrm{sgn}(x) = -1$ if $x < 0$ and $+1$ if $x > 0$ is the algebraic sign (signum) function. As shown in Section B.4 of Appendix B, the integral appearing in this expression can be transformed into a Fresnel integral, which is well tabulated and for which efficient routines exist for numerical calculation.

Putting $a = \sqrt{2k\rho}\cos\frac{\phi}{2}$ in (B.42) and substituting into (5.86) gives:

$$W_0^{(2)}(\rho, \phi) = \frac{1}{\sqrt{\pi}} e^{-jk\rho + j\pi/4} F\left(\sqrt{2k\rho}\cos\frac{\phi}{2}\right) - \vartheta\left(\cos\frac{\phi}{2}\right) e^{jk\rho\cos\phi} \qquad (5.87)$$

where F is the Fresnel integral defined in (B.43). The additional term in (5.87) is exactly cancelled by $W_0^{(1)}(\rho, \phi)$ from (5.78) for all values of ϕ, so we finally have

$$W_0(\rho, \phi) = \frac{1}{\sqrt{\pi}} e^{-jk\rho + j\pi/4} F\left(\sqrt{2k\rho}\cos\frac{\phi}{2}\right) \qquad (5.88)$$

which with (5.68) provides a uniformly valid expression for the total field of the plane wave diffracted by a half-plane. Recall that the angle ϕ in the argument of the Fresnel integral in (5.88) is actually replaced by $\phi \pm \phi_0$ in the expression (5.68) for the potential diffracted by the wedge.

To examine this expression more closely, let us define the angles $\varphi_s = \phi - \phi_0 - \pi$ and $\varphi_r = \phi + \phi_0 - \pi$. Physically, these angles represent the angular distance of the observation direction from the shadow boundary or the reflection boundary, respectively, with positive values of $\varphi = \varphi_{s,r}$ being in the shadow and negative values being in the illuminated region. The angle arguments appearing in (5.68) then become $\varphi + \pi$, and the argument of the Fresnel integral in (5.88) becomes $-\sqrt{2k\rho}\sin(\varphi/2)$. The large argument expansion (B.45) of the Fresnel integral will reduce (5.88) to the GTD result (5.78)/(5.81). From the plot of the magnitude of the Fresnel integral as a function of its argument Figure B.2, we see that this approximation is accurate only if

$$\sqrt{2k\rho}\left|\sin\frac{\varphi}{2}\right| \geq 1.5 \qquad (5.89)$$

Defining normalized Cartesian coordinates

$$X = k\rho\cos\varphi; \qquad Y = k\rho\sin\varphi$$

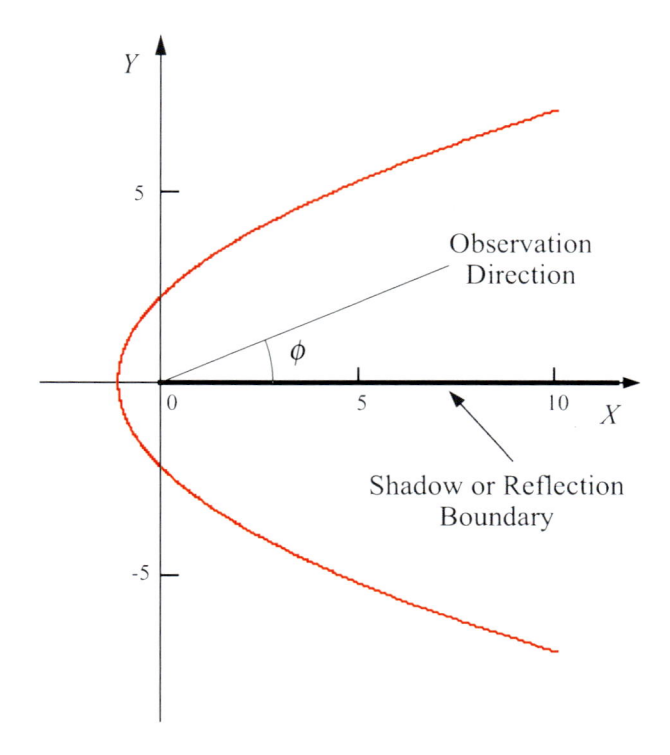

Figure 5.8 Limiting boundary for validity of the GTD approximation.

relative to the shadow or reflection boundary $\varphi = 0$, we can plot the locus of points where equality occurs in (5.89) as shown in Figure 5.8. Outside of this parabolic curve, GTD is valid. The more precise expression (5.88) produces a continuous transition across the shadow or reflection boundary and must be used inside the curve. We can also see from Figure B.2 that for very large values of $k\rho$, the transition from "light" to "shadow" is quite rapid, occurring in an angular range of order $O((k\rho)^{-1/2})$.

For a wedge of arbitrary angle θ between 0 and π (so that $\frac{1}{2} < \nu < 1$), we can make use of the half-plane result to produce an approximate far-field of the wedge that does not break down at the shadow or reflection boundaries. We do this by adding and subtracting the result for the half-plane from that of the arbitrary wedge, such that the difference between the two expressions has no pole in its integrand and can be approximated as in Section 5.3.5, while the uniform transition from light to shadow is taken care of by W_0. The poles of $Q_\theta(t, \pm\phi)$ are located at $jt = \pi \pm \phi + 2\pi p/\nu$, where p is an integer. When $0 < \phi < \pi - \theta$ as in Section 5.3.4, only the poles at $p = 0$ are troublesome. It can be shown that the difference $Q_\theta(t, \phi) - Q_0(t, \phi)$ is bounded (zero, in fact) as $jt \to \pi \pm \phi$, so that we can write

$$
\begin{aligned}
W_\theta^{(2)}(\rho, \phi) &= W_0^{(2)}(\rho, \phi) + [W_\theta^{(2)}(\rho, \phi) - W_0^{(2)}(\rho, \phi)] \\
&= W_0^{(2)}(\rho, \phi) - \frac{1}{2\pi} \int_0^\infty e^{-jk\rho \cosh t} \{Q_\theta(t, \phi) + Q_\theta(t, -\phi) \\
&\quad - Q_0(t, \phi) - Q_0(t, -\phi)\} \, dt \qquad (5.90) \\
&\sim W_0^{(2)}(\rho, \phi) + \frac{e^{-jk\rho - j\pi/4}}{\sqrt{2\pi k\rho}} \left[\frac{\nu \sin \nu\pi}{\cos \nu\pi - \cos \nu\phi} + \frac{1}{2\cos(\phi/2)} \right]
\end{aligned}
$$

by the techniques of the preceding section. Now, the second term in the last line of (5.90) is finite even for $\phi = \pi$, and so the expression for the complete field,

$$
V(\rho, \phi) \sim V_0\{W_0(\rho, \phi - \phi_0) - W_0(\rho, \phi + \phi_0) + \frac{e^{-jk\rho}}{\sqrt{\rho}}[D_\theta(\phi, \phi_0) - D_0(\phi, \phi_0)]\}
$$
$$(5.91)$$

is continuous at the shadow and reflection boundaries. Near these boundaries, the field is approximately the same as for the half-plane, while away from them the terms in W_0 and D_0 combine to give the geometrical optics terms. Other ranges of wedge angles and observation angles can be treated in a similar way; we must ascertain which angles ϕ may give problems, and cancel out the offending poles one by one as done above.

5.4 SPHERICAL HARMONICS

Solutions to the Helmholtz equation in spherical geometries are most conveniently expressed as expansions in spherical harmonics. The Debye potentials v and w in (2.51) will satisfy

$$(\nabla^2 + k^2)v = 0 \qquad (5.92)$$

in a region free of sources, with a similar equation holding true for w. An obvious first step is to use a Fourier series in ϕ as we did for problems with circular cylindrical geometry:

$$
v = \sum_{m=-\infty}^{\infty} e^{jm\phi} v_m(r, \theta)
$$
$$(5.93)$$
$$
v_m(r, \theta) = \frac{1}{2\pi} \int_0^{2\pi} v(r, \theta, \phi) e^{-jm\phi} \, d\phi
$$

This expansion theorem implies the orthogonality relation

$$
\int_0^{2\pi} e^{-jm\phi} e^{jm'\phi} \, d\phi = 2\pi \delta_{mm'} \qquad (5.94)
$$

We now need to find the v_m that satisfy

$$\left[\frac{1}{r^2}\frac{\partial}{\partial r}\left(r^2\frac{\partial}{\partial r}\right) + \frac{1}{r^2\sin\theta}\frac{\partial}{\partial\theta}\left(\sin\theta\frac{\partial}{\partial\theta}\right) - \frac{m^2}{r^2\sin^2\theta} + k^2\right]v_m = 0 \quad (5.95)$$

To make further progress, we need to expand v_m in a set of functions, which separates the variables r and θ in (5.95). If we try the solution

$$v_m = R(r)\Theta(\theta)$$

in (5.95), then we have

$$\frac{1}{R}\left[\frac{d}{dr}\left(r^2\frac{dR}{dr}\right) + k^2r^2R\right] + \frac{1}{\Theta}\left[\frac{1}{\sin\theta}\frac{d}{d\theta}\left(\sin\theta\frac{d\Theta}{d\theta}\right) - \frac{m^2}{\sin^2\theta}\Theta\right] = 0 \quad (5.96)$$

which yields to separation of variables. Calling (with the benefit of some hindsight) the separation constant $n(n+1)$, we have

$$\frac{1}{\sin\theta}\frac{d}{d\theta}\left(\sin\theta\frac{d\Theta}{d\theta}\right) + \left[n(n+1) - \frac{m^2}{\sin^2\theta}\right]\Theta = 0 \quad (5.97)$$

$$\frac{d}{dr}\left(r^2\frac{dR}{dr}\right) + [k^2r^2 - n(n+1)]R = 0 \quad (5.98)$$

Solutions of (5.97) are called *associated Legendre functions* $P_n^m(\cos\theta)$ and $Q_n^m(\cos\theta)$, and their properties are summarized in Section B.5 of Appendix B. By the expansion theorem (B.57), we can further decompose the r from the θ dependence of v and write it as

$$v = \sum_{n=0}^{\infty}\sum_{m=-n}^{n} v_n^m(r)e^{jm\phi}P_n^m(\cos\theta) \quad (5.99)$$

where v_n^m satisfies (5.98). Equations (5.94) and (B.59) can be used with the expansion (5.99) to prove the Parseval identity:

$$\int_0^{2\pi}d\phi\int_0^{\pi}|v|^2\sin\theta\,d\theta = \sum_{n=0}^{\infty}\sum_{m=-n}^{n}|v_n^m(r)|^2\frac{4\pi}{2n+1}\frac{(n+m)!}{(n-m)!} \quad (5.100)$$

Using the substitution $R = (\pi/2kr)^{1/2}f(r)$ in (5.98) gives the equation

$$\frac{1}{r}\frac{d}{dr}\left(r\frac{df}{dr}\right) + \left(k^2 - \frac{(n+\frac{1}{2})^2}{r^2}\right)f = 0 \quad (5.101)$$

which has the same form as Bessel's differential equation encountered earlier. As shown in Appendix B, the general solution for v_n^m is expressed in terms of spherical Bessel functions:

$$v_n^m(r) = B_n^m j_n(kr) + C_n^m y_n(kr) \quad (5.102)$$

where j_n and y_n are defined in (B.28). The constants B_n^m and C_n^m are to be determined by boundary, jump, or radiation conditions appropriate to a specific problem. The complete expansion (5.99), of which the $n = 0$ terms correspond to simple spherical waves, is known as a *spherical harmonic* expansion, or *multipole* expansion. Of the spherical harmonic terms $e^{jm\phi} P_n^m(\cos\theta)$, those with $m = 0$ are called *zonal harmonics*, those with $0 < m < n$ are called *tesseral harmonics*, and those with $m = n$ are called *sectorial harmonics*.

The multipole expansion is used in Appendix C to prove Rellich's theorem, which was used in Chapter 3 to prove a uniqueness theorem for time-harmonic electromagnetic fields. It can also be used like the cylindrical harmonic expansion was used in Section 5.2 to solve radiation and scattering problems from spheres. We will illustrate its use here on the problem of plane wave scattering from a perfectly conducting sphere.

Consider a plane wave

$$\mathbf{E}^{\text{inc}} = \mathbf{u}_x E_0 e^{-jkz} \tag{5.103}$$

and its associated H-field

$$\mathbf{H}^{\text{inc}} = \mathbf{u}_y \frac{E_0}{\zeta} e^{-jkz} \tag{5.104}$$

(where ζ is the wave impedance of the external medium) incident on a perfectly conducting sphere of radius a centered at the origin. The procedure is analogous to that used in Section 5.2 for scattering by a cylinder, the major difference being that the field in the present case will be neither pure TE nor pure TM with respect to the r-direction. We will start by determining the Debye potentials for the incident wave as spherical harmonic expansions, and then enforce boundary conditions to find the scattered field.

The Debye potentials are most easily found from the radial components of the fields. From (5.103) and (5.104), we obtain

$$E_r^{\text{inc}} = \cos\phi \sin\theta E_x^{\text{inc}} = E_0 \frac{\cos\phi}{jkr} \frac{\partial}{\partial\theta} \left(e^{-jkr\cos\theta} \right) \tag{5.105}$$

and

$$H_r^{\text{inc}} = \sin\phi \sin\theta H_y^{\text{inc}} = \frac{E_0}{\zeta} \frac{\sin\phi}{jkr} \frac{\partial}{\partial\theta} \left(e^{-jkr\cos\theta} \right) \tag{5.106}$$

Now the term in parentheses in (5.105) and (5.106) is a plane wave that can be expanded in spherical harmonics using (B.64) from Appendix B:

$$E_r^{\text{inc}} = E_0 \frac{\cos\phi}{jkr} \sum_{n=0}^{\infty} (-j)^n (2n+1) j_n(kr) \frac{\partial P_n(\cos\theta)}{\partial\theta} \tag{5.107}$$

But relation (B.54) gives us

$$
\begin{aligned}
E_r^{\text{inc}} &= E_0 \frac{\cos\phi}{jkr} \sum_{n=1}^{\infty} (-j)^n (2n+1) j_n(kr) P_n^1(\cos\theta) \\
&= E_0 \frac{\cos\phi}{jk^2} \sum_{n=1}^{\infty} (-j)^n \frac{2n+1}{n(n+1)} P_n^1(\cos\theta) \left(\frac{d^2}{dr^2} + k^2 \right) \psi_n(kr) \quad (5.108)
\end{aligned}
$$

where the $n = 0$ term in the sum vanishes because of (B.53), and we have introduced the Riccati-Bessel functions (B.36) and used their governing differential equation (B.37). By (2.53), the component E_r^{inc} relates only to the electric Debye potential v^{inc}, and from the r-component of (2.53) we obtain

$$
v^{\text{inc}} = \frac{E_0}{jk} \cos\phi \sum_{n=1}^{\infty} (-j)^n \frac{2n+1}{n(n+1)} j_n(kr) P_n^1(\cos\theta) \quad (5.109)
$$

and in a similar manner, we get

$$
w^{\text{inc}} = \frac{E_0}{j\omega\mu} \sin\phi \sum_{n=1}^{\infty} (-j)^n \frac{2n+1}{n(n+1)} j_n(kr) P_n^1(\cos\theta) \quad (5.110)
$$

Outside the sphere, the scattered fields can be expressed in terms of the scattered Debye potentials

$$
v^s = \frac{jE_0}{k} \cos\phi \sum_{n=1}^{\infty} (-j)^n \frac{2n+1}{n(n+1)} a_n h_n^{(2)}(kr) P_n^1(\cos\theta) \quad (5.111)
$$

and in a similar manner, we get

$$
w^s = \frac{jE_0}{\omega\mu} \sin\phi \sum_{n=1}^{\infty} (-j)^n \frac{2n+1}{n(n+1)} b_n h_n^{(2)}(kr) P_n^1(\cos\theta) \quad (5.112)
$$

where the scattering coefficients a_n and b_n are to be determined from the boundary conditions, and we have enforced the radiation condition on the scattered field by choosing the outgoing spherical Hankel functions in the expansions (5.111) and (5.112). Expressing the tangential components of the total electric field E_θ and E_ϕ in terms of the Debye potentials using (2.53), we see that for these fields to be zero at at $r = a$, we must impose the boundary conditions

$$
\left. \frac{\partial(rv)}{\partial r} \right|_{r=a} = 0 \quad ; \quad (rw)_{r=a} = 0 \quad (5.113)
$$

on the potentials themselves. Using (5.109)-(5.112) in (5.113), we obtain the following expressions for the scattering coefficients in terms of the Riccati-Bessel functions (B.36):

$$
a_n = \frac{\psi_n'(ka)}{\zeta_n'(ka)} \quad ; \quad b_n = \frac{\psi_n(ka)}{\zeta_n(ka)} \quad (5.114)
$$

and the scattered field can now be obtained from (5.111)-(5.112) and (2.53)-(2.54) by appropriate differentiations.

5.5 PROBLEMS

5–1 Representing the field in terms of Whittaker potentials U and V, obtain these potentials for the case of a horizontal, x-directed electric dipole source located at the point $(0, 0, z_0)$ above the interface between two half-spaces as in Section 5.1. Obtain the far field for these potentials in the upper half-space.

5–2 (a) A tangential surface current distribution $\mathbf{J}_S(x, y)$ is impressed at the interface between air and a grounded dielectric slab of thickness h as shown.

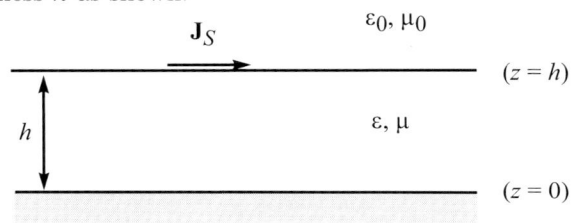

Obtain a formal solution for the electric and magnetic field at any point in space resulting from this current.

(b) If the current density is that of an x-directed Hertz dipole of moment p located at $x = y = 0$, find far-field expressions for $\mathbf{E}$ and $\mathbf{H}$ in $z > h$ as $R_1 \to \infty$, where R_1 is the distance defined in Figure 5.1.

5–3 Consider a z-directed electrostatic charge dipole described by the excess charge density

$$\rho = -\mathbf{p} \cdot \nabla \delta(\mathbf{r})$$

with $\mathbf{p} = p\mathbf{u}_z$ and $\delta(\mathbf{r}) = \delta(x)\delta(y)\delta_{(e)}(z)$, where $\delta_{(e)}(z)$ is a certain Dirac delta-function. As in Section 1.5.5, let this dipole be placed at the interface between two semi-infinite dielectric media of permittivities ϵ_1 and ϵ_2, such that the position-dependent ϵ is given by (1.170).

(a) Find the delta-function and step-function portions (in z) of the field $\mathbf{E}$ and the potential Φ. Assume that Equation (1.172) holds.

(b) Assume the potential Φ goes to zero at infinite distances from the dipole. Use a Fourier transform representation to obtain a solution for this potential everywhere in space, including any singular behavior you may have found in part (a). From this, write out an expression for the electric field everywhere in space, again including the singular behavior found in part (a). Comment on how the value of A (i.e., the relative position of the dipole and the material boundary) affects your solutions.

5–4 Provide the details of the derivation of (5.34).

5–5 Consider the path of the reflected wave in Figure 5.1, but without assuming that ρ_i is chosen so that the incident and reflected angles are equal. Define the electrical path length to be

$$\Theta_r(\rho_i) = k_1 \left(\sqrt{z_0^2 + \rho_i^2} + \sqrt{z^2 + (\rho - \rho_i)^2} \right)$$

Show that $\Theta_r(\rho_i)$ is a minimum for that value of ρ_i for which the incident and reflected angles are equal. This is *Fermat's principle* for the reflected wave.

Do the same for the broken path from source to observation point shown in Figure 5.3, without assuming that ρ_i is chosen so that (5.31) is satisfied. Define the electrical length of this broken path to be

$$\Theta_t(\rho_i) = k_1 \sqrt{z_0^2 + \rho_i^2} + k_2 \sqrt{z^2 + (\rho - \rho_i)^2}$$

Show that $\Theta_t(\rho_i)$ is a minimum for that value of ρ_i for which (5.31) is satisfied. This is Fermat's principle for the transmitted wave.

5–6 (a) A vertical electric dipole (current distribution $\mathbf{J} = j\omega p \mathbf{u}_z \delta(x)\delta(y)\delta(z-z_0)$) is located in the air above a grounded dielectric slab of thickness h as shown ($z_0 > h$).

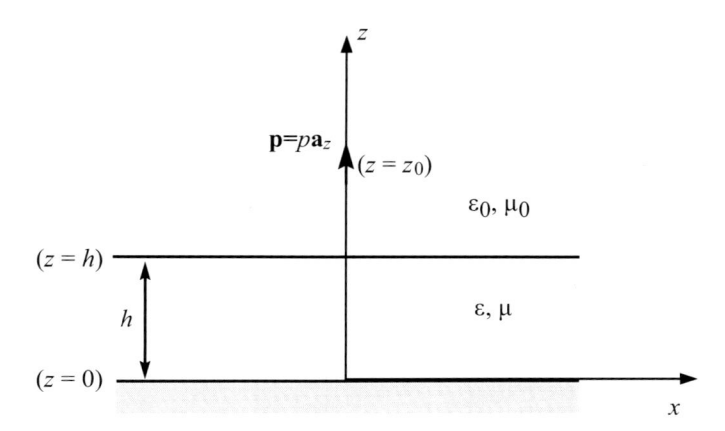

Obtain a formal solution as a Sommerfeld integral for the Whittaker potential V at any point in space resulting from this current. Show that the other Whittaker potential, U, is identically zero.

(b) Obtain expressions for $\mathbf{E}$ and $\mathbf{H}$ at any point in space.

(c) Find far-field expressions for V, $\mathbf{E}$, and $\mathbf{H}$ in $z > h$ as $R_1 \to \infty$, where R_1 is the distance defined in Figure 5.1.

5–7 An infinitely long, circular, perfectly conducting cylinder of radius a is subjected to a *static* incident field corresponding to the scalar potential

$$\Phi^i = -E_0 x = -E_0 \rho \cos \phi$$

The cylinder is grounded ($\Phi = 0$ at $\rho = a$). Find the induced surface charge density ρ_S on the cylinder surface, as well as the scattered static electric field at any point in space, making the assumption that the scattered potential is bounded at infinity.

5–8 A z-polarized plane wave

$$E_z^{inc} = E_0 e^{-jk_1 \rho \cos(\phi - \phi_i)}$$

where $k_1 = \omega \sqrt{\mu_0 \epsilon_1}$, is incident onto a *dielectric* cylinder of radius a as shown.

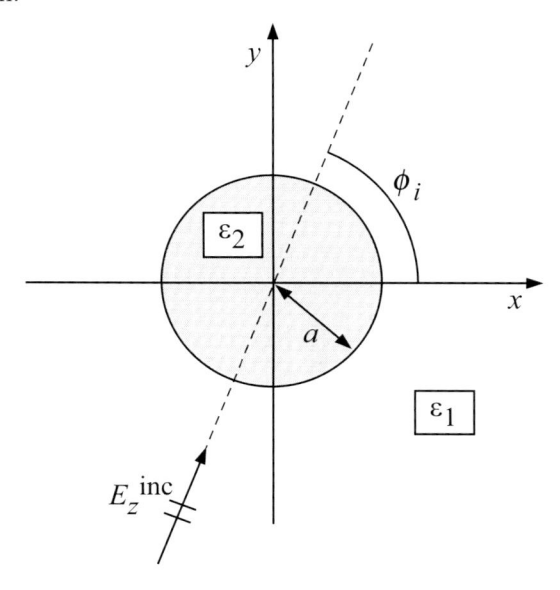

The dielectric has permittivity ϵ_2, while the surrounding medium has permittivity ϵ_1. Both regions have permeability μ_0. Obtain an expression of the form (5.50) for the scattered field in $\rho > a$, and give an expression for the coefficients E_{zm}^s.

5–9 Repeat Problem 5-8, but assuming that $\mu = \mu_1$ outside the cylinder, and μ_2 inside, while $\epsilon = \epsilon_0$ everywhere.

5–10 Obtain the exact expression for the geometrical optics term V^{GO} and the far field (GTD) approximation for the diffracted term V^{DIF} for the wedge scattering problem of Section 5.3 if the wedge angle θ lies between π and $3\pi/2$.

5–11 An impedance cylinder of radius a enforces the boundary condition $\mathbf{E}_{tan}|_{\rho=a} = Z_S \mathbf{u}_\rho \times \mathbf{H}|_{\rho=a}$, for some complex surface impedance Z_S whose real part is nonnegative. A plane wave

$$E_z^{inc} = E_0 e^{-jk_0 \rho \cos(\phi - \phi_i)}$$

in the surrounding free space is incident onto the cylinder. Obtain an expression in the form of (5.50) for the scattered electric field from the cylinder, and give an expression for the coefficients E_{zm}^s. Find the low frequency limit $(k_0 a \ll 1)$ of the radiation pattern $F(\phi)$ for the scattered field analogous to (5.56) and comment on whether it is likely to be smaller or larger than that for the perfectly conducting scatterer.

5–12 An electric Hertz dipole of moment p is located perpendicularly on top of and at a distance ρ_0 from the edge of a perfectly conducting half-plane occupying $y = 0$, $x \geq 0$ as shown.

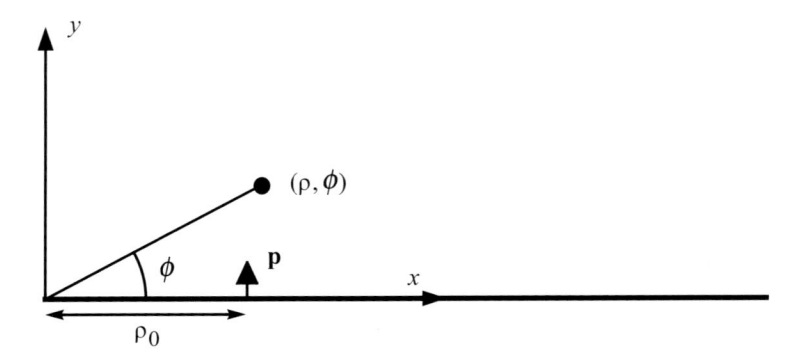

Obtain a formal expression for the Whittaker potentials U and V produced by this source in the presence of the half-plane. The solution has the form of a Fourier series/integral like (5.42) except that the Fourier series is of the type (5.58) suitable for wedge problems.

5–13 Suppose a typical electric or magnetic field component in the vicinity of a *conical* point of a material surface behaves as $O(r^\tau)$ as $r \to 0$ (note that this is r, not ρ). What is the smallest (i.e., most negative) value τ can have if the field is to store finite energy in the neighborhood of the conical point?

5–14 Let a plane wave described by the Whittaker potential $U^{\text{inc}} = U_0 e^{-jk\rho \cos(\phi - \phi_i)}$ be incident onto a perfectly conducting wedge. Express the total potential U that results in terms of the function $W_\theta(\rho, \phi)$.

5–15 An electric Hertz dipole of moment $\mathbf{p} = \mathbf{u}_r p$ is located at $\theta = 0$ on the surface of a perfectly conducting sphere of radius a as shown.

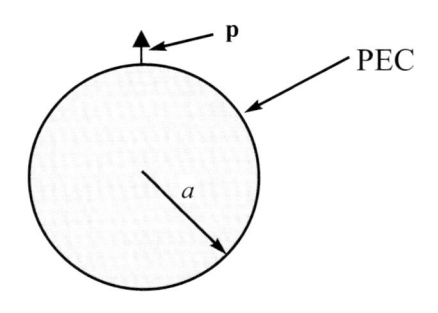

Use Debye potentials to find an expression for the resulting electromagnetic field.

5–16 A perfectly conducting sphere of radius a is set into a uniform, static incident electric field $\mathbf{E}^i = \mathbf{u}_z E_z^i$. Use a spherical harmonic expansion to obtain an expression for the scattered potential Φ^s produced by this object. From this result, find the total charge density on the surface of the sphere, the induced electric dipole moment, and thus obtain an expression for the electric polarizability of the sphere.

5–17 Suppose that the sphere of the previous problem is immersed into a uniform, static incident *magnetic* field $\mathbf{H}^i = \mathbf{u}_z H_z^i$. Find the scattered magnetic field, the induced surface current density on the sphere, the induced magnetic dipole moment, and an expression for the magnetic polarizability of the sphere.

5–18 Using a spherical harmonic expansion, obtain the solution for the scattering of a plane wave in free space from a homogeneous sphere of radius a whose permittivity is ϵ and permeability is μ. This is the so-called *Mie scattering* problem.[6]

---

[6]G. Mie, *Ann. Physik*, ser. 4, vol. 25, pp. 377-445, 1908.

6 Propagation and Scattering in More Complex Regions

6.1 GENERAL CONSIDERATIONS

It is worth asking at this point what it is about the transform representations (Fourier series and Fourier transform) that makes them so useful in obtaining solutions for the fields in various geometrical arrangements of matter. Some consideration shows that the main reason is that these transforms convert a part of the relevant differential operators into algebraic multipliers. For example, in a Fourier series,

$$\frac{\partial^2}{\partial \phi^2} \left\{ \begin{array}{c} \sin m\phi \\ \cos m\phi \\ e^{jm\phi} \end{array} \right\} = -m^2 \left\{ \begin{array}{c} \sin m\phi \\ \cos m\phi \\ e^{jm\phi} \end{array} \right\}$$

If in addition the trigonometric or complex exponential functions obey whatever boundary or periodicity conditions are appropriate to our problem, we say that they are *eigenfunctions* of the operator $\frac{\partial^2}{\partial \phi^2}$. In a similar way, the functions by which the Fourier transform represents a solution are $e^{-j(\alpha x + \beta y)}$, and these are eigenfunctions of the xy part of the Laplacian operator ∇^2.

One important property of the eigenfunctions is that they are orthogonal. For example, in the complex exponential Fourier series we have

$$\int_0^{2\pi} e^{jm\phi} e^{jn\phi} \, d\phi = 0 \quad \text{if } m \neq -n$$

This allows us to determine the coefficient of each term in the Fourier series by multiplying our equation by an eigenfunction and integrating over a suitable range of the variable(s). Another important property is that the eigenfunctions form a *complete set*: any function can be expanded as a linear combination of these eigenfunctions—there are no "missing" functions that might be needed. The eigenfunctions are also linearly independent: there are no "extra" functions that could themselves be expressed as a linear combination of other eigenfunctions. These last two properties taken together mean that the eigenfunctions form a *basis* set for any field, potential or source—the expansion in terms of a basis set is unique.[1]

---

[1] More information on these general issues can be found in:

B. Friedman, *Principles and Techniques of Applied Mathematics*. New York: Wiley, 1956.

Before proceeding to treat more challenging problems by the eigenfunction method, let us get some further insight by looking at the use of a transform to treat a problem that can be solved by much simpler means. Consider a constant time-harmonic current sheet $\mathbf{J}_S = \mathbf{u}_x J_{Sx}$ located in the plane $z = 0$, in an infinite homogeneous medium with parameters μ and ϵ. The resulting electric Hertz potential Π_{ex} obeys

$$\left(\frac{d^2}{dz^2} + k^2\right)\Pi_{ex} = -\frac{J_{Sx}}{j\omega\epsilon}\delta(z) \tag{6.1}$$

The usual solution to this problem is to choose a solution of outgoing plane waves on either side of $z = 0$, and determine their amplitudes by forcing continuity of E_x and the appropriate jump in H_y at $z = 0$. Here, we will instead approach the problem by applying the Fourier transform (4.63).

Applying the transform to (6.1), we obtain the algebraic equation

$$(k^2 - \beta^2)\tilde{\Pi}_{ex}(\beta) = -\frac{J_{Sx}}{j\omega\epsilon}\frac{1}{2\pi} \tag{6.2}$$

It would appear at first glance that the solution to (6.2) should be

$$\tilde{\Pi}_{ex}(\beta) = -\frac{J_{Sx}}{2\pi j\omega\epsilon}\frac{1}{k^2 - \beta^2}$$

However, a closer look [see Equation (A.42)] shows that the solution must in fact be

$$\tilde{\Pi}_{ex}(\beta) = -\frac{J_{Sx}}{2\pi j\omega\epsilon}\text{PV}\left(\frac{1}{k^2 - \beta^2}\right) + C_1\delta(\beta - k) + C_2\delta(\beta + k) \tag{6.3}$$

where C_1 and C_2 are constants to be determined and PV denotes the Cauchy principal value is to be taken when performing the inverse Fourier transform (see Appendix A). Specifically,

$$\text{PV}\left(\frac{1}{k^2 - \beta^2}\right) = \frac{1}{2k}\left[\text{PV}\left(\frac{1}{\beta + k}\right) - \text{PV}\left(\frac{1}{\beta - k}\right)\right] \tag{6.4}$$

Now from entry A.1.11 of Table A.1, $\text{PV}(1/\beta)$ is the Fourier transform of $-\pi j \,\text{sgn}(z)$, so from this together with entry A.1.6, we get

$$\Pi_{ex}(z) = -\frac{J_{Sx}}{4\pi k j\omega\epsilon}\left[-\pi j \,\text{sgn}(z)e^{jkz} + \pi j \,\text{sgn}(z)e^{-jkz}\right] + C_1 e^{jkz} + C_2 e^{-jkz} \tag{6.5}$$

We now choose C_1 and C_2 to ensure outgoing waves (i.e., only e^{-jkz} in $z > 0$ and only e^{jkz} in $z < 0$):

$$\frac{J_{Sx}}{4k\omega\epsilon} + C_1 = 0 \qquad (z > 0)$$

$$\frac{J_{Sx}}{4k\omega\epsilon} + C_2 = 0 \qquad (z < 0) \tag{6.6}$$

We have finally

$$\Pi_{ex}(z) = -\frac{J_{Sx}}{2k\omega\epsilon}\left[\vartheta(z)e^{-jkz} + \vartheta(-z)e^{jkz}\right] = -\frac{J_{Sx}}{2k\omega\epsilon}e^{-jk|z|} \tag{6.7}$$

or with $E_x = k^2\Pi_{ex}$,

$$E_x(z) = -\frac{\zeta J_{Sx}}{2}e^{-jk|z|} \tag{6.8}$$

which is what is found using the more conventional elementary method described before.

Thus, even for this one-dimensional wave problem the transform method is capable of determining the correct solution, despite being far more cumbersome than the elementary method. The main difference between these two techniques (from a methodological point of view) is that with the elementary method, we *assume* the form of the solution as plane waves, whereas in the Fourier method followed above, the form of the solution is obtained as a consequence of the Fourier transform approach—it is a fully constructive method. This is a significant advantage when attempting to solve more complicated problems, where the form of the solution may not be immediately obvious. Notice how the inverse Fourier transform has picked out either e^{-jkz} or e^{jkz} from all possible basis functions $e^{-j\beta z}$ due to the singularities of the delta functions and principal value functions in (6.3)-(6.4). This has been done in such a way as to produce the proper outgoing solution on both sides of the current sheet, even though in much of the range of β employed in the expansion, these basis functions individually violate the outgoing condition (radiation condition).

Some of the difficulty associated with the Fourier method can be removed if we modify our approach using the so-called *principle of limiting absorption*. We temporarily suppose that losses are present in the medium, so that k has a negative imaginary part. Equations (6.3)-(6.4) now reduce to

$$\tilde{\Pi}_{ex}(\beta) = -\frac{J_{Sx}}{2\pi j\omega\epsilon}\frac{1}{k^2 - \beta^2} = \frac{J_{Sx}}{2\pi j\omega\epsilon}\frac{1}{2k}\left(\frac{1}{\beta - k} - \frac{1}{\beta + k}\right) \tag{6.9}$$

The inverse transform of $(\beta - k)^{-1}$ is $-2\pi j\vartheta(z)e^{-jkz}$ by (A.67), and that of $-(\beta + k)^{-1}$ is $-2\pi j\vartheta(-z)e^{jkz}$ by (A.68), so the result (6.7) now follows immediately without the need for intermediate determination of the constants C_1 and C_2, which never appear in this variant of the derivation. We can finally arrive at the result for the lossless case by letting the imaginary part of k approach zero from below. Notice also that in this approach, we never have to explicitly enforce an outgoing radiation condition; those waves are automatically selected in the limiting absorption technique. The method can be used in many problems to simplify the evaluation of fields in various environments, especially those where the medium does not become homogeneous at large distances from the source, and conventional radiation conditions cannot be applied because of the difficulty of ascertaining which solutions are truly the "outgoing" ones.

In this chapter, we will examine several cases where the required eigenfunction expansions are more elaborate than those used in the two preceding chapters. Specifically, we will look at source radiation in a waveguide and in a periodic medium. Although we will limit ourselves for simplicity to consideration of one-dimensional eigenfunction expansions, the methods of this chapter are quite general, and capable of extension to two and three dimensions as well. References to the more general case are given in the footnotes.

6.2   WAVEGUIDES

6.2.1   PARALLEL-PLATE WAVEGUIDE: MODE EXPANSION

In a waveguide, the modes serve as the eigenfunctions that reduce the Maxwell differential equations to ordinary differential equations. Although waveguides of arbitrary configuration find extensive practical use, we will at first limit the discussion to the simplest of all waveguides, the parallel-plate waveguide shown in Figure 6.1, in order to keep the main ideas clear. This guide consists of two perfectly conducting planes, one at $x = 0$ and one at $x = a$. It is assumed that neither sources nor fields have any variation in the y-coordinate. By convention, the direction of propagation has been taken to be z.

Let a z-directed electric line current source I_0 be located at the source position (x_0, z_0). We represent the field in terms of a Whittaker potential $V(x, z)$ as done in Section 4.2.1. We thus have

$$\left(\frac{\partial^2}{\partial x^2} + \frac{\partial^2}{\partial z^2} + k^2\right) V = -\frac{I_0}{j\omega\epsilon}\delta(x - x_0)\delta(z - z_0) \tag{6.10}$$

The boundary conditions on the conducting walls are $V = 0$ at $x = 0$ and $x = a$, and an outgoing wave is required as $z \to \pm\infty$.

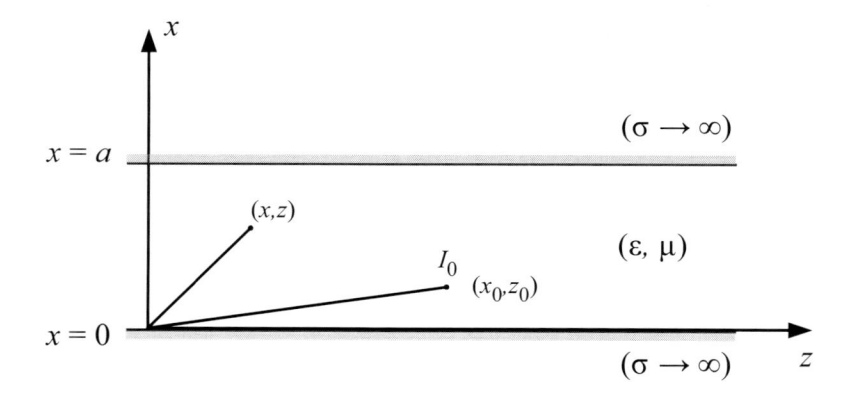

Figure 6.1   Parallel-plate waveguide.

The first method of solution is to expand the potential in a mode series. Here, the modes are proportional to the functions

$$\phi_m(x) = \sin \frac{m\pi x}{a} \qquad m = 1, 2, \ldots \infty \tag{6.11}$$

so the expansion is in fact a Fourier sine series:

$$V(x, z) = \sum_{m=1}^{\infty} V_m(z)\phi_m(x) \tag{6.12}$$

and the Fourier coefficients $V_m(z)$ are found using (A.72):

$$V_m(z) = \frac{2}{a} \int_0^a V(x, z)\phi_m(x)\, dx \tag{6.13}$$

The boundary conditions are satisfied by each of the $\phi_m(x)$ (the modes), so the same is true for (6.12). Finding the Fourier coefficients of each side of (6.10) using (6.13), we obtain an ordinary differential equation for $V_m(z)$:

$$\left(\frac{d^2}{dz^2} + \xi_m^2 \right) V_m(z) = -\frac{I_0}{j\omega\epsilon} \frac{2}{a}\phi_m(x_0)\delta(z - z_0) \tag{6.14}$$

where ξ_m is defined as:

$$\begin{aligned} \xi_m &= \sqrt{k^2 - \left(\frac{m\pi}{a}\right)^2} \qquad \left(k > \frac{m\pi}{a}\right) \\ &= -j\sqrt{\left(\frac{m\pi}{a}\right)^2 - k^2} \qquad \left(k < \frac{m\pi}{a}\right) \end{aligned} \tag{6.15}$$

To solve for the V_m, we consider first the region $z > z_0$. Here the delta function vanishes, and the solution of (6.14) is well-known. We require the solution either to be an outgoing wave, or to decay exponentially as $z \to \infty$. The procedure is similar for $z < z_0$, and we arrive at:

$$\begin{aligned} V_m(z) &= A_m e^{-j\xi_m z} \qquad (z > z_0) \\ &= B_m e^{j\xi_m z} \qquad (z < z_0) \end{aligned} \tag{6.16}$$

It is now clear that the choice of ξ_m in (6.15) was made to ensure the proper behavior of V_m as $z \to \pm\infty$, regardless of the values of k and m. It remains to determine A_m and B_m, which we do by enforcing continuity and jump conditions at $z = z_0$:

$$V_m(z_0^-) = V_m(z_0^+) \tag{6.17}$$

and

$$\left.\frac{dV_m}{dz}\right|_{z=z_0^+} - \left.\frac{dV_m}{dz}\right|_{z=z_0^-} = -\frac{I_0}{j\omega\epsilon}\frac{2}{a}\phi_m(x_0) \tag{6.18}$$

Using (6.16) in (6.17) and (6.18), we find in the usual way that

$$V(x,z) = \frac{I_0}{j\omega\epsilon}\left[-\frac{j}{a}\sum_{m=1}^{\infty}\frac{1}{\xi_m}\phi_m(x)\phi_m(x_0)e^{-j\xi_m|z-z_0|}\right] \tag{6.19}$$

This is the mode expansion for $V(x,z)$, which for the parallel-plate waveguide also happens to be a Fourier sine series expansion. Representation (6.19) is particularly useful in the far-field, when $|z - z_0|$ is sufficiently large that only the modes for which ξ_m are real contribute significantly to the total field; the other modes are cutoff and are exponentially small in the far-field.

There are several points worth emphasizing regarding the mode expansion.

1. Even though we are describing the potential due to a source, the potential of a single mode,

$$\phi_m(x)e^{\mp j\xi_m z}$$

is by definition a solution of the source-free equation governing the potential that has an exponential dependence on the propagation direction. In fact, the mode functions obey the differential equation

$$\left(\frac{d^2}{dx^2} + k^2 - \xi_m^2\right)\phi_m = \left(\frac{d^2}{dx^2} + \alpha_m^2\right)\phi_m = 0 \tag{6.20}$$

where

$$\alpha_m = \sqrt{k^2 - \xi_m^2} = \frac{m\pi}{a} \tag{6.21}$$

The allowed values of α_m and forms of ϕ_m are those that obey both the ordinary differential equation (6.20) and the boundary conditions at the sidewalls:

$$\phi_m(0) = \phi_m(a) = 0 \tag{6.22}$$

Despite the absence of sources in the definition of the eigenfunctions, even at the source point (x_0, z_0), the infinite sum of modes is a correct representation of the potential (in the sense of generalized functions, of course).

2. The mode functions are orthogonal:

$$\int_0^a \phi_{m'}(x)\phi_m(x)\,dx = N_m\delta_{mm'} \tag{6.23}$$

where $\delta_{mm'}$ is the Kronecker delta, and we have defined the *norm* N_m of the mode $\phi_m(x)$ as

$$N_m = \int_0^a \phi_m^2(x)\,dx = \frac{a}{2} \tag{6.24}$$

Relation (6.23) is what allows us to find each of the coefficients in the mode expansion independently of all the others. The proof of (6.23) uses the differential equation (6.20) and boundary conditions (6.22):

$$\int_0^a \phi_{m'}(x)\phi_m(x)\,dx = -\frac{1}{\alpha_{m'}^2}\int_0^a \frac{d^2\phi_{m'}(x)}{dx^2}\phi_m(x)\,dx \qquad (6.25)$$

We integrate by parts twice, use the boundary conditions (6.22) and then the differential equation (6.20):

$$\begin{aligned}
\int_0^a \phi_{m'}(x)\phi_m(x)\,dx &= \frac{1}{\alpha_{m'}^2}\int_0^a \frac{d\phi_{m'}(x)}{dx}\frac{d\phi_m(x)}{dx}\,dx \\
&= -\frac{1}{\alpha_{m'}^2}\int_0^a \phi_{m'}(x)\frac{d^2\phi_m(x)}{dx^2}\,dx \quad (6.26) \\
&= \frac{\alpha_m^2}{\alpha_{m'}^2}\int_0^a \phi_{m'}(x)\phi_m(x)\,dx
\end{aligned}$$

from which (6.23) follows if $m \neq m'$.

3. The modes form a complete set. In other words, we can express a delta function of the transverse variables in terms of the transversely dependent parts of the mode functions. Writing

$$\delta(x - x_0) = \sum_m a_m\phi_m(x)$$

multiplying by $\phi_n(x)$ and integrating from $x = 0$ to $x = a$, we find from (6.23)-(6.24) that

$$\sum_{m=1}^{\infty}\frac{\phi_m(x)\phi_m(x_0)}{N_m} = \delta(x - x_0) \qquad (6.27)$$

By superposition, it follows that a fairly arbitrary transverse field distribution can also be expressed as a mode expansion. For more general waveguides, the completeness property is much more difficult to prove than orthogonality. Indeed, the main challenge is to make sure we have truly found all the possible guided modes of a given waveguide structure before we attempt to expand a general field in terms of them.

Only the three properties above were really essential to our construction of the mode expansion, and with suitable generalizations we can derive mode expansions for other more general types of waveguide as well, including open waveguides, a simple example of which is considered in the next section.

6.2.2 PARALLEL-PLATE WAVEGUIDE: FOURIER EXPANSION

If we use instead a Fourier transform with respect to the z variable in the parallel-plate waveguide problem, then similarly to Section 4.2.1, we arrive at an alternative representation for the potential of the form

$$V(x,z) = \frac{I_0}{j\omega\epsilon} \frac{1}{2\pi} \int_{-\infty}^{\infty} \frac{(\sin\eta x_<)\,\sin[\eta(a-x_>)]}{\eta\sin\eta a} e^{-j\beta(z-z_0)}\,d\beta \tag{6.28}$$

where $x_< = \min(x, x_0)$, $x_> = \max(x, x_0)$, and $\eta = \sqrt{k^2 - \beta^2}$ is as defined in (4.65). Details are left to the reader as an exercise (Problem 6-2). It should be noted that one disadvantage of (6.28) is that the values of β where $\sin\eta a = 0$ cause singularities in the integrand of (6.28) that must be handled by the use of principal values and delta functions with undetermined coefficients as in the previous section, or alternatively using the limiting absorption principle.

The principle of limiting absorption can be used with (6.28) to obtain an image series representation for V. Let k have a negative imaginary part, eventually to be allowed to go to zero. We can write the factor $(\sin\eta a)^{-1}$ of the integrand of (6.28) as

$$\frac{1}{\sin\eta a} = -\frac{je^{-j\eta a}}{2}\frac{1}{1 - e^{-2j\eta a}} = -\frac{je^{-j\eta a}}{2}\sum_{n=0}^{\infty} e^{-2jn\eta a} \tag{6.29}$$

because the imaginary part of η is also negative. If this expansion is substituted into (6.28), and the remaining sines expressed as exponentials, we get

$$V(x,z) = \frac{I_0}{j\omega\epsilon} \frac{1}{4\pi j}\sum_{n=0}^{\infty}\int_{-\infty}^{\infty} e^{-j\eta(2n+1)a}\left(e^{j\eta x_<} - e^{-j\eta x_<}\right)$$

$$\times \left[e^{j\eta(a-x_>)} - e^{-j\eta(a-x_>)}\right]\frac{e^{-j\beta(z-z_0)}}{\eta}\,d\beta \tag{6.30}$$

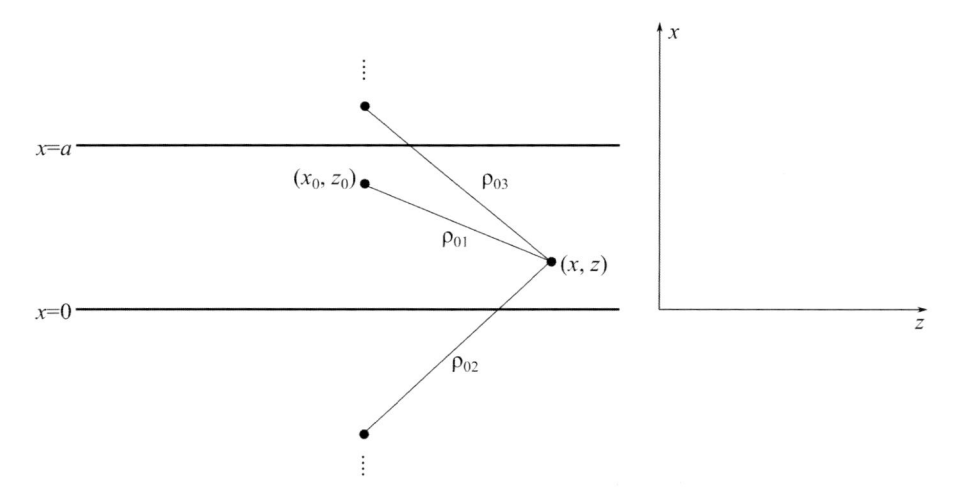

Figure 6.2 Infinite set of images of a line source placed between parallel conducting planes.

Each term in (6.30) can be evaluated in closed form, using (4.45) adapted to the notation of this problem:

$$V(x, z) = \frac{I_0}{j\omega\epsilon}\frac{1}{4j}\sum_{n=0}^{\infty}\left[H_0^{(2)}(k\rho_{n1}) - H_0^{(2)}(k\rho_{n2}) - H_0^{(2)}(k\rho_{n3}) + H_0^{(2)}(k\rho_{n4})\right]$$

(6.31)

where

$$\begin{aligned}
\rho_{n1} &= \sqrt{(|x - x_0| + 2na)^2 + (z - z_0)^2}; \\
\rho_{n2} &= \sqrt{(x + x_0 + 2na)^2 + (z - z_0)^2} \\
\rho_{n3} &= \sqrt{(2a - x - x_0 + 2na)^2 + (z - z_0)^2} \\
\rho_{n4} &= \sqrt{(2a - |x - x_0| + 2na)^2 + (z - z_0)^2}
\end{aligned}$$
(6.32)

The term containing ρ_{01} in (6.31) is the wave emanating from the original line source at (x_0, z_0). Every other term of (6.32) can be identified as the wave emanating from a line source located at an image point: $(x_0 \mp 2na, z_0)$ for terms with ρ_{n1}, $(-x_0 - 2na, z_0)$ for ρ_{n2}, $(-x_0 + 2(n + 1)a, z_0)$ for ρ_{n3}, and $(x_0 \pm 2(n+1)a, z_0)$ for ρ_{n4} (the upper and lower signs refer to the cases $x > x_0$ and $x < x_0$, respectively). The first two images, at $(-x_0, z_0)$ and $(2a - x_0, z_0)$, are directly from the planes $x = 0$ and $x = a$. These in turn are re-imaged repeatedly by alternate planes, producing the infinite series of images, the first few of which are shown in Figure 6.2.

We observe that as a consequence of the results of this section, we have the three-way identity

$$\begin{aligned}
&-\frac{j}{a}\sum_{m=1}^{\infty}\frac{1}{\xi_m}\phi_m(x)\phi_m(x_0)e^{-j\xi_m|z-z_0|} \\
&= \frac{1}{2\pi}\int_{-\infty}^{\infty}\frac{(\sin\eta x_<)\sin[\eta(a - x_>)]}{\eta\sin\eta a}e^{-j\beta(z-z_0)}\,d\beta \\
&= \frac{1}{4j}\sum_{n=0}^{\infty}\left[H_0^{(2)}(k\rho_{n1}) - H_0^{(2)}(k\rho_{n2}) - H_0^{(2)}(k\rho_{n3}) + H_0^{(2)}(k\rho_{n4})\right] \text{ (6.33)}
\end{aligned}$$

It is thus important to note that the mode expansion of the field or potential in a waveguide is by no means the only legitimate representation available to us. Which one to use is dictated by convenience in a particular application. The image representation can be more convenient when $|z - z_0|$ is not large (in the medium-to-near field), when a large number of terms in the mode expansion must be kept for sufficient accuracy, but only a relatively small number of images may be sufficient.

6.2.3 OPEN WAVEGUIDES

Structures such as optical fibers and microstrips, among many others, fall into the category of open waveguides: Their cross-sectional surface extends to infinity in the transverse direction(s). It turns out that such waveguides possess only a finite number of true guided modes, and these are inadequate to represent an arbitrary field on the structure. In order to have a complete set of waveguide modes with which to represent an arbitrary field, we will have to broaden our conception of waveguide modes and include so-called "radiation modes" or "continuous-spectrum modes."[2] These modes allow the representation of fields radiated away from the neighborhood of the waveguide, as well as the portion of the reactive near-field not accounted for by the true guided modes.

To look at how such radiation modes arise, we consider first the case of an electric line current radiating above a single perfectly conducting plane at $x = 0$ (formally, this is the limiting case $a \to \infty$ of the parallel-plate waveguide considered in the previous subsection). Of course, this structure is not really a waveguide any longer (as most people would understand the term): It has no true guided modes that carry finite energy. But we will see below that it does have modes in the extended sense of "radiation modes" that can serve the same purpose as did the modes of the parallel-plate guide in representing the fields and potentials.

The generalized mode expansion can be constructed by starting with the Fourier integral representation of the potential in infinite space, to which we add the analogous expression for the potential of the image current. Setting $I_0 = j\omega\epsilon$ for the rest of this subsection to simplify our expressions, we find by adapting (4.44) that

$$
\begin{aligned}
V(x,z) &= \frac{1}{4\pi j} \int_{-\infty}^{\infty} \left(e^{-j\xi|z-z_0|-j\alpha(x-x_0)} - e^{-j\xi|z-z_0|-j\alpha(x+x_0)} \right) \frac{d\alpha}{\xi} \\
&= \frac{1}{2\pi} \int_{-\infty}^{\infty} e^{-j\xi|z-z_0|-j\alpha x} \sin \alpha x_0 \, \frac{d\alpha}{\xi} \\
&= -\frac{j}{\pi} \int_0^{\infty} e^{-j\xi|z-z_0|} \sin \alpha x \sin \alpha x_0 \, \frac{d\alpha}{\xi}
\end{aligned}
\tag{6.34}
$$

where ξ is given by (4.39). In the last step of (6.34), we have extracted only the even part of the integrand and reduced the integration interval to positive values of α only. The similarity in form of (6.34) with (6.19) is striking. In both cases, the potential is the superposition of a product of two mode functions (one evaluated at the transverse position of the source point, one at that of the

---

[2]For more details on continuous modes, see:

V. V. Shevchenko, *Continuous Transitions in Open Waveguides*. Boulder, CO: Golem Press, 1971.

D. Marcuse, *Theory of Dielectric Optical Waveguides*. New York: Academic Press, 1974.

——, *Light Transmission Optics*. New York: Van Nostrand, 1982.

C. Vassallo, *Théorie des Guides d'Ondes Électromagnétiques*. Paris: Eyrolles, 1985.

observation point) with a propagation factor in the z-direction that involves the "propagation constant" ξ_m or ξ and a normalization factor. In both cases there are some "propagating" modes for which $\alpha < k$, and some attenuating (or evanescent or "cutoff") modes that decay in amplitude as $|z| \to \infty$. The major difference is that in (6.19), the superposition is a sum over a discrete set of modes, while in (6.34) it is an integral over a continuum of radiation modes.

To complete the analogy with the case of a closed waveguide, let us rewrite (6.34) in the notation

$$V(x, z) = -\frac{j}{\pi} \int_0^\infty e^{-j\xi|z - z_0|} \phi(\alpha, x)\phi(\alpha, x_0) \frac{d\alpha}{\xi} \tag{6.35}$$

where $\phi(\alpha, x)$ is a mode eigenfunction defined as

$$\phi(\alpha, x) = \sin \alpha x \tag{6.36}$$

We wish to prove the analogs of the orthogonality relation (6.23) and the completeness relation (6.27). Orthogonality is straightforward; we write (formally)

$$\begin{aligned}
\int_0^\infty \phi(\alpha, x)\phi(\alpha', x)\, dx &= \frac{1}{2} \int_0^\infty [\cos(\alpha - \alpha')x - \cos(\alpha + \alpha')x]\, dx \\
&= \frac{1}{4} \int_{-\infty}^\infty \left[e^{j(\alpha - \alpha')x} - e^{j(\alpha + \alpha')x} \right] dx \\
&= \frac{\pi}{2} [\delta(\alpha - \alpha') - \delta(\alpha + \alpha')]
\end{aligned} \tag{6.37}$$

by virtue of A.1.3 and A.1.8 of Table A.1. Since neither α nor α' can be negative, we can write this in the form

$$\int_0^\infty \phi(\alpha, x)\phi(\alpha', x)\, dx = N(\alpha)\delta(\alpha - \alpha') \tag{6.38}$$

in complete analogy with (6.23), where

$$N(\alpha) = \frac{\pi}{2} \tag{6.39}$$

is called the normalization factor of the mode (in the present case, this factor is a constant, but it may depend on α in general). In a similar way, suppose that the modes form a complete set:

$$\delta(x - x_0) = \int_0^\infty A(\alpha)\phi(\alpha, x)\, d\alpha \tag{6.40}$$

Proceeding as in the derivation of (6.27), using (6.38)-(6.39) leads to the representation of the delta function $\delta(x - x_0)$ in terms of the radiation modes:

$$\delta(x - x_0) = \int_0^\infty \frac{\phi(\alpha, x)\phi(\alpha, x_0)}{N(\alpha)}\, d\alpha \tag{6.41}$$

One important difference with the case of a closed waveguide is that a single radiation mode cannot exist by itself—an infinite amount of power would have to be supplied to maintain it. Only when superposed with all other radiation modes do we get a field that carries finite power. The one thing that does constrain the individual radiation modes is that their fields must be bounded, even if the energy they carry is infinite. Thus, for the general case (when a simple Fourier integral does not suffice to construct the radiation modes), we characterize a radiation mode by the following requirements:

1. The mode satisfies the governing equation (Helmholtz, Maxwell's etc.) with no sources, and is the product of a mode function of the transverse variable(s) and an exponential function of z. Specifically, the mode function obeys

$$\left(\frac{d^2}{dx^2} + \alpha^2 \right) \phi(\alpha, x) = 0 \tag{6.42}$$

2. The mode obeys all boundary conditions on lateral surfaces.
3. The mode is bounded in all transverse directions.

In our example of the half-space bounded by a perfectly conducting plane, criteria (1) and (2) give us the form (6.36) of the mode function, and criterion (3) restricts the variable α to take on only real values. Negative values of α can be excluded because they offer no essentially different functions than those with positive α.

An example of the use of the radiation modes is furnished by the problem of calculating the field produced by a given aperture field $E_{y0}(x)$ in the plane $z = 0$ above a perfectly conducting plane $x = 0$. In this problem,

$$E_y = k^2 V \tag{6.43}$$

so we want to know the potential in the quadrant $x > 0$, $z > 0$ given that $V(x, 0) = E_{y0}(x)/k^2$. We propose the solution

$$V(x, z) = \int_0^\infty A(\alpha) e^{-j\xi z} \phi(\alpha, x) \, d\alpha \tag{6.44}$$

assuming all sources that produce the field in this quadrant lie in $z < 0$. Clearly this solution will obey the governing differential equation and the boundary condition at $x = 0$. We must now determine $A(\alpha)$ so that the boundary condition in the aperture plane $z = 0$ is met. The requirement is:

$$\frac{E_{y0}(x)}{k^2} = \int_0^\infty A(\alpha) \phi(\alpha, x) \, d\alpha \tag{6.45}$$

for all $0 < x < \infty$. We proceed by multiplying both sides of (6.45) by $\phi(\alpha', x)$, integrating from $x = 0$ to ∞ and using the orthogonality property (6.38). The result is

$$A(\alpha) = \frac{1}{k^2 N(\alpha)} \int_0^\infty E_{y0}(x) \phi(\alpha, x) \, dx \tag{6.46}$$

which can be computed if the aperture field as well as the field of an arbitrary radiation mode is known.

6.3 PROPAGATION IN A PERIODIC MEDIUM

Consider the problem of wave propagation in the z direction in a medium where ϵ and μ are real, positive, nonconstant periodic functions of z:

$$\epsilon(z + p) = \epsilon(z); \qquad \mu(z + p) = \mu(z) \tag{6.47}$$

for all z, where p is the period (i.e., the smallest positive value for which (6.47) is true). For a uniform (independent of x and y) electric current sheet source at $z = 0$ polarized in the x direction, the governing differential equations for the field are:

$$\frac{dE_x}{dz} = -j\omega\mu(z)H_y \tag{6.48}$$
$$-\frac{dH_y}{dz} = j\omega\epsilon(z)E_x + J_{Sx}\delta(z)$$

If we eliminate H_y from (6.48), we get

$$\mathcal{L}E_x + \omega^2 E_x = \frac{j\omega}{\epsilon(z)}J_{Sx}\delta(z) \tag{6.49}$$

where the operator $\mathcal{L}$ is defined as

$$\mathcal{L}f(z) \equiv \frac{1}{\epsilon(z)}\frac{d}{dz}\left(\frac{1}{\mu(z)}\frac{df(z)}{dz}\right) \tag{6.50}$$

An attempt to use the ordinary Fourier transform on these equations fails because the dependence of μ and ϵ on z means that $e^{-j\beta z}$ is not an eigenfunction of these equations. In fact, taking the Fourier transform of the product $\mu(z)H_y(z)$ in (6.48) gives a convolution between the Fourier transforms of the two factors (see (A.61)). The resulting transformed equations are, if anything, more difficult to solve than the originals. Moreover, just because a field exists in a periodic structure, this does not mean the field itself will be periodic with the same period as the structure. Thus a Fourier series does not achieve the kind of transformation we are seeking. We require a different kind of transform that truly simplifies the differential equation. Finally, a periodic medium does not stabilize to a homogeneous one at large distances. This means that we are not entitled to use a conventional radiation condition to select the unique solution to our problem. An alternative means will therefore have to be found.

6.3.1 GEL'FAND'S LEMMA

For problems involving periodic structures, such a transform can be constructed with the aid of Gel'fand's lemma.[3] The essential idea behind this

---

[3]I. M. Gel'fand, *Collected Papers*, vol. 1 (S. G. Gindikin, et al., eds.). Berlin: Springer, 1987, pp. 401-404.

lemma is to take a reasonably arbitrary function $f(z)$ (not necessarily periodic) with suitable properties and make a new periodic function from it. Specifically, define the Gel'fand transform of f to be

$$\tilde{f}(z;\beta) \equiv \sum_{n=-\infty}^{\infty} f(z+np)e^{j\beta(np+z)} = e^{j\beta z} \sum_{n=-\infty}^{\infty} f(z+np)e^{j\beta np} \quad (6.51)$$

By direct calculation, we obtain

$$\tilde{f}(z+p;\beta) = \sum_{n=-\infty}^{\infty} f[z+(n+1)p]e^{j\beta[(n+1)p+z]}$$

$$= \sum_{m=-\infty}^{\infty} f(z+mp)e^{j\beta(mp+z)} = \tilde{f}(z;\beta) \quad (6.52)$$

having renamed the index from n to $m = n+1$. Thus $\tilde{f}(z;\beta)$ is periodic in z. To recover f from $\tilde{f}$, we need to perform an operation that will extract only the $n = 0$ term from (6.51). This can be done by carrying out an integral similar to that which determines the coefficients of a Fourier series (cf. (A.70)):

$$\int_{-\pi/p}^{\pi/p} \tilde{f}(z;\beta)e^{-j\beta z}\,d\beta = \sum_{n=-\infty}^{\infty} \left(\int_{-\pi/p}^{\pi/p} e^{j\beta np}\,d\beta \right) f(z+np) = \frac{2\pi}{p}f(z) \quad (6.53)$$

because the integral inside the summation is zero if $n \neq 0$. The result (6.53) is known as Gel'fand's lemma. This inversion integral resembles the inverse Fourier transform, but here β is confined to the first *Brillouin zone*[4] $-\pi/p < \beta < \pi/p$.

Hence, the inverse transform

$$f(z) = \frac{p}{2\pi} \int_{-\pi/p}^{\pi/p} \tilde{f}(z;\beta)e^{-j\beta z}\,d\beta \quad (6.54)$$

goes along with (6.51) to form a *Gel'fand transform pair*. This transform has many properties; we will have particular need of two (both of which are easily proved). First, the Gel'fand transform of $f'(z)$ is:

$$\left(\frac{\partial}{\partial z} - j\beta \right) \tilde{f}(z;\beta) \quad (6.55)$$

where we use a partial derivative on $\tilde{f}$ because it is a function of two variables, not one. Second, if $A(z)$ is a periodic function of period p, then the Gel'fand transform of $A(z)f(z)$ is

$$A(z)\tilde{f}(z;\beta) \quad (6.56)$$

---

[4]L. Brillouin, *Wave Propagation in Periodic Structures*. New York: Dover, 1953. The other Brillouin zones are shifted from the first by integer multiples of $2\pi/p$.

6.3.2 BLOCH WAVE MODES AND THEIR PROPERTIES

The integrand

$$\tilde{f}(z; \beta)e^{-j\beta z}$$

of (6.54), where $\tilde{f}$ is periodic in z, is known as a Bloch wave, or Bloch-Floquet wave.[5] We wish to use Bloch waves as eigenfunctions (Bloch wave modes) that will convert the differential operator $\mathcal{L}$ in (6.49) into an algebraic multiplier. To this end, let an eigenfunction $\phi_l(z; \beta)$ for an integer index l be a solution of

$$\mathcal{L}\phi_l + \omega_l^2(\beta)\phi_l = 0 \tag{6.57}$$

along with the quasi-periodicity condition

$$\phi_l(z + p; \beta) = e^{-j\beta p}\phi_l(z; \beta) \qquad \text{for any } z \tag{6.58}$$

and the conditions that $\phi_l(z; \beta)$ and $\frac{1}{\mu(z)}\phi_l'(z; \beta)$ are continuous (which correspond to the continuity of the tangential fields E_x and H_y in the original problem). Here the prime denotes differentiation with respect to z and the eigenvalue $\omega_l^2(\beta)$ is a constant (that is, independent of z). We note that at this stage the eigenvalue is $\omega_l^2(\beta)$, and that $\omega_l(\beta)$ itself is only determined to within a $\pm$ sign.

Alternatively, we can define eigenfunctions $\tilde{\phi}_l$ that are periodic in z by

$$\tilde{\phi}_l(z; \beta) = \phi_l(z; \beta)e^{j\beta z} \tag{6.59}$$

and which obey

$$L(\beta)\tilde{\phi}_l(z; \beta) + \omega_l^2(\beta)\tilde{\phi}_l(z; \beta) = 0 \tag{6.60}$$

where the operator $L(\beta)$ is defined by

$$L(\beta)f = \frac{1}{\epsilon}\left(\frac{d}{dz} - j\beta\right)\left[\frac{1}{\mu}\left(\frac{df}{dz} - j\beta f\right)\right] \tag{6.61}$$

and $\tilde{\phi}_l(z; \beta)$ and $\frac{1}{\mu(z)}\left(\frac{d}{dz} - j\beta\right)\phi_l(z; \beta)$ are continuous.

---

[5]Floquet discovered these solutions in the one-dimensional case:

G. Floquet, *Ann. Sci. École Norm. Sup.*, ser. 2, vol. 12, pp. 47-88, 1883.

and Bloch later found their three-dimensional generalization:

F. Bloch, *Z. Phys.*, vol. 52, pp. 555-600, 1928.

but it was not until Gel'fand's work that the use of these functions as basis functions was made sound theoretically. For more examples, see

Brillouin, *loc. cit.*

D. A. Watkins, *Topics in Electromagnetic Theory*. New York: Wiley, 1958, Chapter 1.

W. Kohn, *Phys. Rev.*, vol. 115, pp. 809-821, 1959.

E. H. Lee and W. H. Yang, *SIAM J. Appl. Math.*, vol. 25, pp. 492-499, 1973.

C. Elachi, *Proc. IEEE*, vol. 64, pp. 1666-1698, 1976.

In order to make use of the Bloch wave modes to expand an arbitrary function of z, we will need some further properties of the eigenfunctions and eigenvalues. From (6.59) we see that the definitions of the eigenfunctions and eigenvalues can be extended outside of the first Brillouin zone by putting:

$$\omega_l^2\left(\beta + \frac{2\pi}{p}\right) = \omega_l^2(\beta) \qquad \text{and} \qquad \phi_l\left(z; \beta + \frac{2\pi}{p}\right) = \phi_l(z; \beta) \qquad (6.62)$$

or equivalently,

$$\tilde{\phi}_l\left(z; \beta + \frac{2\pi}{p}\right) = \tilde{\phi}_l(z; \beta)\, e^{j\frac{2\pi z}{p}} \qquad (6.63)$$

First, let us multiply (6.57) by $\epsilon(z)\phi_m^*(z; \beta)$ (where $*$ denotes the complex conjugate) and integrate over a period:[6]

$$\int_0^p \phi_m^*(z; \beta)\frac{d}{dz}\left(\frac{1}{\mu(z)}\frac{d\phi_l(z; \beta)}{dz}\right)\, dz + \omega_l^2(\beta)\int_0^p \epsilon(z)\phi_m^*(z; \beta)\phi_l(z; \beta)\, dz = 0 \qquad (6.64)$$

Integrating the first integral of (6.64) by parts and using the quasi-periodicity condition (6.58) we obtain

$$\int_0^p \frac{1}{\mu(z)}\frac{d\phi_m^*(z; \beta)}{dz}\frac{d\phi_l(z; \beta)}{dz}\, dz = \omega_l^2(\beta)\int_0^p \epsilon(z)\phi_m^*(z; \beta)\phi_l(z; \beta)\, dz \qquad (6.65)$$

If $m = l$, we obtain from (6.65) a representation of the eigenvalue, which clearly shows that it is real and nonnegative:

$$\omega_l^2(\beta) = \frac{\int_0^p \frac{1}{\mu(z)}\left|\frac{d\phi_l(z; \beta)}{dz}\right|^2\, dz}{\int_0^p \epsilon(z)\left|\phi_l(z; \beta)\right|^2\, dz} \geq 0 \qquad (6.66)$$

Thus $\omega_l(\beta)$ itself is always real. From (6.60)-(6.61), it is then clear that we have

$$\omega_l^2(-\beta) = \omega_l^2(\beta) \qquad (6.67)$$

and we can take

$$\phi_l(z; -\beta) = \phi_l^*(z; \beta) \qquad (6.68)$$

Taking $m \leftrightarrow l$ and the complex conjugate in (6.65) gives

$$\int_0^p \frac{1}{\mu(z)}\frac{d\phi_l(z; \beta)}{dz}\frac{d\phi_m^*(z; \beta)}{dz}\, dz = \omega_m^2(\beta)\int_0^p \epsilon(z)\phi_l(z; \beta)\phi_m^*(z; \beta)\, dz \qquad (6.69)$$

Subtracting (6.69) from (6.65) results in the orthogonality relation

$$\int_0^p \epsilon(z)\phi_l(z; \beta)\phi_m^*(z; \beta)\, dz = 0 \qquad \text{if } \omega_l^2(\beta) \neq \omega_m^2(\beta) \qquad (6.70)$$

---

[6]We show this period as $[0, p]$, but in (6.64) and all subsequent integrals over a spatial period, any interval of length p along the z-axis may be used.

We denote the norm of a Bloch mode as

$$N_l(\beta) = \int_0^p \epsilon(z)\phi_l(z;\beta)\phi_l^*(z;\beta)\,dz = \int_0^p \epsilon(z)|\phi_l(z;\beta)|^2\,dz = N_l(-\beta) \quad (6.71)$$

where the last equality follows from (6.68). From (6.66), we also have

$$\int_0^p \frac{1}{\mu(z)} \left| \frac{d\phi_l(z;\beta)}{dz} \right|^2 dz = \omega_l^2(\beta)N_l(\beta) \quad (6.72)$$

Equations (6.70)-(6.71) are the properties required to use Bloch wave modes as an eigenfunction expansion, which we do in the next subsection. Bloch wave eigenfunction expansions can also be constructed for fields in two and three dimensions in a similar way. Moreover, they can be constructed so that they satisfy boundary conditions on a periodic surface such as a grating.

Except in a few simple cases, the most tedious part of the solution process is the construction of the eigenfunctions themselves, along with the computation of the corresponding eigenvalues, which often requires a numerical procedure. One way of doing this is to expand the periodic eigenfunction $\tilde{\phi}_l$ into a standard Fourier series:

$$\tilde{\phi}_l(z;\beta) = \sum_{n=-\infty}^{\infty} e^{jn\frac{2\pi z}{p}} c_{n,l}(\beta) \quad (6.73)$$

where the Fourier coefficients are given as usual by

$$c_{n,l}(\beta) = \frac{1}{p} \int_0^p e^{-jn\frac{2\pi z}{p}} \tilde{\phi}_l(z;\beta)\,dz \quad (6.74)$$

Substituting (6.73) into the differential equation (6.60), multiplying by $e^{-j2\pi mz/p}$, and integrating the result from $z = 0$ to p leads to an infinite matrix equation for the coefficients $c_{n,l}$:

$$p\omega_l^2(\beta)c_{m,l}(\beta) = \sum_{n=-\infty}^{\infty} \left(\beta - \frac{2\pi n}{p} \right)^2 N_{mn}c_{n,l}(\beta) - j \sum_{n=-\infty}^{\infty} \left(\beta - \frac{2\pi n}{p} \right) M_{mn}c_{n,l}(\beta) \quad (6.75)$$

for $m = -\infty, \ldots, -1, 0, 1, \ldots, +\infty$, where

$$N_{mn} = \int_0^p \frac{e^{j(n-m)\frac{2\pi z}{p}}}{\mu(z)\epsilon(z)}\,dz\,, \qquad M_{mn} = \int_0^p \frac{\mu'(z)e^{j(n-m)\frac{2\pi z}{p}}}{\mu^2(z)\epsilon(z)}\,dz \quad (6.76)$$

The determinant of the system (6.75) (*Hill's determinant*) is set equal to zero to determine the eigenvalues.[7] In practice, the infinite system must be

---

[7]See:

Brillouin, *loc. cit.*

S. Solimeno, B. Crosignani, and P. DiPorto, *Guiding, Diffraction, and Confinement of Optical Radiation*. Orlando, FL: Academic Press, 1986, pp. 191-196.

truncated to a finite number of terms, sufficiently many to achieve a desired accuracy.

The eigenfunctions ϕ_l will be useful for representing the electric field E_x of our original problem (6.48). It is convenient to introduce additional eigenfunctions to represent the magnetic field. This will in addition enable us to fix the sign of $\omega_l(\beta)$. Considering (6.48), we define

$$\psi_l(z;\beta) \equiv -\frac{1}{j\omega_l(\beta)\mu(z)}\frac{d\phi_l(z;\beta)}{dz} \tag{6.77}$$

so that the set of the electric and magnetic field Bloch wave eigenfunctions $\{\phi_l(z;\beta), \psi_l(z;\beta)\}$ obeys the system of equations

$$\frac{d\phi_l(z;\beta)}{dz} = -j\omega_l(\beta)\mu(z)\psi_l(z;\beta)$$

$$\frac{d\psi_l(z;\beta)}{dz} = -j\omega_l(\beta)\epsilon(z)\phi_l(z;\beta) \tag{6.78}$$

Evidently the $\psi_l(z;\beta)$ must also be continuous functions of z. We easily derive the additional relations

$$\psi_l(z;-\beta) = \psi_l^*(z;\beta) \tag{6.79}$$

and from (6.72),

$$N_l(\beta) = \int_0^p \mu(z)|\psi_l(z;\beta)|^2\,dz \tag{6.80}$$

Analogously to the electric field eigenfunctions, we can introduce magnetic field eigenfunctions that are periodic in z:

$$\psi_l(z;\beta) = \tilde{\psi}_l(z;\beta)e^{-j\beta z} \tag{6.81}$$

and obey

$$\tilde{\psi}_l\left(z;\beta + \frac{2\pi}{p}\right) = \tilde{\psi}_l(z;\beta)\,e^{j\frac{2\pi z}{p}} \tag{6.82}$$

By (6.78), the periodic eigenfunctions obey the system

$$\frac{d\tilde{\phi}_l(z;\beta)}{dz} - j\beta\tilde{\phi}_l(z;\beta) = -j\omega_l(\beta)\mu(z)\tilde{\psi}_l(z;\beta)$$

$$\frac{d\tilde{\psi}_l(z;\beta)}{dz} - j\beta\tilde{\psi}_l(z;\beta) = -j\omega_l(\beta)\epsilon(z)\tilde{\phi}_l(z;\beta) \tag{6.83}$$

It is clear from (6.78) that for every set of eigenfunctions $\{\phi_l(z;\beta), \psi_l(z;\beta)\}$ with an eigenvalue $\omega_l(\beta)$, there will also be a set $\{\phi_l(z;\beta), -\psi_l(z;\beta)\}$ with eigenvalue $-\omega_l(\beta)$. These are analogous to forward and backward traveling plane waves in a homogeneous medium. The product $\phi_l(z;\beta)\psi_l^*(z;\beta)$ represents the complex Poynting vector for each set of eigenfunctions, and its real part will be positive if the time-average power flow of the mode is in the

$+z$-direction, or negative if power flows in the $-z$-direction.[8] It can be shown (Problem 6-5) that

$$p \operatorname{Re}\left[\phi_l(z;\beta)\psi_l^*(z;\beta)\right] = v_{gl}(\beta)N_l(\beta) \tag{6.84}$$

where

$$v_{gl}(\beta) = \frac{d\omega_l(\beta)}{d\beta} \tag{6.85}$$

is the group velocity of the lth Bloch wave eigenfunction, to be contrasted with the phase velocity

$$v_{pl}(\beta) = \frac{\omega_l(\beta)}{\beta} \tag{6.86}$$

of this eigenfunction, both of which have the usual physical meanings. Since $N_l > 0$, we see that time-average power flows in the $+z$-direction if $v_{gl}(\beta) > 0$, and in the $-z$-direction if $v_{gl}(\beta) < 0$. Indeed, since N_l/p is proportional to average stored energy, (6.84) implies that the group velocity of the eigenfunction is equal to the velocity with which energy is carried in the z-direction—a result that is true for very general situations. Note that it is possible that $\omega_l(\beta) < 0$ for a mode that carries power in the $+z$-direction; such modes are called *backward-wave* modes. By convention, we will define $\omega_l(\beta) > 0$ regardless of the sign of the group velocity.

For certain eigenfunctions it is useful to consider their average over a period of the medium. Define

$$\langle f \rangle = \frac{1}{p} \int_0^p f(z)\, dz \tag{6.87}$$

for any function $f(z)$. Taking the average of (6.83) and using the periodicity of $\tilde{\phi}_l$ and $\tilde{\psi}_l$, we obtain

$$\begin{aligned} \beta\langle\tilde{\phi}_l\rangle &= \omega_l(\beta)\langle\mu\tilde{\psi}_l\rangle \\ \beta\langle\tilde{\psi}_l\rangle &= \omega_l(\beta)\langle\epsilon\tilde{\phi}_l\rangle \end{aligned} \tag{6.88}$$

If further we have $\langle\tilde{\phi}_l\rangle \neq 0$ and $\langle\tilde{\psi}_l\rangle \neq 0$ for a given eigenfunction, we can define an *effective* permittivity and permeability for that mode by means of

$$\epsilon_{\text{eff},l}(\beta) \equiv \frac{\langle\epsilon\tilde{\phi}_l\rangle}{\langle\tilde{\phi}_l\rangle}; \qquad \mu_{\text{eff},l}(\beta) \equiv \frac{\langle\mu\tilde{\psi}_l\rangle}{\langle\tilde{\psi}_l\rangle} \tag{6.89}$$

Then (6.88) becomes

$$\begin{aligned} \beta\langle\tilde{\phi}_l\rangle &= \omega_l(\beta)\mu_{\text{eff},l}(\beta)\langle\tilde{\psi}_l\rangle \\ \beta\langle\tilde{\psi}_l\rangle &= \omega_l(\beta)\epsilon_{\text{eff},l}(\beta)\langle\tilde{\phi}_l\rangle \end{aligned} \tag{6.90}$$

---

[8]By the complex Poynting theorem, the time-average power flow of a Bloch mode (i.e., $\operatorname{Re}\left[\phi_l(z;\beta)\psi_l^*(z;\beta)\right]$) in this lossless medium must be independent of z

whence

$$\beta^2 = \omega_l^2(\beta)\mu_{\mathrm{eff},l}(\beta)\epsilon_{\mathrm{eff},l}(\beta) \tag{6.91}$$

and

$$\frac{\langle \tilde{\phi}_l \rangle}{\langle \tilde{\psi}_l \rangle} = \sqrt{\frac{\mu_{\mathrm{eff},l}(\beta)}{\epsilon_{\mathrm{eff},l}(\beta)}} \equiv \zeta_{\mathrm{eff},l}(\beta) \tag{6.92}$$

which serves as a kind of effective wave impedance for the Bloch wave mode.

It turns out that for any given functions $\mu(z)$ and $\epsilon(z)$ there is an infinite set of eigenfunctions and eigenvalues, which we number in such a way that

$$0 \leq \omega_0^2(\beta) \leq \omega_1^2(\beta) \leq \ldots \tag{6.93}$$

within the first Brillouin zone. However, there is at most one eigenvalue that can become zero in this problem, which we denote with the index $l = 0$ and refer to as the fundamental Bloch mode. Other periodic structures may not have any eigenvalues that can vanish, and in such cases, we number the Bloch modes beginning with $l = 1$. The vanishing of $\omega_0(\beta)$ can happen only when $\beta = 0$, and its eigenfunctions in that case are found from (6.78) to be:

$$\phi_0(z;0) = \tilde{\phi}_0(z;0) = \phi_{00}; \qquad \psi_0(z;0) = \tilde{\psi}_0(z;0) = \psi_{00} \qquad [\text{if } \omega_0(0) = 0] \tag{6.94}$$

where ϕ_{00} and ψ_{00} are constants. The effective permittivity and permeability in this limit are

$$\epsilon_{\mathrm{eff},0}(0) = \langle \epsilon \rangle = \frac{1}{p}\int_0^p \epsilon(z)\,dz; \qquad \mu_{\mathrm{eff},0}(0) = \langle \mu \rangle = \frac{1}{p}\int_0^p \mu(z)\,dz \tag{6.95}$$

from (6.89) (note that this result is specific to our one-dimensional problem, and is not true for more general periodic media). The constants ϕ_{00} and ψ_{00} are related to the norm N_0 of this eigenfunction by

$$N_0 = p\langle \epsilon \rangle|\phi_{00}|^2 = p\langle \mu \rangle|\psi_{00}|^2 \tag{6.96}$$

Typical behavior of the first few Bloch modes is shown in Figure 6.3(a). In this figure (showing the case when an $l = 0$ mode exists), the constant c_0 is an effective wave velocity characteristic of the low-frequency behavior of the periodic medium:

$$c_0 = \frac{1}{\sqrt{\mu_{\mathrm{eff},0}(0)\epsilon_{\mathrm{eff},0}(0)}} \tag{6.97}$$

Such graphs of $\omega_l(\beta)$ versus β are called *Brillouin diagrams* or *ω-β diagrams*. Because of properties (6.62)-(6.63) and (6.82), only the behavior of $\omega_l(\beta)$ over the first Brillouin zone is needed to completely characterize its behavior for all β. It is common to indicate this by the use of a reduced Brillouin diagram as shown in Figure 6.3(b). Figure 6.3 illustrates some of the typical behavior that occurs in a periodic structure. Portions of the real ω-axis are such that one or more values of β produce a real eigenvalue $\omega_l(\beta)$ in that range. These

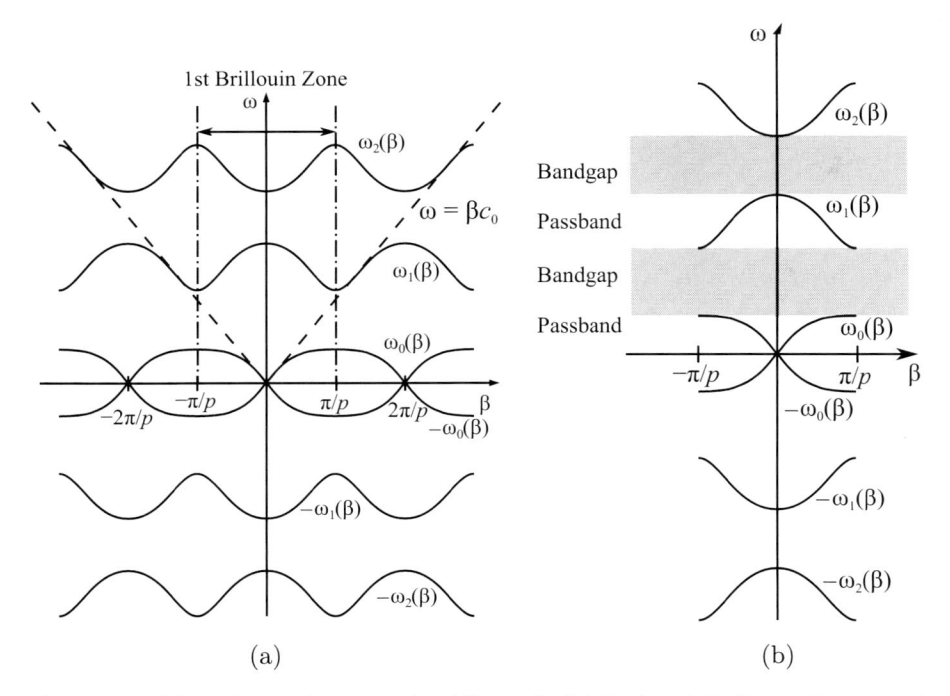

Figure 6.3 (a) Brillouin diagram of $\omega_l(\beta)$ vs. β; (b) Reduced Brillouin diagram of $\omega_l(\beta)$ vs. β.

ranges of ω are called *passbands* of the periodic medium, in which waves can propagate without attenuation. There are, however, portions of the real ω-axis where no real value of β can produce an eigenvalue $\omega_l(\beta)$. Such ranges are known as *bandgaps* or *stopbands*. Nevertheless, there can be *complex* values of β such that $\omega_l(\beta)$ lies in a stopband, and this situation is much like that of a cutoff mode in a waveguide; a wave can exist, but will be attenuated along the z-direction.

6.3.3　THE BLOCH WAVE EXPANSION

We now propose to represent the Gel'fand transform of any function $f(z)$ as an expansion in the Bloch wave eigenfunctions. This requires that the Bloch wave modes form a complete set, so that an arbitrary Gel'fand transform $\tilde{f}(z;\beta)$, periodic in z, can always be written in the form

$$\tilde{f}(z;\beta) = \sum_l F_l(\beta)\tilde{\phi}_l(z;\beta) \tag{6.98}$$

from which the function $f(z)$ itself can be reconstructed using (6.54):

$$f(z) = \frac{p}{2\pi} \int_{-\pi/p}^{\pi/p} \sum_l F_l(\beta)\phi_l(z;\beta)\,d\beta \tag{6.99}$$

where $F_l(\beta)$ are coefficients to be determined. The sums in (6.98) and (6.99) are taken over all the Bloch-Floquet modes, from $l = 0$ or $l = 1$ to ∞ depending on whether or not an $l = 0$ mode exists. An expansion of this kind is called a Bloch wave expansion[9] and serves the same purpose as a Fourier series or Fourier transform (to both of which it has some similarities) in the solution of field problems in periodic structures. It requires a superposition of all Bloch wave eigenfunctions ($l = 0$ or 1 to $+\infty$) but only over the first Brillouin zone $(-\pi/p < \beta < \pi/p)$.

In order to make use of this expansion, we must be able to compute the coefficients $F_l(\beta)$ for a given function $f(z)$; that is, we need to obtain the second equation of a transform pair with (6.99). This is accomplished by multiplying both sides of (6.98) by $\epsilon(z)\tilde{\phi}_m^*(z; \beta)$, integrating from $z = 0$ to p, and using (6.70)-(6.71), with the result that:

$$F_l(\beta) = \frac{1}{N_l(\beta)} \int_0^p \tilde{f}(z; \beta)\epsilon(z)\tilde{\phi}_l^*(z; \beta)\, dz \qquad (6.100)$$

Now we substitute for the Gel'fand transform from (6.51) and get

$$F_l(\beta) = \frac{1}{N_l(\beta)} \sum_{n=-\infty}^{\infty} \int_0^p e^{j\beta(z+np)} f(z + np)\epsilon(z)\tilde{\phi}_l^*(z; \beta)\, dz \qquad (6.101)$$

or

$$F_l(\beta) = \frac{1}{N_l(\beta)} \int_{-\infty}^{\infty} f(z)\epsilon(z)\phi_l^*(z; \beta)\, dz \qquad (6.102)$$

the latter form following from the periodicity of $\tilde{\phi}_l$. Thus (6.99) and (6.102) serve as a Bloch wave transform pair. Evidently, application of this transform to a specific problem requires that the Bloch wave eigenfunctions and

---

[9]See:

F. Odeh and J. B. Keller, *J. Math. Phys.*, vol. 5, pp. 1499-1504, 1964.

C. H. Wilcox, *J. Analyse Math.*, vol. 33, pp. 146-167, 1978.

A. Bensoussan, J.-L. Lions and G. Papanicolaou, *Asymptotic Analysis for Periodic Structures*. Amsterdam: North-Holland, 1978, Chapter 4, Sections 3 and 4.

J. Sanchez-Hubert and E. Sanchez-Palencia, *Vibration and Coupling Analysis of Continuous Systems*. Berlin: Springer, 1989, Chapter 4, Section 5.

F. Santosa and W. W. Symes, *SIAM J. Appl. Math.*, vol. 51, pp. 984-1005, 1991.

A. Oster and N. Turbé, *RAIRO Modél. Math. Anal. Numér.*, vol. 27, pp. 481-496, 1993.

C. Conca, J. Planchard and M. Vanninathan, *Fluids and Periodic Structures*. Chichester, UK: Wiley, 1995, Chapter 3.

C. Conca and M. Vanninathan, *SIAM J. Appl. Math.*, vol. 57, pp. 1639-1659, 1997.

G. Allaire, C. Conca and M. Vanninathan, *ESAIM: Proceedings*, vol. 3, pp. 65-84, 1998.

H.-Y. D. Yang and D. R. Jackson, *IEEE Trans. Ant. Prop.*, vol. 48, pp. 556-564, 2000.

D. Sjöberg et al., *Multiscale Model. Simul.*, vol. 4, pp. 149-171, 2005.

especially for proofs of the completeness property of the Bloch wave modes and extensions to multiple dimensions, among other results. A number of these references also present specific applications to Maxwell's equations.

eigenvalues be calculated in advance, just as waveguide modes and their propagation constants need to be calculated in order to expand an arbitrary field in terms of them.

Note that the explicit Gel'fand transform has disappeared from (6.99) and (6.102) and never appeared in Equations (6.57)-(6.72), which are obeyed by the eigenfunctions and eigenvalues. It was only needed as an intermediate step in deriving the Bloch wave transform, and is not used when applying the Bloch wave eigenfunction expansion to a specific problem. The Gel'fand representation is in fact a Fourier series expansion for the Bloch wave modes, which is often used as a numerical method for their determination.

6.3.4 SOLUTION FOR THE FIELD OF A CURRENT SHEET IN TERMS OF BLOCH MODES

We now return to the solution of Equation (6.49). We assume a Bloch wave expansion for E_x of the form (6.99):

$$E_x(z) = \frac{p}{2\pi} \int_{-\pi/p}^{\pi/p} \sum_{l=0}^{\infty} E_{xl}(\beta)\tilde{\phi}_l(z;\beta)e^{-j\beta z}\,d\beta \tag{6.103}$$

Multiplying both sides of (6.49) by $\epsilon(z)\tilde{\phi}_l^*(z;\beta)e^{j\beta z}$, integrating from $z = -\infty$ to $+\infty$, and using (6.57) and (6.102) results in

$$\left[\omega^2 - \omega_l^2(\beta)\right]N_l(\beta)E_{xl}(\beta) = j\omega J_{Sx}\tilde{\phi}_l^*(0;\beta) \tag{6.104}$$

We could solve (6.104) by a method analogous to that of Section 6.1:

$$E_{xl}(\beta) = \frac{j\omega J_{Sx}\tilde{\phi}_l^*(0;\beta)}{N_l(\beta)}\text{PV}\frac{1}{\omega^2 - \omega_l^2(\beta)} + C_1\delta\left[\omega_l(\beta) - \omega\right] + C_2\delta\left[\omega_l(\beta) + \omega\right] \tag{6.105}$$

where C_1 and C_2 are constants eventually to be determined by the requirement that only outgoing waves be present on each side of the current sheet, analogously to (6.6). However, as has already been mentioned, identifying exactly what constitutes an outgoing wave in this periodic medium is not obvious. Therefore, we choose instead to use the principle of limiting absorption, in which the medium is temporarily assigned some small losses (in the form of negative imaginary parts of $\epsilon(z)$ and/or $\mu(z)$), which are allowed to approach zero in the final solution. In that case, we have instead of (6.105)

$$E_{xl}(\beta) = \frac{j\omega J_{Sx}\tilde{\phi}_l^*(0;\beta)}{N_l(\beta)}\frac{1}{\omega^2 - \omega_l^2(\beta)} \tag{6.106}$$

without the need for principal value integrals or delta functions, analogous to (6.9). Substituting (6.106) into (6.103), we obtain the Bloch wave expansion of $E_x(z)$:

$$E_x(z) = \frac{j\omega J_{Sx}p}{2\pi} \int_{-\pi/p}^{\pi/p} \sum_{l=0}^{\infty} \frac{\phi_l^*(0;\beta)\phi_l(z;\beta)}{N_l(\beta)\left[\omega^2 - \omega_l^2(\beta)\right]}\,d\beta \tag{6.107}$$

in the limit as the losses approach zero.

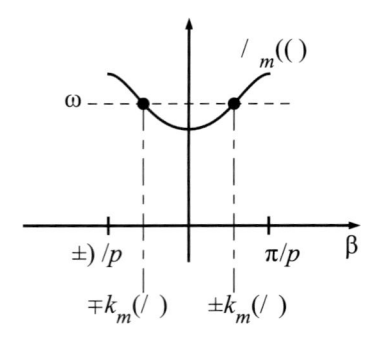

Figure 6.4   Solutions of (6.108) in the first Brillouin zone.

At the operating frequency ω (here assumed positive), there can be at most one pair of solutions $\pm k_m(\omega)$ of

$$\omega = \omega_m[k_m(\omega)] \qquad (6.108)$$

located on one of the branches $l = m$ (where m depends on ω) as shown in Figure 6.4. There are no real solutions for $k_m(\omega)$ in the bandgaps, but if there does exist such a real $k_m(\omega)$ for some m, then in the first Brillouin zone the right side of (6.106) will have singularities at $\beta = \pm k_m(\omega)$. We will defer until later the question of whether $k_m(\omega)$ is positive or negative. Using Taylor series expansions about $\beta = \pm k_m$, we have

$$\omega_m(\beta) = \omega \pm [\beta \mp k_m(\omega)]v_{gm}[k_m(\omega)] + \ldots \qquad (6.109)$$

where the group velocity is given by (6.85), and we have used the fact that $v_{gm}[-k_m(\omega)] = -v_{gm}[k_m(\omega)]$. Equations 6.109 can be combined into the single expression

$$\omega_m^2(\beta) = \omega^2 + [\beta^2 - k_m^2(\omega)]v_{pm}[k_m(\omega)]v_{gm}[k_m(\omega)] + \ldots \qquad (6.110)$$

valid near both singularities, where the phase velocity is given by (6.86).

It will now be seen that if real solutions to (6.108) do exist, then the major contribution to the Bloch wave expansion (6.107) of E_x will come only from the term $l = m$ in the summation, and only in the neighborhood of the points $\beta = \pm k_m(\omega)$ in the integral. We can in this case make an approximation, analogous to that which led to the stationary-phase approximation of integrals in Section 4.3.3, which we will call the *single-mode* approximation. Its physical sense is analogous to that of using only a single mode to represent the field in a waveguide if all higher-order modes are cut off. Accordingly, we keep only the $l = m$ term in (6.107), and extend the limits of the remaining integral to

$\pm\infty$ since the integrand decays rapidly as $\beta \to \pm\infty$ (this will not be accurate if k_m is near a band edge $\pm\pi/p$). We now have

$$E_x(z) \simeq \frac{j\omega J_{Sx}p}{2\pi} \int_{-\infty}^{\infty} \frac{\phi_m^*(0;\beta)\phi_m(z;\beta)}{N_m(\beta)\left[\omega^2 - \omega_m^2(\beta)\right]} \, d\beta \qquad (6.111)$$

We further put $N_m(\beta) \simeq N_m[k_m(\omega)]$ and use (6.110) to approximate

$$\omega^2 - \omega_m^2(\beta) \simeq [k_m^2(\omega) - \beta^2]v_{pm}[k_m(\omega)]v_{gm}[k_m(\omega)] \qquad (6.112)$$

so our approximation becomes

$$E_x(z) \simeq \frac{j\omega J_{Sx}p}{2\pi N_m[k_m(\omega)]v_{pm}[k_m(\omega)]v_{gm}[k_m(\omega)]} \int_{-\infty}^{\infty} \frac{\phi_m^*(0;\beta)\phi_m(z;\beta)}{k_m^2(\omega) - \beta^2} \, d\beta \qquad (6.113)$$

We are now very close to the Fourier transform result obtained in Section 6.1 for the field created by a current sheet in a homogeneous medium (cf. (6.9)). What remains is to choose $k_m(\omega)$ so that it has a negative imaginary part when losses are added to the periodic medium. Since ω is real, and $\omega_m(\beta)$ has a positive imaginary part ω_{mi} if β is real when losses are present (this is shown in Problem 6-6), $k_m = k_{mr} + jk_{mi}$ will now have a non-zero imaginary part. If this imaginary part is small, it can be determined by expanding (6.108) into a Taylor series in the imaginary part of k_m:

$$\omega = \omega_m(k_{mr} + jk_{mi}) = \omega_m(k_{mr}) + jk_{mi}\omega_m'(k_{mr}) + \dots \qquad (6.114)$$

Taking the imaginary part of (6.114), we obtain to leading order in k_{mi}

$$0 = \omega_{mi}(k_{mr}) + k_{mi}v_{gm}(k_{mr}) \qquad (6.115)$$

whence $k_{mi} < 0$ if we choose $k_m(\omega)$ to be the solution of (6.108) for which the group velocity $v_{gm}[k_m(\omega)]$ is positive. Note that this is *not* necessarily the solution with $k_m(\omega) > 0$.

We can now finish the derivation by following the same method used in Section 6.1. The eigenfunctions $\phi_m(z;\beta)$ are approximated near $\beta = \pm k_m(\omega)$ as

$$\phi_m(z;\beta) \simeq \tilde{\phi}_m[z;\pm k_m(\omega)]e^{-j\beta z} \qquad (6.116)$$

and the single-mode approximation (6.113) finally becomes

$$\begin{aligned} E_x(z) \quad \simeq \quad & \frac{pJ_{Sx}}{2N_m[k_m(\omega)]v_{gm}[k_m(\omega)]} \\ & \times \left[\vartheta(z)\tilde{\phi}_m^*[0;k_m(\omega)]\tilde{\phi}_m[z;k_m(\omega)]e^{-jk_m(\omega)z} \right. \\ & \left. + \vartheta(-z)\tilde{\phi}_m^*[0;-k_m(\omega)]\tilde{\phi}_m[z;-k_m(\omega)]e^{+jk_m(\omega)z} \right] \end{aligned} \qquad (6.117)$$

Our choice of sign for k_m has led to waves whose group velocity (and therefore energy velocity) is directed away from the current sheet source on either side, regardless of how the phase terms $e^{\mp jk_m(\omega)z}$ are varying. This general result is sometimes referred to as *Mandel'shtam's energy radiation principle*:[10] The unique solution of a time-harmonic electromagnetic field in a passive environment is that whose time-average power flow is outgoing at infinite distances from the source. It, as well as the principle of limiting absorption, apply to more general situations than do the Sommerfeld or Silver-Müller radiation conditions.

If the period p is small compared to a wavelength ($\omega p/c_0 \ll 1$), we see that $m = 0$ and from (6.91) and (6.108),

$$k_0(\omega) = \omega\sqrt{\mu_{\mathrm{eff},0}[k_0(\omega)]\epsilon_{\mathrm{eff},0}[k_0(\omega)]} \tag{6.118}$$

which for ω sufficiently small is approximately

$$k_0(\omega) \simeq \omega\sqrt{\langle\mu\rangle\langle\epsilon\rangle} \tag{6.119}$$

by (6.95). In this same limit, the eigenfunctions $\tilde{\phi}_0$ and $\tilde{\psi}_0$ become constants by (6.94). As a result, the single-mode approximation becomes

$$E_x(z) \simeq \frac{\zeta_{\mathrm{eff},0}(0)}{2} J_{Sx} e^{-jk_0(\omega)|z|} \tag{6.120}$$

where we have used (6.96) and (6.92). Formula (6.120) is analogous to the pair of plane waves obtained in (6.8) for a homogeneous medium. The approximation (6.120) is known as an *effective medium* or *homogenization* approximation.[11] The fine details of the periodic medium can only be "seen" by the field (6.117) through the factors $\tilde{\phi}_m[z; \pm k_m(\omega)]$, which vary rapidly (on the scale of the period p) with respect to z. When $m = 0$, for low enough frequencies that p is small compared to the effective wavelength $2\pi/k_0(\omega)$ in the medium, $\tilde{\phi}_0$ and $\tilde{\psi}_0$ in the present case are essentially constant. Our solutions have

---

[10]L. I. Mandel'shtam, *Lektsii po Optike, Teorii Otnocitel'nosti i Kvantovoi Mekhanike.* Moscow: Nauka, 1972, pp. 431-437.

[11]See:

Bensoussan et al., *loc. cit.*

E. Sanchez-Palencia, *Non-Homogeneous Media and Vibration Theory.* Berlin: Springer-Verlag, 1980.

N. S. Bakhvalov and G. P. Panasenko, *Homogenisation: Averaging Processes in Periodic Media.* Dordrecht: Kluwer, 1989.

J. Sanchez-Hubert and E. Sanchez-Palencia, *Introduction aux Méthodes Asymptotiques et à l'Homogénéisation.* Paris: Masson, 1992.

Conca et al., *loc. cit..*

D. I. Bardzokas and A. I. Zobnin, *Mathematical Modelling of Physical Processes.* Moscow: URSS, 2005.

become plane waves (6.120) propagating in an effective medium characterized by the material parameters $\epsilon_{\text{eff},0}(k_{\text{eff}}(\omega)) \simeq \langle \epsilon \rangle$ and $\mu_{\text{eff},0}(k_{\text{eff}}(\omega)) \simeq \langle \mu \rangle$. If the fine-scale variations of $\tilde{\phi}_0$ and $\tilde{\psi}_0$ can be neglected (for example, because the available measuring instruments average over them), then the effective-medium description may be sufficient for applications.

6.4 PROBLEMS

6–1 A *static* line current I flowing in the y-direction is located at $(x, z) = (a/2, 0)$ halfway between two perfectly conducting planes $x = 0, a$ as shown.

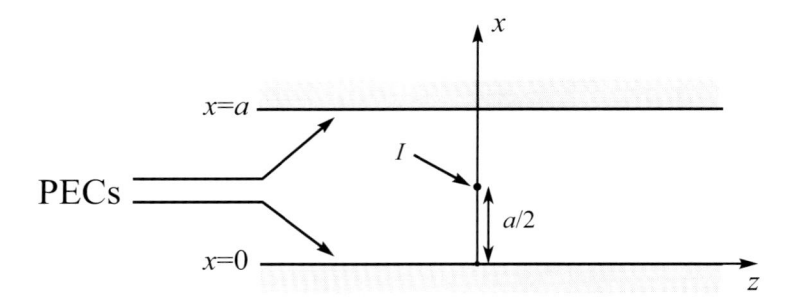

Recall that the normal component of **B** is zero at a perfect conductor, and assume that $\mathbf{B} \to 0$ as $z \to \pm\infty$ to eliminate the possibility of infinite currents on the conducting planes. Find two expressions for the magnetic field everywhere in the region between the parallel plates: one corresponding to (6.19), and the other to (6.28). For extra credit, obtain a closed-form expression for this field by evaluating the sums or integrals.

6–2 Furnish the details of the derivation of (6.28).

6–3 Consider an open waveguide consisting of the half-space $(x > 0, -\infty < z < \infty)$ in which (6.10) holds for the potential (with $I_0 = j\omega\epsilon$), and a reactive surface at $x = 0$, on which the boundary condition

$$\left[\frac{\partial V}{\partial x} + b_S V\right]_{x=0} = 0$$

applies, where b_S (a normalized surface susceptibility) is a real constant. Under what condition on b_S can there exist a discrete mode (called a surface wave mode) that decays exponentially as $x \to \infty$? What is the propagation constant ξ_0 of this mode? Obtain the eigenfunction expansion analogous to (6.35) for V, keeping in mind that in addition to the continuous mode spectrum there may be a discrete surface wave mode as well.

6–4 A periodic array of magnetodielectric slabs is arranged so that

$$\epsilon(z) = \epsilon_1; \qquad \mu(z) = \mu_1$$

for $0 < z < a$, and

$$\epsilon(z) = \epsilon_2; \qquad \mu(z) = \mu_2$$

for $a < z < p$. The pattern is periodically repeated for other z between $-\infty$ and ∞:

$$\epsilon(z+p) = \epsilon(z); \qquad \mu(z+p) = \mu(z)$$

for all z.

Find the Floquet-Bloch modes of the differential equation

$$\frac{1}{\epsilon}\frac{d}{dz}\left(\frac{1}{\mu}\frac{d\phi_l(z;\beta)}{dz}\right) + \omega_l^2(\beta)\phi_l(z;\beta) = 0$$

such that

$$\phi_l(z;\beta) = \tilde{\phi}_l(z;\beta)e^{-j\beta z}$$

β is a real number between $-\pi/p$ and π/p, and $\tilde{\phi}_l$ (and thus also its derivative) is a periodic function of z:

$$\tilde{\phi}_l(z+p;\beta) = \tilde{\phi}_l(z;\beta)$$

The function ϕ_l, which represents the electric field E_x, must be continuous at all values of z, while

$$\frac{1}{\mu}\frac{d\phi_l}{dz}$$

which is proportional to H_y, must also be continuous. You will not be able to find ω_l in closed form; obtain the transcendental equation that it obeys.

6–5 Prove (6.84). Hint: Consider the imaginary part of the quantity

$$\frac{\partial}{\partial \beta}\int_0^p \frac{d}{dz}[\phi_l(z;\beta)\psi_l^*(z;\beta)]\,dz$$

and use the fact that $\mathrm{Re}\,[\phi_l(z;\beta)\psi_l^*(z;\beta)]$ is independent of z.

6–6 Suppose that losses are present in the periodic medium of Section 6.3, so that $\epsilon(z) = \epsilon'(z) - j\epsilon''(z)$ and $\mu(z) = \mu'(z) - j\mu''(z)$, where $\epsilon' > 0$, $\mu' > 0$, $\epsilon'' \geq 0$ and $\mu'' \geq 0$, with at least one of ϵ'' and μ'' not identically equal to zero. Prove the identity (a one-dimensional version of the complex Poynting theorem)

$$\frac{d}{dz}[\phi_l(z;\beta)\psi_l^*(z;\beta)] = j\omega_l^*(\beta)\epsilon^*(z)\,|\phi_l(z;\beta)|^2 - j\omega_l(\beta)\mu(z)\,|\psi_l(z;\beta)|^2$$

if β is real and $\omega_l(\beta) = \omega_{lr}(\beta) + j\omega_{li}(\beta)$ is now complex. From this result, show that

$$\frac{\omega_{li}(\beta)}{\omega_{lr}(\beta)} > 0$$

7 Integral Equations in Scattering Problems

7.1 GREEN'S THEOREM AND GREEN'S FUNCTIONS

7.1.1 SCALAR PROBLEMS

The Lorentz reciprocity theorem is representative of a type of reciprocity relationship that holds for scalar, vector, or dyadic functions satisfying certain kinds of governing differential equations. For example, let f and g be some scalar functions (or generalized functions) of $\mathbf{r}$. Consider the expression

$$\nabla \cdot (f\nabla g - g\nabla f) = f\nabla^2 g - g\nabla^2 f \tag{7.1}$$

which is true by elementary vector identities. Integrating this over a volume V bounded by a closed surface S (Figure 7.1) and applying the divergence theorem in the usual way, we obtain *Green's theorem*:

$$\oint_S (f\nabla g - g\nabla f) \cdot \mathbf{u}_n \, dS = \int_V (f\nabla^2 g - g\nabla^2 f) \, dV \tag{7.2}$$

Suppose, moreover, that the functions f and g satisfy the scalar Helmholtz equations with "sources" s_f and s_g:

$$(\nabla^2 + k^2)f = -s_f$$

$$(\nabla^2 + k^2)g = -s_g \tag{7.3}$$

then (7.2) becomes the analog of Lorentz reciprocity for f and g:

$$\oint_S (f\nabla g - g\nabla f) \cdot \mathbf{u}_n \, dS = \int_V (gs_f - fs_g) \, dV \tag{7.4}$$

An important application of Green's theorem is for the case when one of the functions corresponds to a delta-function source term, such as were studied in Chapter 4. We will define a scalar *Green's function* $G(\mathbf{r}, \mathbf{r}_0)$ to be a solution of

$$(\nabla^2 + k^2)G(\mathbf{r}, \mathbf{r}_0) = -\delta(\mathbf{r} - \mathbf{r}_0) \tag{7.5}$$

The Green's function is not uniquely defined unless boundary conditions are imposed upon it, and although at this point we do not need to say precisely what these conditions are, we will restrict what type of conditions they can be

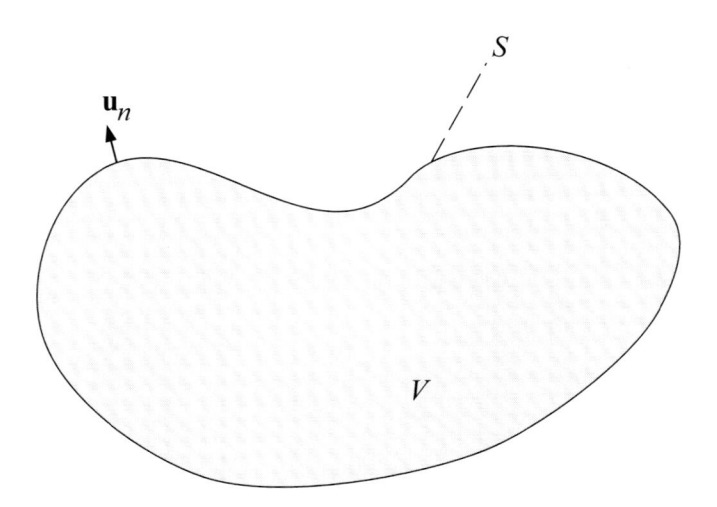

Figure 7.1   Volume and surface used in Green's theorem.

in order to guarantee that the Green's function will possess certain properties. For example, let $f = G(\mathbf{r}, \mathbf{r}_1)$ and $g = G(\mathbf{r}, \mathbf{r}_2)$ in (7.4). Then if $\mathbf{r}_1$ and $\mathbf{r}_2$ are both in V, we have

$$G(\mathbf{r}_1, \mathbf{r}_2) - G(\mathbf{r}_2, \mathbf{r}_1) = \oint_S \left[G(\mathbf{r}, \mathbf{r}_1) \frac{\partial G(\mathbf{r}, \mathbf{r}_2)}{\partial n} - G(\mathbf{r}, \mathbf{r}_2) \frac{\partial G(\mathbf{r}, \mathbf{r}_1)}{\partial n} \right] dS \quad (7.6)$$

If one of the following types of boundary conditions holds:

$$G(\mathbf{r}, \mathbf{r}_0) = 0 \quad \text{for } \mathbf{r} \in S$$

$$\frac{\partial G(\mathbf{r}, \mathbf{r}_0)}{\partial n} = 0 \quad \text{for } \mathbf{r} \in S$$

$$\frac{\partial G(\mathbf{r}, \mathbf{r}_0)}{\partial n} + h(\mathbf{r})G(\mathbf{r}, \mathbf{r}_0) = 0 \quad \text{for } \mathbf{r} \in S$$

where h is some given function on S, or (on a portion of S that recedes to infinity) the radiation condition (3.36) holds, then the surface integral in (7.6) vanishes, and we have that

$$G(\mathbf{r}_1, \mathbf{r}_2) = G(\mathbf{r}_2, \mathbf{r}_1) \quad (7.7)$$

which is to say that the Green's function is symmetric in its arguments. This is also known as the reciprocity property of the Green's function G.

It is traditional to use the notation $\mathbf{r}$ to denote an "observation point," i.e., a point at which we wish to observe the value of a function or field.

On the other hand, $\mathbf{r}'$ will denote a "source point" over which we may be integrating or at which a source is located, as in the Green's function $G(\mathbf{r}, \mathbf{r}')$. The symbol ∇' will be used to denote the del operator in "primed" coordinates:

$$\nabla' = \mathbf{u}_x \frac{\partial}{\partial x'} + \mathbf{u}_y \frac{\partial}{\partial y'} + \mathbf{u}_z \frac{\partial}{\partial z'}$$

and so, by (7.5) and (7.7), we have the properties

$$\nabla' G(\mathbf{r}, \mathbf{r}') = -\nabla G(\mathbf{r}, \mathbf{r}') \tag{7.8}$$

and

$$(\nabla'^2 + k^2) G(\mathbf{r}, \mathbf{r}') = -\delta(\mathbf{r} - \mathbf{r}') \tag{7.9}$$

Using primed variables as integration variables in (7.4), and putting $g(\mathbf{r}') = G(\mathbf{r}, \mathbf{r}')$, we arrive at

$$f(\mathbf{r}) = \int_V G(\mathbf{r}, \mathbf{r}') s_f(\mathbf{r}') \, dV' + \oint_S \left[G(\mathbf{r}, \mathbf{r}') \frac{\partial f(\mathbf{r}')}{\partial n'} - f(\mathbf{r}') \frac{\partial G(\mathbf{r}, \mathbf{r}')}{\partial n'} \right] dS' \tag{7.10}$$

if $\mathbf{r} \in V$. Since the left side of (7.10) evaluates to zero if $\mathbf{r}$ is outside of V, we can combine both cases into the single expression:

$$[1 - \vartheta_S(n)] f(\mathbf{r}) = \int_V G(\mathbf{r}, \mathbf{r}') s_f(\mathbf{r}') \, dV'$$
$$+ \oint_S \left[G(\mathbf{r}, \mathbf{r}') \frac{\partial f(\mathbf{r}')}{\partial n'} - f(\mathbf{r}') \frac{\partial G(\mathbf{r}, \mathbf{r}')}{\partial n'} \right] dS' \tag{7.11}$$

where ϑ_S is the unit step function defined in (1.106). If we suppose further that V approaches unbounded infinite space, by letting $S \to \infty$, and if the radiation condition (3.36) holds for both G and f, then (7.10) takes the simpler form:

$$f(\mathbf{r}) = \int G(\mathbf{r}, \mathbf{r}') s_f(\mathbf{r}') \, dV' \tag{7.12}$$

for all $\mathbf{r}$, where the integral is carried out over all points where the source function s_f is nonzero. In physical terms, this result states that the field f due to a general source distribution s_f is formed by the superposition of point source fields, weighted by the density of the actual source. Note that the solution of (7.5) that satisfies the radiation condition is (see (4.22)):

$$G(\mathbf{r}, \mathbf{r}') = \frac{e^{-jk|\mathbf{r} - \mathbf{r}'|}}{4\pi |\mathbf{r} - \mathbf{r}'|} \tag{7.13}$$

An application of (7.12) with $k = 0$ is the proof of Helmholtz's theorem, which is carried out in Appendix D.

7.1.2 VECTOR PROBLEMS

Vector versions of Green's theorem also exist, along with both vector Green's functions and dyadic (or tensor) Green's functions, which are suitable for application to Maxwell's equations. Analogously to (7.1), we evaluate

$$\nabla \cdot (\mathbf{P} \times \nabla \times \mathbf{Q} - \mathbf{Q} \times \nabla \times \mathbf{P}) = \mathbf{Q} \cdot \nabla \times \nabla \times \mathbf{P} - \mathbf{P} \cdot \nabla \times \nabla \times \mathbf{Q} \quad (7.14)$$

for any vector functions $\mathbf{P}$ and $\mathbf{Q}$. Integrating (7.14) over a volume V gives the vector Green's theorem

$$\oint_S (\mathbf{P} \times \nabla \times \mathbf{Q} - \mathbf{Q} \times \nabla \times \mathbf{P}) \cdot \mathbf{u}_n \, dS$$
$$= \int_V (\mathbf{Q} \cdot \nabla \times \nabla \times \mathbf{P} - \mathbf{P} \cdot \nabla \times \nabla \times \mathbf{Q}) \, dV \quad (7.15)$$

Now suppose that $\mathbf{P}$ is replaced by, let us say, the electric field $\mathbf{E}$ that is produced by a known electric current distribution $\mathbf{J}$ and a known magnetic current distribution $\mathbf{M}$ in a homogeneous region of space. Then from Maxwell's equations we have

$$-\nabla \times \nabla \times \mathbf{E} + k^2 \mathbf{E} = j\omega\mu\mathbf{J} + \nabla \times \mathbf{M} \quad (7.16)$$

in that region. As $\mathbf{Q}$, let us take a vector Green's function $\mathbf{G}^c(\mathbf{r}, \mathbf{r}_0)$ which is the electric field of the electric dipole source

$$\mathbf{J} = -\mathbf{u}_c \frac{\delta(\mathbf{r} - \mathbf{r}_0)}{j\omega\mu}$$

oriented along a given direction c (where c is either x, y or z):

$$-\nabla \times \nabla \times \mathbf{G}^c + k^2 \mathbf{G}^c = -\mathbf{u}_c \delta(\mathbf{r} - \mathbf{r}_0) \quad (7.17)$$

By comparing this problem with (4.18)-(4.24) and (2.33), we can construct one such vector Green's function[1] as

$$\mathbf{G}^c(\mathbf{r}, \mathbf{r}_0) = \mathbf{u}_c G(\mathbf{r}, \mathbf{r}_0) + \frac{1}{k^2} \nabla\nabla \cdot [\mathbf{u}_c G(\mathbf{r}, \mathbf{r}_0)] \quad (7.18)$$

where G is a solution of (7.9). For infinite free space, we can use the G in (7.13).

Applying (7.15) to these functions, we obtain

$$[1 - \vartheta_S(n_0)] \, \mathbf{u}_c \cdot \mathbf{E}(\mathbf{r}_0) = \quad (7.19)$$
$$-j\omega\mu \int_V \mathbf{G}^c(\mathbf{r}', \mathbf{r}_0) \cdot \mathbf{J}(\mathbf{r}') \, dV' - \int_V \mathbf{G}^c(\mathbf{r}', \mathbf{r}_0) \cdot \nabla' \times \mathbf{M}(\mathbf{r}') \, dV'$$
$$+ \oint_S [\mathbf{G}^c(\mathbf{r}', \mathbf{r}_0) \times \nabla' \times \mathbf{E}(\mathbf{r}') - \mathbf{E}(\mathbf{r}') \times \nabla' \times \mathbf{G}^c(\mathbf{r}', \mathbf{r}_0)] \cdot \mathbf{u}_n' \, dS'$$

---

[1] Not all solutions of (7.17) will have this form; in particular, if certain types of boundary condition are imposed.

The term involving $\mathbf{M}$ can be transformed by integration by parts:

$$-\int_V \mathbf{G}^c(\mathbf{r}', \mathbf{r}_0) \cdot \nabla' \times \mathbf{M}(\mathbf{r}') \, dV'$$

$$= \int_V \{\nabla' \cdot [\mathbf{G}^c(\mathbf{r}', \mathbf{r}_0) \times \mathbf{M}(\mathbf{r}')] - \mathbf{M}(\mathbf{r}') \cdot \nabla' \times \mathbf{G}^c(\mathbf{r}', \mathbf{r}_0)\} \, dV'$$

$$= \oint_S \mathbf{u}'_n \cdot [\mathbf{G}^c(\mathbf{r}', \mathbf{r}_0) \times \mathbf{M}(\mathbf{r}')] \, dS' - \int_V \mathbf{M}(\mathbf{r}') \cdot \nabla' \times \mathbf{G}^c(\mathbf{r}', \mathbf{r}_0) \, dV'$$

so that (7.19) becomes

$$[1 - \vartheta_S(n_0)] \, \mathbf{u}_c \cdot \mathbf{E}(\mathbf{r}_0) = \quad\quad\quad (7.20)$$
$$- j\omega\mu \int_V \mathbf{G}^c(\mathbf{r}', \mathbf{r}_0) \cdot \mathbf{J}(\mathbf{r}') \, dV' - \int_V \mathbf{M}(\mathbf{r}') \cdot \nabla' \times \mathbf{G}^c(\mathbf{r}', \mathbf{r}_0) \, dV'$$
$$+ \oint_S [\mathbf{G}^c(\mathbf{r}', \mathbf{r}_0) \times [\nabla' \times \mathbf{E}(\mathbf{r}') + \mathbf{M}(\mathbf{r}')]$$
$$- \mathbf{E}(\mathbf{r}') \times \nabla' \times \mathbf{G}^c(\mathbf{r}', \mathbf{r}_0)] \cdot \mathbf{u}'_n \, dS'$$

Choosing the particular case of $\mathbf{E}(\mathbf{r}) = \mathbf{G}^d(\mathbf{r}, \mathbf{r}_1)$ (so that $\mathbf{J}$ is a Hertz dipole and $\mathbf{M}$ is zero), and putting $\mathbf{r}_0 = \mathbf{r}_2$, we obtain that

$$\mathbf{u}_c \cdot \mathbf{G}^d(\mathbf{r}_2, \mathbf{r}_1) - \mathbf{u}_d \cdot \mathbf{G}^c(\mathbf{r}_1, \mathbf{r}_2) \quad\quad\quad (7.21)$$
$$= \oint_S [\mathbf{G}^c(\mathbf{r}', \mathbf{r}_2) \times \nabla' \times \mathbf{G}^d(\mathbf{r}', \mathbf{r}_1) - \mathbf{G}^d(\mathbf{r}', \mathbf{r}_1) \times \nabla' \times \mathbf{G}^c(\mathbf{r}', \mathbf{r}_2)] \cdot \mathbf{u}'_n \, dS'$$

if both $\mathbf{r}_1$ and $\mathbf{r}_2$ lie inside S. Appropriately chosen boundary conditions on $\mathbf{G}^{c,d}$ will cause the surface integral in (7.21) to vanish. If this is done (which we will assume from here on), then we have the property

$$\mathbf{u}_c \cdot \mathbf{G}^d(\mathbf{r}_2, \mathbf{r}_1) = \mathbf{u}_d \cdot \mathbf{G}^c(\mathbf{r}_1, \mathbf{r}_2) \quad\quad\quad (7.22)$$

which is the analog of (7.7), and is a direct consequence of Lorentz reciprocity.

7.1.3 DYADIC GREEN'S FUNCTIONS

Since only one component of the electric field is represented by (7.20), we must add up three such terms to get the full vector $\mathbf{E}$. This is most conveniently done using dyadic notation (Appendix D). Define a *dyadic Green's function*[2]

---

[2]See, for example,

R. Zich, in *URSI Symposium on Electromagnetic Waves*, Rome, Italy, 1969, pp. 30-34.

C.-T. Tai, *Dyadic Green's Functions in Electromagnetic Theory*. Scranton, PA: Intext, 1971.

W. C. Chew, *Waves and Fields in Inhomogeneous Media*. New York: Van Nostrand Reinhold, 1990, Chapter 7.

$\overset{\leftrightarrow}{\mathbf{G}}$ as

$$
\begin{aligned}
\overset{\leftrightarrow}{\mathbf{G}}(\mathbf{r},\mathbf{r}') &= \mathbf{u}_x \mathbf{G}^x(\mathbf{r},\mathbf{r}') + \mathbf{u}_y \mathbf{G}^y(\mathbf{r},\mathbf{r}') + \mathbf{u}_z \mathbf{G}^z(\mathbf{r},\mathbf{r}') \\
&= \mathbf{G}_x(\mathbf{r},\mathbf{r}')\mathbf{u}_x + \mathbf{G}_y(\mathbf{r},\mathbf{r}')\mathbf{u}_y + \mathbf{G}_z(\mathbf{r},\mathbf{r}')\mathbf{u}_z \qquad (7.23)
\end{aligned}
$$

Taking the dot product on the right of (7.23) with a unit vector $\mathbf{u}_c$ and using (7.22) gives

$$
\overset{\leftrightarrow}{\mathbf{G}}(\mathbf{r}',\mathbf{r}) \cdot \mathbf{u}_c = \mathbf{G}^c(\mathbf{r},\mathbf{r}') = \mathbf{G}_c(\mathbf{r}',\mathbf{r}) \qquad (7.24)
$$

Its posterior components $\mathbf{G}^x$, $\mathbf{G}^y$, and $\mathbf{G}^z$ by definition, and its anterior components $\mathbf{G}_x$, $\mathbf{G}_y$, and $\mathbf{G}_z$ by (7.24) thus represent, respectively, the electric field due to elementary electric dipole sources oriented in the x, y, and z directions (although the source and observation points are interchanged in the latter). This property is succinctly expressed as the reciprocity relation

$$
\overset{\leftrightarrow}{\mathbf{G}}^T(\mathbf{r},\mathbf{r}') = \overset{\leftrightarrow}{\mathbf{G}}(\mathbf{r}',\mathbf{r}) \qquad (7.25)
$$

We then have

$$
\begin{aligned}
[1 - \vartheta_S(n)]\,\mathbf{E}(\mathbf{r}) = \oint_S &\left\{ \overset{\leftrightarrow}{\mathbf{G}}(\mathbf{r}',\mathbf{r}) \times [\nabla' \times \mathbf{E}(\mathbf{r}') + \mathbf{M}(\mathbf{r}')] \right. \\
&\left. + \left[\nabla' \times \overset{\leftrightarrow}{\mathbf{G}}^T(\mathbf{r}',\mathbf{r}) \right]^T \times \mathbf{E}(\mathbf{r}') \right\} \cdot \mathbf{u}'_n\, dS' \qquad (7.26) \\
-j\omega\mu \int_V &\overset{\leftrightarrow}{\mathbf{G}}(\mathbf{r}',\mathbf{r}) \cdot \mathbf{J}(\mathbf{r}')dV' - \int_V \left[\nabla' \times \overset{\leftrightarrow}{\mathbf{G}}^T(\mathbf{r}',\mathbf{r}) \right]^T \cdot \mathbf{M}(\mathbf{r}')dV'
\end{aligned}
$$

or, by (7.25),

$$
\begin{aligned}
[1 - \vartheta_S(n)]\,\mathbf{E}(\mathbf{r}) = \oint_S &\left\{ \overset{\leftrightarrow}{\mathbf{G}}^T(\mathbf{r},\mathbf{r}') \times [\nabla' \times \mathbf{E}(\mathbf{r}') + \mathbf{M}(\mathbf{r}')] \right. \\
&\left. + \left[\nabla' \times \overset{\leftrightarrow}{\mathbf{G}}(\mathbf{r},\mathbf{r}') \right]^T \times \mathbf{E}(\mathbf{r}') \right\} \cdot \mathbf{u}'_n\, dS' \qquad (7.27) \\
-j\omega\mu \int_V &\overset{\leftrightarrow}{\mathbf{G}}^T(\mathbf{r},\mathbf{r}') \cdot \mathbf{J}(\mathbf{r}')dV' - \int_V \left[\nabla' \times \overset{\leftrightarrow}{\mathbf{G}}(\mathbf{r},\mathbf{r}') \right]^T \cdot \mathbf{M}(\mathbf{r}')dV'
\end{aligned}
$$

If we use the vector Green's functions of (7.18) in (7.23), we obtain a particular dyadic Green's function given by

$$
\overset{\leftrightarrow}{\mathbf{G}} = \overset{\leftrightarrow}{\mathbf{I}}\,G + \frac{1}{k^2}\nabla\nabla G = \overset{\leftrightarrow}{\mathbf{I}}\,G + \frac{1}{k^2}\nabla'\nabla'G \qquad (7.28)
$$

However, if boundary conditions are imposed on $\overset{\leftrightarrow}{\mathbf{G}}$ at a finite surface, such a representation may not be possible. Once again, for infinite free space, we can use the G in (7.13), and we have

$$
\overset{\leftrightarrow}{\mathbf{G}} = \overset{\leftrightarrow}{\mathbf{I}}\,\frac{e^{-jk|\mathbf{r}-\mathbf{r}'|}}{4\pi|\mathbf{r}-\mathbf{r}'|} + \frac{1}{k^2}\nabla\nabla\frac{e^{-jk|\mathbf{r}-\mathbf{r}'|}}{4\pi|\mathbf{r}-\mathbf{r}'|} \qquad (7.29)
$$

As an example, consider an electric current density $\mathbf{J}$ of bounded extent radiating in an infinite homogeneous medium, and suppose we employ the dyadic Green's function (7.29) in (7.27). Letting the surface S recede to infinity, and because the fields satisfy the radiation condition, we can show that the surface integrals in (7.27) are zero (show this!). Since magnetic currents are absent, we then have:

$$\mathbf{E}(\mathbf{r}) = -j\omega\mu \int \frac{e^{-jk|\mathbf{r}-\mathbf{r}'|}}{4\pi|\mathbf{r}-\mathbf{r}'|}\mathbf{J}(\mathbf{r}')\,dV' - \frac{j\omega\mu}{k^2} \int \nabla\nabla \frac{e^{-jk|\mathbf{r}-\mathbf{r}'|}}{4\pi|\mathbf{r}-\mathbf{r}'|}\cdot\mathbf{J}(\mathbf{r}')\,dV' \quad (7.30)$$

where the integrals are carried out over all space (or at any rate, over the extent of the sources). While the first integral in (7.30) is well defined, the second is singular enough that it must be considered as a generalized function. Its singularity is of the same type as considered in Example 20 of the Appendix, so we find that (7.30) must be interpreted as:

$$\begin{aligned}
\mathbf{E}(\mathbf{r}) &= -j\omega\mu \int \frac{e^{-jk|\mathbf{r}-\mathbf{r}'|}}{4\pi|\mathbf{r}-\mathbf{r}'|}\mathbf{J}(\mathbf{r}')\,dV' - \frac{j\omega\mu}{k^2}\mathrm{PV}_{V_0}\int \nabla\nabla \frac{e^{-jk|\mathbf{r}-\mathbf{r}'|}}{4\pi|\mathbf{r}-\mathbf{r}'|} \\
&\quad \times \mathbf{J}(\mathbf{r}')\,dV' + \frac{j\omega\mu}{k^2}\overset{\leftrightarrow}{\mathbf{L}}_{V_0}\cdot\mathbf{J}(\mathbf{r})
\end{aligned} \quad (7.31)$$

for a given principal volume V_0, where $\overset{\leftrightarrow}{\mathbf{L}}_{V_0}$ is the depolarization dyadic associated with V_0. It is instructive to rewrite (7.31) as:

$$\begin{aligned}
\mathbf{E}_{\mathrm{local}}(\mathbf{r}) &= \mathbf{E}(\mathbf{r}) + \frac{1}{j\omega\epsilon}\overset{\leftrightarrow}{\mathbf{L}}_{V_0}\cdot\mathbf{J}(\mathbf{r}) = -j\omega\mu \int \frac{e^{-jk|\mathbf{r}-\mathbf{r}'|}}{4\pi|\mathbf{r}-\mathbf{r}'|}\mathbf{J}(\mathbf{r}')\,dV' \\
&\quad - \frac{j\omega\mu}{k^2}\mathrm{PV}_{V_0}\int \nabla\nabla \frac{e^{-jk|\mathbf{r}-\mathbf{r}'|}}{4\pi|\mathbf{r}-\mathbf{r}'|}\cdot\mathbf{J}(\mathbf{r}')\,dV'
\end{aligned} \quad (7.32)$$

Here $\mathbf{E}_{\mathrm{local}}$ is the so-called *local field* at the point $\mathbf{r}$. Its meaning is this: imagine that we cut a tiny volume V_0 of the current density $\mathbf{J}$ from around the observation point V_0. We now observe the field in a source-free region, rather than at a point where currents are actually present. Now despite the very small size of the region of currents we have removed, a possibly significant change occurs in the field now seen at $\mathbf{r}$, even in the limit as V_0 shrinks to zero. The difference occurs because when we cut out the small portion of $\mathbf{J}$ but leave the rest of the current density intact, we generally cause a surface charge density to appear at the surface S_0 that bounds V_0, and this surface density in turn results in a quasistatic electric field inside V_0 that does not approach zero there as $V_0 \to 0$. This new field, $\mathbf{E}_{\mathrm{local}}$, is what would be "seen" by a small particle such as an atom or molecule placed at $\mathbf{r}$ in a "gap" between the moving charged particles that make up $\mathbf{J}$. It differs from the macroscopic field $\mathbf{E}$, which could in principle never be measured without disturbing the current distribution that is causing it. The local field is important in deriving the macroscopic behavior of materials based on microscopic effects of fields on polarizable particles.

7.1.4   RELATION TO EQUIVALENCE PRINCIPLE

Performing a slight rearrangement of (7.27) by means of Maxwell's equations and the vector triple product law, we obtain the following version of this representation:

$$[1 - \vartheta_S(n)] \mathbf{E}(\mathbf{r}) = \oint_S \left\{ \overset{\leftrightarrow}{\mathbf{G}}{}^T(\mathbf{r}, \mathbf{r}') \cdot [j\omega\mu \mathbf{u}'_n \times \mathbf{H}(\mathbf{r}')] \right.$$
$$\left. + \left[\nabla' \times \overset{\leftrightarrow}{\mathbf{G}}(\mathbf{r}, \mathbf{r}') \right]^T \cdot \mathbf{E}(\mathbf{r}') \times \mathbf{u}'_n \right\} dS' \qquad (7.33)$$
$$- j\omega\mu \int_V \overset{\leftrightarrow}{\mathbf{G}}{}^T(\mathbf{r}, \mathbf{r}') \cdot \mathbf{J}(\mathbf{r}') \, dV' - \int_V \left[\nabla' \times \overset{\leftrightarrow}{\mathbf{G}}(\mathbf{r}, \mathbf{r}') \right]^T \cdot \mathbf{M}(\mathbf{r}') \, dV'$$

Aside from the fact that the first integrals are over the surface S while the second are over the volume V, both terms on the right side of (7.33) involve either $\overset{\leftrightarrow}{\mathbf{G}}{}^T$ or $\left[\nabla' \times \overset{\leftrightarrow}{\mathbf{G}} \right]^T$ dotted into a vector on the right. It is as if there were surface electric currents given by

$$\mathbf{J}_{\mathrm{eq}} = -\mathbf{u}_n \times \mathbf{H}\delta_S(n)$$

and surface magnetic currents given by

$$\mathbf{M}_{\mathrm{eq}} = -\mathbf{E} \times \mathbf{u}_n\delta_S(n)$$

at the surface S, in addition to the volume densities of $\mathbf{J}$ and $\mathbf{M}$. But this is precisely the content of Love's equivalence theorem stated in Sections 1.4.3 and 3.2.4, if we remember that there the unit normal was chosen pointing *into* the region where the observation point was located, while in this case, $\mathbf{u}_n$ points away from this region (V).

In a case where the dyadic Green's function has the form (7.28), then (7.33) can be expressed in terms of the Lorenz vector potential $\mathbf{A}$ and its dual $\mathbf{F}$ as in (2.22)-(2.23), with

$$\mathbf{A} = \mu \int [\mathbf{J}_V(\mathbf{r}') + \mathbf{J}_{\mathrm{eq}}(\mathbf{r}')] G(\mathbf{r}, \mathbf{r}') \, dV'$$
$$\mathbf{F} = \epsilon \int [\mathbf{M}_V(\mathbf{r}') + \mathbf{M}_{\mathrm{eq}}(\mathbf{r}')] G(\mathbf{r}, \mathbf{r}') \, dV' \qquad (7.34)$$

where $\mathbf{J}_V$ and $\mathbf{M}_V$ are respectively equal to $\mathbf{J}$ and $\mathbf{M}$ within V and equal to zero outside of V.

7.2   INTEGRAL EQUATIONS FOR SCATTERING BY A PERFECT CONDUCTOR

It is not normally the case that we know the fields completely on a boundary surface S. Until we do, it is not possible to use any of the foregoing versions

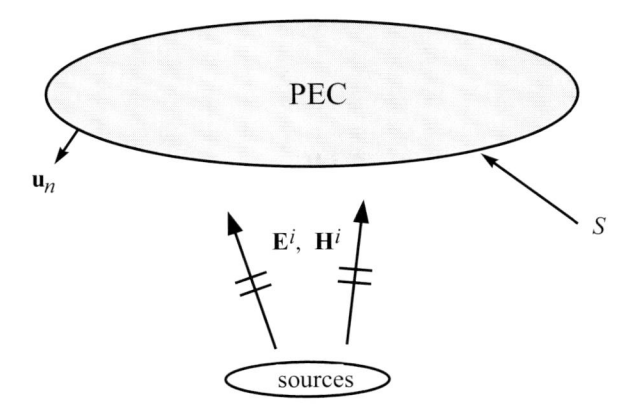

Figure 7.2 Scattering by a perfect conductor.

of Green's theorem to calculate the field at an arbitrary point in space. In a specific problem, we will need to subject the surface field to some appropriate boundary condition in order to fully determine it. In this section, we will study the formulation of integral equations for this purpose.[3] We do this by means of a number of examples.

7.2.1 ELECTRIC-FIELD INTEGRAL EQUATION (EFIE)

Consider a perfectly conducting, arbitrarily shaped scatterer whose surface is S, and whose outward unit normal is $\mathbf{u}_n$ as shown in Figure 7.2. An incident wave $\mathbf{E}^i$, $\mathbf{H}^i$ impinges upon the scatterer. This field is what would be produced in infinite empty space by a set of given sources $\mathbf{J}$, $\mathbf{M}$.

Let $G = G(\mathbf{r}, \mathbf{r}')$ be the scalar time-harmonic Green's function (7.13) for infinite space. Let V denote the region *exterior* to the scatterer, so that its boundary surface is S plus the surface S_∞ at infinity. We apply (2.22)-(2.23) and (7.34), and note that

$$\mathbf{u}'_n \times \mathbf{E}(\mathbf{r}')|_S = 0 \qquad (7.35)$$

[3]For further background on integral equations in electromagnetics, see:

D. Colton and R. Kress, *Integral Equation Methods in Scattering Theory.* New York: Wiley, 1983.

D. S. Jones, *Methods in Electromagnetic Wave Propagation.* Oxford, UK: Clarendon Press, 1987, Chapter 6.

since S is a perfect conductor:

$$\mathbf{E} = -j\omega\mu \left(1 + \frac{1}{k^2}\nabla\nabla\cdot\right) \int_V G\mathbf{J}(\mathbf{r}')\,dV' \tag{7.36}$$

$$-\nabla \times \int_V G\mathbf{M}(\mathbf{r}')\,dV' - j\omega\mu \left(1 + \frac{1}{k^2}\nabla\nabla\cdot\right) \oint_S G\mathbf{J}_S(\mathbf{r}')\,dS'$$

(observe that $\mathbf{u}'_n$ appearing in (7.33) is the outward normal *away from* V, and thus must be taken here as $-\mathbf{u}'_n$ as shown in Figure 7.2). Here,

$$\mathbf{J}_S(\mathbf{r}') = \mathbf{u}'_n \times \mathbf{H}(\mathbf{r}')|_S \tag{7.37}$$

is the (yet unknown) surface current induced on the scatterer. There is also an associated induced surface charge density ρ_S related to $\mathbf{J}_S$ by

$$\rho_S = -\frac{\nabla \cdot \mathbf{J}_S}{j\omega} \tag{7.38}$$

Now, the first two terms on the right side of (7.36) represent the electric field that would be produced by $\mathbf{J}$ and $\mathbf{M}$ if the scatterer were absent; i.e., the incident field:

$$\mathbf{E}^i = -j\omega\mu \left(1 + \frac{1}{k^2}\nabla\nabla\cdot\right) \int_V G\mathbf{J}(\mathbf{r}')\,dV' - \nabla \times \int_V G\mathbf{M}(\mathbf{r}')\,dV' \tag{7.39}$$

It is usually more convenient (and more common) to specify this incident $\mathbf{E}$-field directly than to specify the sources $\mathbf{J}$ and $\mathbf{M}$. The last term on the right side of (7.36) is the scattered field $\mathbf{E}^s$; this is what would be produced by the yet unknown induced currents $\mathbf{J}_S$, radiating as equivalent currents in empty space (with the scatterer removed). Then (7.36) can be written as:

$$\mathbf{E} = \mathbf{E}^i - j\omega\mu \left(1 + \frac{1}{k^2}\nabla\nabla\cdot\right) \oint_S G\mathbf{J}_S(\mathbf{r}')\,dS' \tag{7.40}$$

Equation (7.40) can be rewritten using (7.38) along with the types of manipulation employed in Section 7.1; the result is:

$$\mathbf{E} = \mathbf{E}^i(\mathbf{r}) - j\omega\mu \oint_S G\mathbf{J}_S(\mathbf{r}')\,dS' - \frac{1}{\epsilon}\nabla \oint_S G\rho_S(\mathbf{r}')\,dS' \tag{7.41}$$

Taking the curl of (7.40) or (7.41) and using

$$\nabla \times (\mathbf{E} - \mathbf{E}^i) = -j\omega\mu(\mathbf{H} - \mathbf{H}^i) \tag{7.42}$$

which follows from Maxwell's equations, we get

$$\mathbf{H}(\mathbf{r}) = \mathbf{H}^i(\mathbf{r}) + \nabla \times \oint_S G\mathbf{J}_S(\mathbf{r}')\,dS' \tag{7.43}$$

The unknown $\mathbf{J}_S$ (and the corresponding surface charge ρ_S) is determined in principle by bringing the observation point to the surface of the conductor ($\mathbf{r} \in S$) in (7.40) or (7.43) and enforcing the boundary condition (7.35) on the field at that surface. In general, this results in what is usually called an *integral equation*[4] for $\mathbf{J}_S$, which must be solved either by analytical means or (most often, in practice) using numerical techniques.

If we put $\mathbf{r} \in S$ in (7.40)—that is, we enforce the boundary condition on the electric field—we have

$$\mathbf{u}_n \frac{\rho_S}{\epsilon} = \mathbf{E}^i(\mathbf{r}) - j\omega\mu\left(1 + \frac{1}{k^2}\nabla\nabla\cdot\right)\oint_S G\mathbf{J}_S(\mathbf{r}')\,dS'; \quad (\mathbf{r} \in S) \qquad (7.44)$$

It turns out that enforcement of the normal component of (7.44) at S is usually redundant (see below for the exceptions to this), so the remaining tangential components give

$$0 = \mathbf{E}^i_{\tan}(\mathbf{r}) - j\omega\mu\left\{\left(1 + \frac{1}{k^2}\nabla\nabla\cdot\right)\oint_S G\mathbf{J}_S(\mathbf{r}')\,dS'\right\}_{\tan}; \quad (\mathbf{r} \in S) \qquad (7.45)$$

This equation, together with any necessary edge conditions on $\mathbf{J}_S$ at sharp edges,[5] is sufficient to determine $\mathbf{J}_S$ over all of S (except at resonant frequencies of the interior problem as discussed below). Equation (7.45) (which is obtained by imposing a boundary condition on the electric field) is known as an *electric field integral equation* (EFIE).

Some simplification of the EFIE occurs if the problem is two-dimensional; that is, if the incident wave is independent of one coordinate (z, say) and S is the surface of a cylinder whose axis is parallel to this direction. In that case, $\mathbf{J}_S(\mathbf{r}') = \mathbf{J}_S(\boldsymbol{\rho}')$ is independent of z', and the surface integral in (7.45) can be written

$$\oint_S G\mathbf{J}_S(\mathbf{r}')\,dS' = \oint_C \mathbf{J}_S(\boldsymbol{\rho}')\int_{-\infty}^{\infty} G(\mathbf{r},\mathbf{r}')\,dz'\,dl' \qquad (7.46)$$

where C is the boundary of the cylinder cross-section and l' is an arc length along it (Figure 7.3). But

$$\int_{-\infty}^{\infty} G(\mathbf{r},\mathbf{r}')\,dz' = \frac{1}{4\pi}\int_{-\infty}^{\infty}\frac{e^{-jk\sqrt{|\boldsymbol{\rho}-\boldsymbol{\rho}'|^2+(z-z')^2}}}{\sqrt{|\boldsymbol{\rho}-\boldsymbol{\rho}'|^2+(z-z')^2}}\,dz'$$

$$= -\frac{j}{4}H_0^{(2)}(k|\boldsymbol{\rho}-\boldsymbol{\rho}'|) \equiv G_0(\boldsymbol{\rho},\boldsymbol{\rho}') \qquad (7.47)$$

---

[4]More properly, it is an *integro-differential equation* because differentiation as well as integration of the unknown is involved.

[5]The component of $\mathbf{J}_S$ at an edge that is perpendicular to that edge must be zero. Otherwise, conservation of surface charge would require a buildup of charge at the edge (for example, a line charge), and this in turn would produce a radial electric field that blows up more rapidly as the edge is approached than is permitted by the edge condition—see Section 5.3.2.

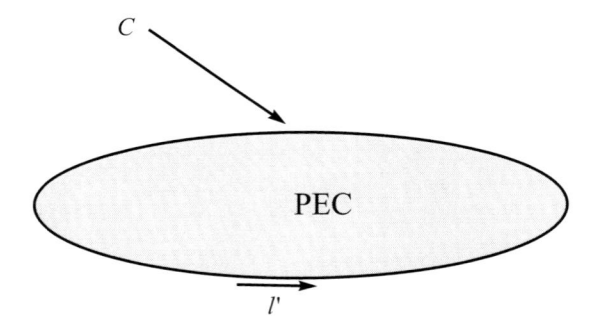

Figure 7.3 Cross section of cylindrical scatterer.

by virtue of the Fourier transform relation inverse to (4.68). Hence (7.46) in (7.45) yields

$$0 = \mathbf{E}_{\text{tan}}^i(\boldsymbol{\rho}) - j\omega\mu \left\{ \left(1 + \frac{1}{k^2}\nabla_t\nabla_t\cdot\right) \oint_C G_0 \mathbf{J}_S(\boldsymbol{\rho}') \, dl' \right\}_{\text{tan}} \qquad (7.48)$$

If $\mathbf{E}^i$ is polarized in the z-direction only, then so must $\mathbf{J}_S$ be as well, and (7.48) reduces to a pure *integral equation*:

$$0 = E_z^i(\boldsymbol{\rho}) - j\omega\mu \oint_C G_0 J_{Sz}(\boldsymbol{\rho}') \, dl' \qquad (7.49)$$

On the other hand, when $\mathbf{E}^i$ is polarized parallel to the xy-plane, no significant simplification occurs.

7.2.2 MAGNETIC-FIELD INTEGRAL EQUATION (MFIE)

This same scattering problem can be formulated in an alternate manner as a *magnetic field integral equation* (MFIE), by imposing the boundary condition on the magnetic field at S. By letting $\mathbf{r} \to S$ in (7.43) and enforcing the boundary condition on $\mathbf{H}$ there, we have:

$$\mathbf{J}_S(\mathbf{r}) \times \mathbf{u}_n = \mathbf{H}^i(\mathbf{r}) + \nabla \times \oint_S G\mathbf{J}_S(\mathbf{r}') \, dS'; \quad (\mathbf{r} \in S) \qquad (7.50)$$

Again, apart from some exceptional circumstances noted below, the normal components of Equation (7.50) are redundant, and so we get the MFIE in the form

$$\mathbf{J}_S(\mathbf{r}) \times \mathbf{u}_n = \mathbf{H}_{\text{tan}}^i(\mathbf{r}) + \left\{ \nabla \times \oint_S G\mathbf{J}_S(\mathbf{r}') \, dS' \right\}_{\text{tan}} ; \quad (\mathbf{r} \in S) \qquad (7.51)$$

A few comments are in order about (7.45) and (7.51). Equation (7.45) is known as an integral equation of the *first kind*, since the unknown $\mathbf{J}_S$ appears

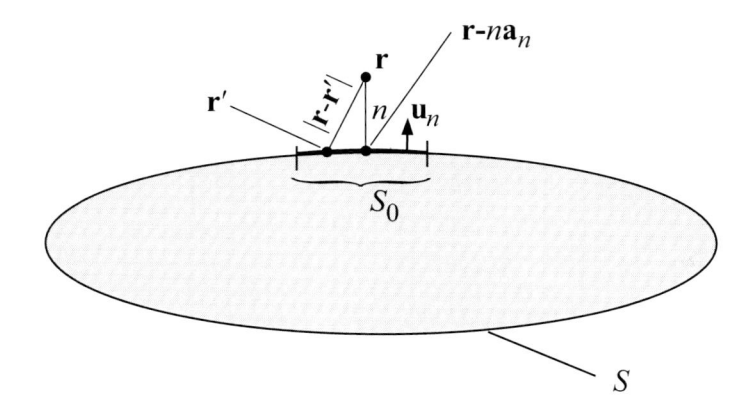

Figure 7.4 Calculation of integral with removable singularity on a surface.

only under the integral sign, whereas (7.51) is known as an integral equation of the *second kind*, because the unknown function appears both inside and outside of the integral sign. Moreover, the differentiations appearing in (7.45) and (7.51) are to be carried out *after* the integrations over dS', if they are to be regarded in the sense of ordinary functions. In practical calculations this may be inconvenient, and the following technique must be used to allow the differentiation to be moved under the integral sign.

Consider $\mathbf{r} \in V$, but not on S, and examine the integral

$$\nabla \oint_S f(\mathbf{r}')G(\mathbf{r}, \mathbf{r}')\, dS'$$

where G is given by (7.13). Let the perpendicular distance n of the point $\mathbf{r}$ from the surface S be small, and consider a small portion S_0 of S located below this point. Let the "diameter" d of S_0 (the maximum distance between any two points of S_0) be small enough compared to the radii of curvature of S_0 such that it is locally plane (i.e., flat) underneath $\mathbf{r}$, and let $k|\mathbf{r} - \mathbf{r}'| \ll 1$ for $\mathbf{r}' \in S_0$ (see Figure 7.4). We finally assume that $n \leq d$, so that an observer located at $\mathbf{r}$ looking down at S_0 also sees a nearly planar surface. Then we have for such a fixed $n > 0$:

$$\nabla \int_{S_0} f(\mathbf{r}')G(\mathbf{r}, \mathbf{r}')\, dS' \simeq \nabla \left\{ \frac{f(\mathbf{r} - n\mathbf{u}_n)}{4\pi} \int_{S_0} \frac{dS'}{|\mathbf{r} - \mathbf{r}'|} \right\} \tag{7.52}$$

But

$$\nabla \left(\frac{1}{|\mathbf{r} - \mathbf{r}'|} \right) = \nabla_t \left(\frac{1}{|\mathbf{r} - \mathbf{r}'|} \right) + \mathbf{u}_n \frac{\partial}{\partial n} \left(\frac{1}{|\mathbf{r} - \mathbf{r}'|} \right)$$

$$= -\nabla'_t \left(\frac{1}{|\mathbf{r} - \mathbf{r}'|} \right) + \mathbf{u}_n \frac{\partial}{\partial n} \left(\frac{1}{|\mathbf{r} - \mathbf{r}'|} \right) \tag{7.53}$$

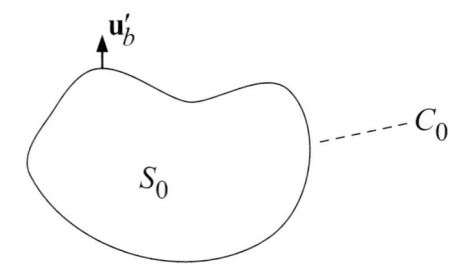

Figure 7.5 Detail of the excluded surface S_0.

where ∇_t denotes the components of ∇ tangent to the surface S, so carrying out the derivatives indicated in (7.53), we have

$$
\nabla \int_{S_0} f(\mathbf{r}')G(\mathbf{r},\mathbf{r}')\, dS'
$$

$$
\simeq \frac{\nabla f(\mathbf{r}-n\mathbf{u}_n)}{4\pi} \int_{S_0} \frac{dS'}{|\mathbf{r}-\mathbf{r}'|}
$$

$$
- \frac{f(\mathbf{r}-n\mathbf{u}_n)}{4\pi} \left[\int_{S_0} \nabla_t' \left(\frac{1}{|\mathbf{r}-\mathbf{r}'|} \right) dS' + \mathbf{u}_n \int_{S_0} \frac{n\, dS'}{(|\boldsymbol{\rho}-\boldsymbol{\rho}'|^2 + n^2)^{3/2}} \right]
$$

$$
= \frac{\nabla f(\mathbf{r}-n\mathbf{u}_n)}{4\pi} \int_{S_0} \frac{dS'}{|\mathbf{r}-\mathbf{r}'|}
$$

$$
- \frac{f(\mathbf{r}-n\mathbf{u}_n)}{4\pi} \left[\oint_{C_0} \frac{\mathbf{u}_b'\, dl'}{|\mathbf{r}-\mathbf{r}'|} + \mathbf{u}_n \int_{S_0} \frac{n\, dS'}{(|\boldsymbol{\rho}-\boldsymbol{\rho}'|^2 + n^2)^{3/2}} \right] \qquad (7.54)
$$

where $\boldsymbol{\rho}' = \mathbf{r}'$ is the source point lying in S_0, $\boldsymbol{\rho} = \mathbf{r} - n\mathbf{u}_n$ is the projection of the observation point $\mathbf{r}$ onto S_0, C_0 is the contour bounding S_0, and $\mathbf{u}_b'$ is the unit normal vector to C_0 lying tangential to the surface S and pointing towards the exterior of S_0 (see Figure 7.5). We next take the limit as $n \to 0^+$ ($\mathbf{r} \to S$ from above), and obtain

$$
\nabla \int_{S_0} f(\mathbf{r}')G(\mathbf{r},\mathbf{r}')\, dS' \simeq \frac{\nabla f(\mathbf{r})}{4\pi} \int_{S_0} \frac{dS'}{|\boldsymbol{\rho}-\boldsymbol{\rho}'|} - \frac{f(\mathbf{r})}{4\pi} \oint_{C_0} \frac{\mathbf{u}_b'\, dS'}{|\boldsymbol{\rho}-\boldsymbol{\rho}'|} - \frac{\mathbf{u}_n f(\mathbf{r})}{2}
$$

$$
(7.55)
$$

using the result of Problem A-11. Some careful calculations will show that the first two terms of the right side of (7.55) approach 0 as the diameter d of S_0 shrinks to zero.

The remainder of the surface integral over $S - S_0$ is an ordinary one, and the gradient operator can be taken inside of it without special treatment. Combining this with (7.55) and only now taking the limit as S_0 shrinks to zero, we can thus write

$$\nabla \oint_S f(\mathbf{r}')G(\mathbf{r},\mathbf{r}')\,dS' \tag{7.56}$$

$$= \lim_{d\to 0}\left\{-\int_{S-S_0} f(\mathbf{r}')\nabla'G(\mathbf{r},\mathbf{r}')\,dS'\right\} - \frac{1}{2}\mathbf{u}_n f(\mathbf{r})$$

$$= -\oint_S f(\mathbf{r}')\nabla'G(\mathbf{r},\mathbf{r}')\,dS' - \frac{\mathbf{u}_n f(\mathbf{r})}{2};\quad (\mathbf{r}\in S)$$

The limit of the integral appearing in the second line of (7.56) is independent of the shape of S_0 and the manner in which it shrunk to zero; the integral has only a removable singularity. Numerically, the final integral in (7.56) can be computed by excluding a small enough surface S_0 from the integration, thereby avoiding the apparent singular behavior of the integral. We could thus write (7.51) as

$$\frac{1}{2}\mathbf{J}_S(\mathbf{r})\times\mathbf{u}_n = \mathbf{H}^i_{\text{tan}}(\mathbf{r}) - \left\{\oint_S \nabla'G\times\mathbf{J}_S(\mathbf{r}')\,dS'\right\}_{\text{tan}};\quad (\mathbf{r}\in S) \tag{7.57}$$

while (7.45) could be rewritten using (7.41) as:

$$0 = \mathbf{E}^i_{\text{tan}}(\mathbf{r}) - j\omega\mu\left\{\oint_S G\mathbf{J}_S(\mathbf{r}')\,dS'\right\}_{\text{tan}} + \frac{1}{\epsilon}\left\{\oint_S \nabla'G\rho_S(\mathbf{r}')\,dS'\right\}_{\text{tan}} \tag{7.58}$$

for $\mathbf{r}\in S$, with ρ_S given by (7.38). Note that no additional term results in this EFIE when differentiation is brought inside the integral sign.

As with the EFIE, there is a two-dimensional version of the MFIE for general cylindrical scatterers. It follows from (7.57) by using (7.46)-(7.47):

$$\frac{1}{2}\mathbf{J}_S(\boldsymbol{\rho}) = \mathbf{u}_n\times\mathbf{H}^i(\boldsymbol{\rho}) - \mathbf{u}_n\times\oint_C \nabla'_t G_0\times\mathbf{J}_S(\boldsymbol{\rho}')\,dl';\quad (\boldsymbol{\rho}\in C) \tag{7.59}$$

For an incident $\mathbf{H}$-field polarized transverse to z, this reduces to

$$\frac{1}{2}J_{Sz}(\boldsymbol{\rho}) = \mathbf{u}_l\cdot\mathbf{H}^i(\boldsymbol{\rho}) + \oint_C \frac{\partial G_0}{\partial n'}\times J_{Sz}(\boldsymbol{\rho}')\,dl' \tag{7.60}$$

while for an incident magnetic field purely in the z-direction,

$$\frac{1}{2}H_z(\boldsymbol{\rho}) = H^i_z(\boldsymbol{\rho}) + \oint_C \frac{\partial G_0}{\partial n'}\times H_z(\boldsymbol{\rho}')\,dl' \tag{7.61}$$

7.2.3 NONUNIQUENESS AND OTHER DIFFICULTIES

Unfortunately, unlike the formulation of a scattering problem via differential equations together with the appropriate boundary conditions, the solutions to the EFIE or MFIE may not be unique. If, for example, there is a solution to the homogeneous equation

$$\left[(\nabla\nabla \cdot + k^2) \oint_S G\mathbf{J}_S(\mathbf{r}')\,dS' \right]_{\text{tan}} = 0 \quad (\mathbf{r} \in S)$$

then the solution to (7.45) may not be unique. Although no nonzero solution to the source-free exterior scattering problem can exist, an integral equation of the same form as (7.45) also results from the *interior* scattering problem for the same surface S. Namely, if the sources of the incident field are *inside* a cavity V bounded by the perfect conductor S, the same type of integral equation results. Unlike the exterior problem, however, the source-free interior problem *does* possess nontrivial solutions at certain values of ω. These are just the resonant cavity modes of the volume interior to S. Although there is no direct *physical* relationship between the exterior and interior problems, they are mathematically related by sharing a common integral equation formulation. This is a problem not only when ω is exactly equal to one of these so-called "resonant" values for the interior problem, but (in approximate numerical solutions) even for values of ω merely in the neighborhood of such a resonance. A similar nonuniqueness appears in the MFIE formulation (7.51).

Several ways to eliminate the nonuniqueness in the solution of the traditional EFIE and MFIE have been suggested. We mention here only a particular one[6] that augments the tangential components of (7.44) or (7.50) with the normal components. In all but a few cases of high symmetry (certain resonant frequencies for the case when S is a sphere), these additional equations make the solution to the EFIE or MFIE unique.

Finally, some caution must be exercised when the scatterer reduces to an infinitesimally thin surface (a "shell"). For the EFIE, it is possible to combine the currents from both sides of the shell into a "total" ("top" side plus "bottom" side) current, which is the only unknown in the equation; if desired, splitting of the current into top and bottom parts individually can be done after the solution of the EFIE is obtained by substituting the total current into (7.43) and taking the limits of $\mathbf{H}_{\text{tan}}$ as the top side and bottom side of the shell are approached. The MFIE on the other hand does not permit this simplification, and in fact results in a pair of degenerate equations as the scatterer approaches a shell. Modifications to the MFIE must be made to enable its use in shell problems.[7]

7.3 VOLUME INTEGRAL EQUATIONS FOR SCATTERING BY A DIELECTRIC BODY

If instead of a perfect conductor, the scatterer is a dielectric body, with a (possibly inhomogeneous) permittivity $\epsilon(\mathbf{r})$ occupying a volume V, we can

---

[6] A. D. Yaghjian, *Radio Science*, vol. 16, pp. 987-1001, 1981.

[7] R. Mittra et al., *Radio Science*, vol. 8, pp. 869-875, 1973.

formulate the scattering problem as a volume integral equation in the following way. Let $\mathbf{E}^i$ and $\mathbf{H}^i$ denote the incident field, which is produced in a homogeneous "background medium" of permittivity ϵ_b and permeability μ_b by the impressed sources $\mathbf{J}$. This field obeys

$$\nabla \times \mathbf{E}^i = -j\omega\mu_b\mathbf{H}^i; \qquad \nabla \times \mathbf{H}^i = j\omega\epsilon_b\mathbf{E}^i + \mathbf{J} \tag{7.62}$$

The total field $\mathbf{E}, \mathbf{H}$ obeys

$$\nabla \times \mathbf{E} = -j\omega\mu_b\mathbf{H}; \qquad \nabla \times \mathbf{H} = j\omega\epsilon_b\mathbf{E} + \mathbf{J}_{\text{eq}} \tag{7.63}$$

where the equivalent current density

$$\mathbf{J}_{\text{eq}} = \mathbf{J} + j\omega\left[\epsilon(\mathbf{r}) - \epsilon_b\right]\mathbf{E}$$

is the sum of the impressed current and a (not yet known, since $\mathbf{E}$ is unknown) polarization current in the volume V. That is, the actual field can be regarded as produced *in the homogeneous background medium* by the equivalent current density $\mathbf{J}_{\text{eq}}$.

Pretending for the moment that $\mathbf{J}_{\text{eq}}$ is known, the formulas of Section 7.1.4 allow us to write the total electric field as:

$$\mathbf{E}(\mathbf{r}) = \mathbf{E}^i(\mathbf{r}) + \left(\nabla\nabla \cdot + k_b^2\right) \int_V \frac{\epsilon(\mathbf{r}') - \epsilon_b}{\epsilon_b}\mathbf{E}(\mathbf{r}')G\,dV' \tag{7.64}$$

where $k_b = \omega\sqrt{\mu_b\epsilon_b}$ is the wavenumber in the background medium, and

$$G = \frac{e^{-jk_b|\mathbf{r}-\mathbf{r}'|}}{4\pi|\mathbf{r}-\mathbf{r}'|}$$

is the scalar time-harmonic Green's function for the infinite background medium. Setting $\mathbf{r} \in V$ in (7.64), we obtain a consistency condition in the form of a volume integro-differential equation for $\mathbf{E}$. Using the result (7.31), we can write (7.64) as a pure integral equation, involving a principal value integral (see (A.83)):

$$\mathbf{E}(\mathbf{r}) \quad + \quad \frac{\epsilon(\mathbf{r}) - \epsilon_b}{\epsilon_b}\overset{\leftrightarrow}{\mathbf{L}}_{V_0} \cdot \mathbf{E}(\mathbf{r}) = \mathbf{E}^i(\mathbf{r}) + k_b^2 \int_V \frac{\epsilon(\mathbf{r}') - \epsilon_b}{\epsilon_b}\mathbf{E}(\mathbf{r}')G\,dV'$$

$$+ \quad \text{PV}_{V_0} \int_V \frac{\epsilon(\mathbf{r}') - \epsilon_b}{\epsilon_b}\mathbf{E}(\mathbf{r}') \cdot \nabla\nabla G\,dV' \tag{7.65}$$

where V_0 is a conveniently chosen principal volume.

7.4 INTEGRAL EQUATIONS FOR STATIC "SCATTERING" BY CONDUCTORS

In this section we will again consider the geometry of Section 7.2, but now with a *static* field "incident" upon the scatterer (i.e., the incident field is what would be produced by the impressed sources in the absence of the conductor

S).[8] In the static limit, the electric and magnetic problems decouple, and we consider them separately (see Section E.2).

7.4.1 ELECTROSTATIC SCATTERING

Let a static incident electric field $\mathbf{E}^i(\mathbf{r}) = -\nabla\Phi^i(\mathbf{r})$ be characterized by means of a potential $\Phi^i(\mathbf{r})$, which is the result of some static charge distribution $\rho_{\text{ext}}(\mathbf{r})$ located external to S. When a perfect conductor[9] whose surface is S is introduced into this incident field, a redistribution of surface charge takes place on S, resulting in the surface charge density $\rho_S(\mathbf{r})$. If G is the static Green's function (D.51):

$$G = \frac{1}{4\pi|\mathbf{r} - \mathbf{r}'|}$$

for unbounded space, and

$$\Phi(\mathbf{r}) = \Phi^i(\mathbf{r}) + \Phi^s(\mathbf{r}) \tag{7.66}$$

is the total potential resulting from the action of ρ_{ext} acting in the presence of the conductor S, then putting $f = \Phi^s$ (and $k = 0$) in (7.10) gives

$$\Phi^s(\mathbf{r}) = -\oint_S \left[G\frac{\partial\Phi^s(\mathbf{r}')}{\partial n'} - \Phi^s(\mathbf{r}')\frac{\partial G}{\partial n'} \right] dS' \tag{7.67}$$

for $\mathbf{r}$ outside of S, since $\nabla^2\Phi^s = 0$ there, and assuming that the surface integral analogous to the right side of (7.67) taken over the sphere S_∞ at infinity is zero (an assumption that we must eventually verify). Note once again the reversal of sign due to the fact that the surface normal $\mathbf{u}_n$ points *into* our volume of integration, rather than away from it as in the original statement (7.2) of Green's theorem. Now $\Phi^s = \Phi - \Phi^i$, and using (7.2) we have that

$$\oint_S \left[G\frac{\partial\Phi^i(\mathbf{r}')}{\partial n'} - \Phi^i(\mathbf{r}')\frac{\partial G}{\partial n'} \right] dS' = \int_{V_{\text{in}}} [G\nabla'^2\Phi^i - \Phi^i\nabla'^2 G]\,dV' = 0 \tag{7.68}$$

where V_{in} is the region *interior* to S, since both $\nabla^2\Phi^i$ and $\nabla^2 G$ vanish in that region (recall that $\mathbf{r}$ is located outside of S). Thus, we have

$$\Phi^s(\mathbf{r}) = -\oint_S \left[G\frac{\partial\Phi(\mathbf{r}')}{\partial n'} - \Phi(\mathbf{r}')\frac{\partial G}{\partial n'} \right] dS' \tag{7.69}$$

Now, on S, the potential Φ will be equal to some constant Φ_S. Moreover, $\partial\Phi(\mathbf{r}')/\partial n' = -\mathbf{u}'_n \cdot \mathbf{E}(\mathbf{r}') = -\rho_S(\mathbf{r}')/\epsilon$ on S. Thus (7.69) gives

$$\Phi^s(\mathbf{r}) = \frac{1}{\epsilon}\oint_S G\rho_S(\mathbf{r}')\,dS' + \Phi_S\oint_S \frac{\partial G}{\partial n'}\,dS' = \frac{1}{\epsilon}\oint_S G\rho_S(\mathbf{r}')\,dS' \tag{7.70}$$

---

[8]Van Bladel, Sections 4.7 and 6.11.

[9]The conductor is, as in Section E.2, assumed to be connected—all of one piece—to avoid additional complexity in the formulation.

where we have used the divergence theorem to convert the surface integral of $\partial G/\partial n'$ to an integral of $\nabla'^2 G = 0$ over V_{in}.

The far field of the scattered potential can be investigated by the usual means:

$$
\begin{aligned}
\Phi^s(\mathbf{r}) &= \frac{1}{4\pi\epsilon} \oint_S \frac{\rho_S(\mathbf{r}')}{|\mathbf{r}-\mathbf{r}'|}\,dS' \\
&= \frac{1}{4\pi\epsilon} \oint_S \rho_S(\mathbf{r}') \left[\frac{1}{r} + \frac{\mathbf{r}'\cdot\mathbf{r}}{r^3} + \cdots \right] dS' \\
&= \frac{\oint_S \rho_S(\mathbf{r}')\,dS'}{4\pi\epsilon r} + \frac{\mathbf{r}\cdot\oint_S \mathbf{r}'\rho_S(\mathbf{r}')\,dS'}{4\pi\epsilon r^3} + \cdots
\end{aligned}
\tag{7.71}
$$

Thus $\Phi^s = O(r^{-1})$ and $\partial\Phi^s/\partial r = O(r^{-2})$ as $r \to \infty$, and our earlier assumption about the vanishing of the surface integral over S_∞ is justified.

Letting $\mathbf{r} \to S$ from the outside, and using the boundary condition on the total potential Φ at S, we arrive at an integral equation for the charge distribution ρ_S on S:

$$
\Phi_S - \Phi^i(\mathbf{r}) = \frac{1}{4\pi\epsilon} \oint_S \frac{\rho_S(\mathbf{r}')}{|\mathbf{r}-\mathbf{r}'|}\,dS'; \quad (\mathbf{r} \in S)
\tag{7.72}
$$

By superposition, the solution to (7.72) is expressible as the sum of the solutions to two simpler problems. First, suppose that $\Phi^i \equiv 0$, but that the total charge $\oint_S \rho_S(\mathbf{r}')\,dS'$ is not zero, and therefore there must be a nonzero potential Φ_S at the conductor S (this follows because the field is nonzero, at least in the far field, by (7.71), and (E.14) demands that Φ_S not vanish). This situation is the *capacitance problem*, whose solution (denoted by ρ_{Sc}, corresponding to a conductor potential Φ_{Sc}) is unique by the results of Section E.2:

$$
\Phi_{Sc} = \frac{1}{4\pi\epsilon} \oint_S \frac{\rho_{Sc}(\mathbf{r}')}{|\mathbf{r}-\mathbf{r}'|}\,dS'; \quad (\mathbf{r} \in S)
\tag{7.73}
$$

On the other hand, if Φ^i is not identically zero, but there is no net charge on the conductor, we have a generalized *polarizability problem*:

$$
\Phi_{Sp} - \Phi^i(\mathbf{r}) = \frac{1}{4\pi\epsilon} \oint_S \frac{\rho_{Sp}(\mathbf{r}')}{|\mathbf{r}-\mathbf{r}'|}\,dS'; \quad (\mathbf{r} \in S)
\tag{7.74}
$$

The conductor potential Φ_{Sp} is determined by the requirement that the total charge $\oint_S \rho_{Sp}(\mathbf{r}')\,dS'$ be zero, and with this condition the solution is again unique by the results of Section E.2.

In the generalized polarizability problem (7.74), it is often possible to choose the arbitrary constant additive term in Φ^i by inspection (using symmetry, for example) so that $\Phi_{Sp} = 0$, but if not, we can use (E.22) to compute it, if the capacitance problem has been solved. Thus, we solve the auxiliary integral equation

$$
-\Phi^i(\mathbf{r}) = \frac{1}{4\pi\epsilon} \oint_S \frac{\tilde{\rho}_{Sp}(\mathbf{r}')}{|\mathbf{r}-\mathbf{r}'|}\,dS'; \quad (\mathbf{r} \in S)
\tag{7.75}
$$

and finally obtain the solution to the generalized polarizability problem as:

$$\rho_{Sp} = \tilde{\rho}_{Sp} + \frac{\Phi_{Sp}}{\Phi_{Sc}}\rho_{Sc} \tag{7.76}$$

We would proceed by solving (7.75) for $\tilde{\rho}_{Sp}$, and then determining Φ_{Sp} by integrating (7.76) over S, and using (E.16) along with the fact that the total charge due to ρ_{Sp} on S is zero to get:

$$\Phi_{Sp} = -\Phi_{Sc}\frac{\oint_S \tilde{\rho}_{Sp}(\mathbf{r}')\,dS'}{\oint_S \tilde{\rho}_{Sc}(\mathbf{r}')\,dS'} = -\frac{\oint_S \tilde{\rho}_{Sp}(\mathbf{r}')\,dS'}{C}$$

which is then inserted into (7.76) to completely specify ρ_{Sp}.

The leading-order behavior of the far field also serves to characterize the polarizability problem by means of a single parameter. From (7.71), since the total charge on the conductor is zero, the scattered potential is

$$\Phi^s(\mathbf{r}) \sim \frac{\mathbf{r} \cdot \mathbf{p}}{4\pi\epsilon r^3} \tag{7.77}$$

where

$$\mathbf{p} = \oint_S \mathbf{r}'\rho_S(\mathbf{r}')\,dS'$$

is the induced electric dipole moment on the conductor. By linearity, it must evidently be proportional to the strength of the incident electric *field* (and not to the incident potential, which may have an arbitrary constant added to it without affecting the fields in the problem). If $\mathbf{E}^i$ is constant in space, we define the electric polarizability dyadic $\overset{\leftrightarrow}{\boldsymbol{\alpha}}_E$ by means of (E.24).

In two-dimensional problems where Φ^i is independent of z and the conductor S is a cylinder whose axis is parallel to the z-axis, with cross-sectional boundary C, the above treatment must be slightly modified. The difficulty lies in the assumption that $\Phi^s \to 0$ as $r \to \infty$. This is impossible in a two-dimensional problem. Indeed, a circular conducting cylinder of radius a whose total linear charge density (charge per unit length in the z-direction) is Q_l will have an electric field of

$$\mathbf{E} = \mathbf{u}_\rho\frac{Q_l}{2\pi\epsilon\rho} \tag{7.78}$$

in the capacitance problem, and hence a potential of

$$\Phi = \frac{Q_l}{2\pi\epsilon}\ln\frac{\rho_0}{\rho} = \frac{\Phi_{Sc}}{\ln(\rho_0/a)}\ln\frac{\rho_0}{\rho} \tag{7.79}$$

where $\Phi_{Sc} = \Phi|_{\rho=a}$, and ρ_0 is some arbitrary constant. No matter what value we take for ρ_0, Φ will not go to zero as $\rho \to \infty$ (it will always become infinite, in fact). It is thus impossible to define a capacitance per unit length in such a case because the voltage of the conductor with respect to infinity (for a finite linear charge density) is infinite.

Let us nevertheless take

$$G(\boldsymbol{\rho}, \boldsymbol{\rho}') = \frac{1}{2\pi} \ln \frac{\rho_0}{|\boldsymbol{\rho} - \boldsymbol{\rho}'|}$$

as our two-dimensional static Green's function (cf. Problem 7–1), with ρ_0 some arbitrary constant. Applying the two-dimensional analog of Green's theorem to the region exterior to C, we obtain

$$\Phi^s(\boldsymbol{\rho}) = \frac{1}{2\pi\epsilon} \oint_C \rho_S(\boldsymbol{\rho}') \ln \frac{\rho_0}{|\boldsymbol{\rho} - \boldsymbol{\rho}'|} \, dl' \qquad (7.80)$$

if we assume that the integral over the infinite circle C_∞ vanishes, which must yet be verified. We may proceed as with (7.71) to compute the far field of this potential:

$$
\begin{aligned}
\Phi^s(\boldsymbol{\rho}) &= \frac{1}{2\pi\epsilon} \oint_C \rho_S(\boldsymbol{\rho}') \left[\ln \frac{\rho_0}{\rho} + \frac{\boldsymbol{\rho} \cdot \boldsymbol{\rho}'}{\rho^2} + \cdots \right] dl' \\
&\simeq \frac{1}{2\pi\epsilon} \left[\oint_C \rho_S(\boldsymbol{\rho}') \, dl' \ln \frac{\rho_0}{\rho} + \frac{\oint_C \rho_S(\boldsymbol{\rho}') \boldsymbol{\rho}' \, dl' \cdot \boldsymbol{\rho}}{\rho^2} \right]
\end{aligned}
\qquad (7.81)
$$

A somewhat more careful computation than in the three-dimensional case verifies that the integral over C_∞ discarded above does indeed vanish. Note that if ρ is large enough that only the first term on the right side of (7.81) is needed:

$$\Phi^s(\boldsymbol{\rho}) \sim \frac{1}{2\pi\epsilon} \oint_C \rho_S(\boldsymbol{\rho}') \, dl' \ln \frac{\rho_0}{\rho}$$

then this potential is approximately zero at $\rho = \rho_0$. Thus, if for all points on the boundary C we have $\rho' \ll \rho_0$, our representation of the potential is also an approximate representation in the case where C is located not in infinite space, but near the center of a large circular cylindrical conducting shield of radius ρ_0 which is held at zero potential.

Again, we distinguish two distinct cases: the capacitance problem where the incident potential is zero, and the polarizability problem where the net charge per unit length on the cylinder vanishes. In the capacitance problem, the surface charge density ρ_{Sc} satisfies

$$\Phi_{Sc} = \frac{1}{2\pi\epsilon} \oint_C \rho_{Sc}(\boldsymbol{\rho}') \ln \frac{\rho_0}{|\boldsymbol{\rho} - \boldsymbol{\rho}'|} \, dl'; \quad (\boldsymbol{\rho} \in C) \qquad (7.82)$$

with a total charge per unit length of

$$Q_l = \oint_C \rho_{Sc}(\boldsymbol{\rho}') \, dl' \qquad (7.83)$$

In the case where C is located in infinite space, we cannot define a capacitance per unit length for the reasons discussed above, but we *can* introduce the idea

of an equivalent radius by demanding that the leading term in the far field of
the charged cylinder C be the same as that of a circular cylinder of equivalent
radius a_{eq} held at the same potential Φ_{Sc}, as given by (7.79). We have

$$\ln \frac{\rho_0}{a_{eq}} = \frac{2\pi\epsilon\Phi_{Sc}}{Q_l} \tag{7.84}$$

If, on the other hand, we regard the problem as an approximate formulation
of the case where C is surrounded by a large circular conducting shield of
radius ρ_0, then the capacitance per unit length of this system is well defined,
and is given by

$$C_l = \frac{Q_l}{\Phi_{Sc}} \tag{7.85}$$

For the polarizability problem no special treatment is necessary for the
two-dimensional case. The integral equation is now

$$\Phi_{Sp} - \Phi^i(\rho) = \frac{1}{2\pi\epsilon} \oint_C \rho_{Sp}(\rho') \ln \frac{\rho_0}{|\rho - \rho'|} \, dl'; \quad (\rho \in C) \tag{7.86}$$

where the constant Φ_{Sp} is chosen as in the three-dimensional case to make the
total charge per unit length on C vanish. The dipole moment per unit length
is

$$\mathbf{p}_l = \oint_C \rho_{Sp}(\rho')\rho' \, dl' \tag{7.87}$$

which is used to define a polarizability per unit length

$$\mathbf{p}_l = \epsilon \overset{\leftrightarrow}{\alpha}_{El} \cdot \mathbf{E}^i \tag{7.88}$$

in the case of an arbitrary uniform transverse incident field $\mathbf{E}^i$.

7.4.2 MAGNETOSTATIC SCATTERING

The corresponding magnetostatic problems can also be formulated. It is possible to use the zero-frequency limit of the MFIE from Section 7.2, but this
has the disadvantage that it leads to a two-component vector equation in the
(two-component) unknown surface current density, as compared to the single scalar equation we obtained for the electrostatic problem. Additionally,
the MFIE is not directly usable for thin shell conductors, as was pointed out
above. Thus, we will employ a formulation based on the dual scalar potential
Ψ rather than on the vector potential $\mathbf{A}$. We will make the assumption that
the conductor may have a hole (a surface spanning this hole is called, as in
Section E.2, S_f). Then, as with the electrostatic case, the magnetostatic scattering problem can be decomposed into two distinct cases—the inductance
problem and the general magnetic polarizability problem, and the general
solution is a superposition of the two.

For the inductance problem, we follow the formulation (E.33) ff., where
the potential Ψ_0 is to be determined. If the stream potential $\mathbf{T}_e$ chosen for

(E.33) is further subjected to the condition $\nabla \cdot (\mu \mathbf{T}_e) = 0$ (as would be true for an actual magnetic field), then Ψ_0 will obey Laplace's equation. We now apply Green's theorem to Ψ_0 and the static free-space Green's function on the region V_e exterior to the surface S of the conductor; we get (7.67) with Φ^s replaced by Ψ_0. From (E.36), we have

$$\Psi_0(\mathbf{r}) = - \oint_S G T_{en}(\mathbf{r}') \, dS' + \oint_S \Psi_0(\mathbf{r}') \frac{\partial G}{\partial n'} \, dS' \tag{7.89}$$

We take the gradient of (7.89), let $\mathbf{r} \to S$, form the dot product with the unit vector $\mathbf{u}_n$ at S and use the boundary condition (E.36) one more time to get:

$$\mathbf{u}_n \cdot \nabla \nabla \cdot \oint_S \Psi_0(\mathbf{r}') G \mathbf{u}'_n \, dS' = -T_{en}(\mathbf{r}) - \mathbf{u}_n \cdot \nabla \oint_S G T_{en}(\mathbf{r}') \, dS' \tag{7.90}$$

for $\mathbf{r} \in S$. This integro-differential equation for the auxiliary potential Ψ_0 is a scalar equation in a scalar unknown. Once (7.90) is solved for Ψ_0, we can use (E.37) and some manipulations with vector identities to get:

$$L = \frac{1}{I^2} \left[\int_{V_e} \mu \mathbf{T}_e \cdot \mathbf{T}_e \, dV + \oint_S \mu \Psi_0 T_{en} \, dS \right] \tag{7.91}$$

For the general magnetic polarizability problem, we use the formulation of (E.38) ff. From Green's theorem applied to Ψ_p^s we now have:

$$\begin{aligned} \Psi_p^s(\mathbf{r}) &= - \oint_S G H_n^i(\mathbf{r}') \, dS' + \oint_S \Psi_p^s(\mathbf{r}') \frac{\partial G}{\partial n'} \, dS' = \oint_S G \nabla' \Psi^i(\mathbf{r}') \cdot \mathbf{u}'_n \, dS' \\ &\quad + \oint_S \left[\Psi_p(\mathbf{r}') - \Psi^i(\mathbf{r}') \right] \frac{\partial G}{\partial n'} \, dS' \end{aligned} \tag{7.92}$$

But since there are no sources for the incident field *inside* S,

$$\oint_S \left[G \nabla' \Psi^i(\mathbf{r}') - \Psi^i(\mathbf{r}') \nabla' G \right] \cdot \mathbf{u}'_n \, dS' = 0 \tag{7.93}$$

so we can rewrite (7.92) in terms of the *total* potential $\Psi_p = \Psi^i + \Psi_p^s$:

$$\Psi_p^s(\mathbf{r}) = \oint_S \Psi(\mathbf{r}') \frac{\partial G}{\partial n'} \, dS' \tag{7.94}$$

Proceeding as with the inductance problem, we obtain the following integro-differential equation for $\Psi_p(\mathbf{r})$:

$$\mathbf{u}_n \cdot \nabla \nabla \cdot \oint_S \Psi_p(\mathbf{r}') G \mathbf{u}'_n \, dS' = -H_n^i(\mathbf{r}) \tag{7.95}$$

for points $\mathbf{r} \in S$.

Once Ψ_p has been found, the far field of the scattered potential can be found from (7.94):

$$
\begin{aligned}
\Psi_p^s(\mathbf{r}) &= -\nabla \cdot \oint_S \frac{\Psi_p(\mathbf{r}')\mathbf{u}_n' \, dS'}{4\pi|\mathbf{r} - \mathbf{r}'|} \\
&\simeq -\frac{1}{4\pi}\nabla\left(\frac{1}{r}\right) \cdot \oint_S \Psi_p(\mathbf{r}')\mathbf{u}_n' \, dS'
\end{aligned}
\tag{7.96}
$$

Using (E.45) shows that the surface integral at the end of (7.96) is expressible in terms of the induced magnetic dipole moment of the current on the conductor. Thus the far scattered field (7.96) reduces to:

$$
\Psi_p^s(\mathbf{r}) \simeq -\frac{\mathbf{m} \cdot \mathbf{r}}{4\pi r^3}
\tag{7.97}
$$

If the incident field is uniform, we further have an expression for $\mathbf{m}$ in terms of $\mathbf{H}^i$ from (E.47), so that much of the information in the solution to this problem is distilled into the single parameter $\overleftrightarrow{\boldsymbol{\alpha}}_M$.

In the two-dimensional case, with only a transverse magnetic field $\mathbf{H} = \mathbf{H}_t$, the magnetic dipole moment per unit length induced on a z-directed cylinder of cross-sectional boundary C can be shown to be (see especially Problem 1-2 in this connection):

$$
\mathbf{m}_l = \int_{S_e} \mathbf{H}_t \, dS
\tag{7.98}
$$

where S_e is the surface exterior to C. The magnetic polarizability per unit length is then defined from

$$
\mathbf{m}_l = \overleftrightarrow{\boldsymbol{\alpha}}_{Ml} \cdot \mathbf{H}_t^i
\tag{7.99}
$$

by analogy with the three-dimensional case.

7.5   ELECTROSTATICS OF A THIN CONDUCTING STRIP

We consider now an electrostatic problem for which an exact, closed-form solution can be obtained.[10] Let the conductor be in the form of a thin (in the limit that we consider, infinitesimal) strip of width $2w$ occupying $y = 0$, $|x| < w$ as shown in Figure 7.6. Both the capacitance and polarizability problems reduce to the solution of an integral equation of the form

$$
\int_{-w}^{w} f(x') \ln \frac{\rho_0}{|x - x'|} \, dx' = g(x); \quad (|x| < w)
\tag{7.100}
$$

---

[10]See B. Noble, in *Electromagnetic Waves* (R. E. Langer, ed.), U. of Wisconsin Press, Madison, 1962, pp. 323-360; also C. M. Butler and D. R. Wilton, *IEEE Trans. Ant. Prop.*, vol. 28, pp. 42-48, 1980.

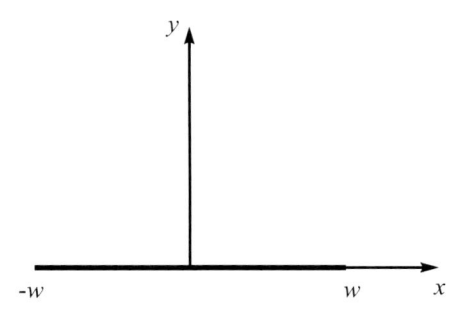

Figure 7.6 Geometry of a thin conducting strip.

where $g(x)$ is a given function on the strip (here either $2\pi\epsilon\Phi_{Sc}$ or $2\pi\epsilon[\Phi_{Sp} - \Phi^i(x,0)]$, while $f(x')$ is the unknown *total* charge density on the strip (we sum the top side and bottom side charge densities to get f).

The *kernel* of this integral equation, $\ln\frac{\rho_0}{|x-x'|}$, can be expressed as a bilinear series of orthogonal functions, each term of which is the product of a function of x and the same function of x', which will allow us to find a solution to the integral equation in closed form. We proceed as follows: Starting with the known Taylor series for $\ln(1+x)$, we have

$$\ln(1 - e^{j\theta}) = -\sum_{m=1}^{\infty} \frac{e^{jm\theta}}{m} \tag{7.101}$$

and take the real part of both sides:

$$\frac{1}{2}\ln 2|1 - \cos\theta| = -\sum_{m=1}^{\infty} \frac{\cos m\theta}{m} \tag{7.102}$$

Let $\theta = u + u'$ and $u - u'$ by turns, and add the resulting expansions together:

$$\ln 2|\cos u - \cos u'| = -2\sum_{m=1}^{\infty} \frac{\cos mu \cos mu'}{m} \tag{7.103}$$

Finally, the changes of variable $\cos u = x/w$ and $\cos u' = x'/w$ give

$$\begin{aligned}
\ln\frac{\rho_0}{|x - x'|} &= \ln\frac{2\rho_0}{w} + 2\sum_{m=1}^{\infty} \frac{\cos m(\cos^{-1}\frac{x}{w}) \cos m(\cos^{-1}\frac{x'}{w})}{m} \\
&= \ln\frac{2\rho_0}{w} + 2\sum_{m=1}^{\infty} \frac{T_m(\frac{x}{w})T_m(\frac{x'}{w})}{m} \tag{7.104}
\end{aligned}$$

where $T_m(z) \equiv \cos(m\cos^{-1}z)$ is the mth order *Chebyshev polynomial* of the first kind.

Using this bilinear series, we can solve the integral equation as follows. Putting $x = w \cos u$ in (7.100) gives

$$w \int_0^\pi [F(u') \sin u'] \ln \frac{\rho_0/w}{|\cos u - \cos u'|} \, du' = G(u); \quad (0 < u < \pi) \quad (7.105)$$

where $F(u') = f(w \cos u')$ and $G(u) = g(w \cos u)$. But the kernel can be replaced by the bilinear expansion (7.104), transforming (7.105) into

$$G(u) = w \ln \frac{2\rho_0}{w} \int_0^\pi [F(u') \sin u'] \, du' \quad (7.106)$$

$$+ 2w \sum_{m=1}^\infty \frac{\cos mu}{m} \int_0^\pi [F(u') \sin u'] \cos mu' \, du' \quad (7.107)$$

which is none other than the Fourier cosine series for $G(u)$. But from the theory of such series,

$$G(u) = \sum_{m=0}^\infty G_m \cos mu \quad (7.108)$$

where the coefficients are given by:

$$G_0 = \frac{1}{\pi} \int_0^\pi G(u) \, du \quad (7.109)$$

$$G_m = \frac{2}{\pi} \int_0^\pi G(u) \cos mu \, du \quad (m \geq 1)$$

Term-by-term comparison of (7.106) and (7.108) gives:

$$w \ln \frac{\rho_0}{w} \int_0^\pi [F(u') \sin u'] \, du' = \frac{1}{\pi} \int_0^\pi G(u) \, du$$

and

$$\frac{2w}{m} \int_0^\pi [F(u') \sin u'] \cos mu' \, du' = \frac{2}{\pi} \int_0^\pi G(u) \cos mu \, du \quad (m \geq 1)$$

which is to say that $[F(u) \sin u]$ itself can be expressed as a Fourier cosine series:

$$F(u) \sin u = \frac{1}{\pi^2 w \ln(2\rho_0/w)} \int_0^\pi G(u') \, du'$$

$$+ \sum_{m=1}^\infty \frac{2m}{\pi^2 w} \cos mu \int_0^\pi G(u') \cos mu' \, du' \quad (7.110)$$

Changing back to the original variables x and x' gives

$$f(x) = \frac{1}{\pi^2 \sqrt{w^2 - x^2}} \left\{ \frac{1}{\ln(2\rho_0/w)} \int_{-w}^w \frac{g(x') \, dx'}{\sqrt{w^2 - x'^2}} \right.$$

$$\left. + 2 \sum_{m=1}^\infty m T_m \left(\frac{x}{w} \right) \int_{-w}^w \frac{g(x') T_m(\frac{x'}{w}) \, dx'}{\sqrt{w^2 - x'^2}} \right\} \quad (7.111)$$

which is the exact, general solution to (7.100).

We should take note here of an exceptional case, which although it does not arise in the context of the physical problem we are considering, is of some mathematical importance. If $\rho_0 = w/2$, we see that the solution given in (7.111) is invalid (it becomes formally infinite). Such a small value of ρ_0 could not correspond to a grounded circular cylinder surrounding the strip because it would intersect it. However, it is an indication of the fact that under this condition, our integral equation does not have a solution at all: A solution to (7.100) with $g(x) \neq 0$ does not exist. If we consider instead the homogeneous version of (7.100) for this case,

$$\int_{-w}^{w} f(x') \ln \frac{w/2}{|x - x'|} \, dx' = 0; \quad (|x| < w), \tag{7.112}$$

and follow the solution procedure above, we arrive at

$$f(x) = \frac{A}{\sqrt{w^2 - x^2}} \tag{7.113}$$

where A is *any* constant. This is an example of the Fredholm alternative, to be discussed further in Section 8.2. For the present, we will assume in what follows that this exceptional circumstance does not occur.

For the case of the capacitance problem,

$$g(x) = G(u) = G_0 \equiv 2\pi\epsilon\Phi_{Sc} \tag{7.114}$$

so that $G_m \equiv 0$ for $m \geq 1$. Equation (7.110) or (7.111) then gives simply

$$f(x) = \frac{G_0}{\pi \ln(2\rho_0/w)\sqrt{w^2 - x^2}} \tag{7.115}$$

The corresponding total charge Q_l per unit length of the strip is

$$Q_l = \int_{-w}^{w} f(x) \, dx = \frac{G_0}{\ln(2\rho_0/w)} \tag{7.116}$$

Comparing (7.114) with (7.116), we have

$$2\pi\epsilon\Phi_{Sc} = Q_l \ln \frac{2\rho_0}{w}$$

By (7.85), the capacitance per unit length of this strip when surrounded by a large circular cylindrical shield of radius ρ_0 centered at the origin is

$$C_l = \frac{2\pi\epsilon}{\ln(2\rho_0/w)} \tag{7.117}$$

Alternatively, we can use (7.84) to infer an equivalent radius of $a_{eq} = w/2 = (2w)/4$ for this strip. The expression

$$C_l = \frac{2\pi\epsilon}{\ln(b/a)}$$

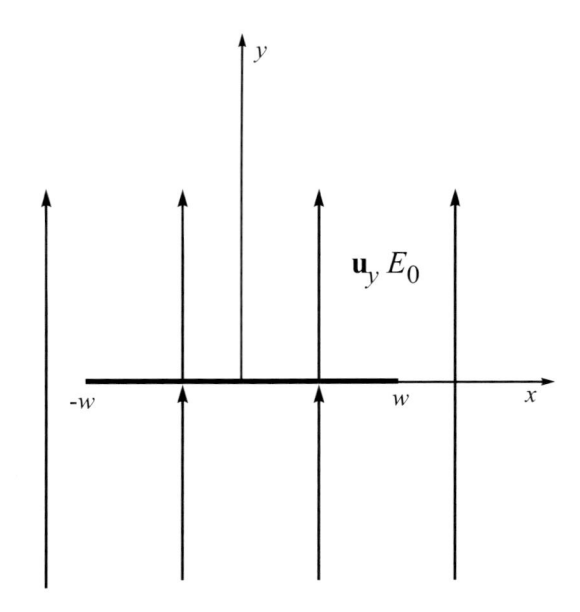

Figure 7.7   Uncharged strip in a perpendicular incident static field.

for the capacitance per unit length of two circular coaxial cylinders with inner radius a and outer radius b reduces to (7.117) when we put $a = a_{\text{eq}}$ and $b = \rho_0$, illustrating the usefulness of the equivalent radius concept in computing capacitance.

For the polarizability problem, let the incident field be $\mathbf{E}^i = \mathbf{u}_x E_0$ or $\mathbf{u}_y E_0$, where E_0 is a constant. Considerations of symmetry show that we may take $\Phi^i - \Phi_{Sp}$ equal to $-xE_0$ or $-yE_0$ respectively in order to assure that the solution of the integral equation will have zero total charge. The second of these choices—that of the y-polarized incident field—produces a net total charge density on the strip of exactly zero; the bottom charge density is equal and opposite to that on the top side of the strip (Figure 7.7). On the other hand, for $\mathbf{E}^i = \mathbf{u}_x E_0$,

$$g(x) = 2\pi \epsilon E_0 x \tag{7.118}$$

or

$$G(u) = G_1 \cos u \tag{7.119}$$

where $G_1 = 2\pi \epsilon E_0 w$. Then (7.110) or (7.111) gives

$$f(x) = \frac{2\epsilon E_0 x}{\sqrt{w^2 - x^2}} \tag{7.120}$$

since $T_1(z) = z$. This function is therefore the total (top plus bottom) charge density corresponding to ρ_{Sp} in (7.86). Since it is an odd function of x, the

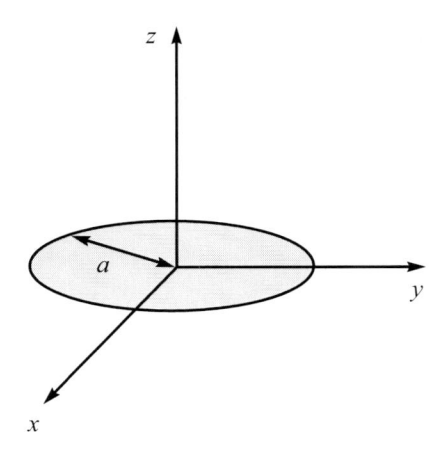

Figure 7.8   Circular conducting disk.

total charge on the strip is zero as required in the polarizability problem. The dipole moment per unit length of the strip is

$$\mathbf{p}_l = \mathbf{u}_x \int_{-w}^{w} x' f(x')\, dx' = \mathbf{u}_x \pi \epsilon w^2 E_0 \tag{7.121}$$

and so by (7.88) the polarizability tensor is

$$\overset{\leftrightarrow}{\boldsymbol{\alpha}}_{El} = \mathbf{u}_x \mathbf{u}_x \pi w^2 \tag{7.122}$$

Since the polarizability tensor for a circular cylinder of radius a is $\overset{\leftrightarrow}{\boldsymbol{\alpha}}_{El} = \mathbf{u}_x \mathbf{u}_x \pi a^2 + \mathbf{u}_y \mathbf{u}_y \pi a^2$, it will be realized that the concept of equivalent radius as we have defined it applies only to capacitance problems, and not to problems of polarizability.

7.6   ELECTROSTATICS OF A THIN CONDUCTING CIRCULAR DISK

Another important problem that can be solved exactly is that for the electrostatic field in the presence of a perfectly conducting circular disk of radius a lying in the xy-plane as shown in Figure 7.8. Putting $f(\boldsymbol{\rho}) = \rho_S^{tot}(\boldsymbol{\rho})$, the total (top plus bottom) charge density, and either

$$g(\boldsymbol{\rho}) = 4\pi\epsilon\Phi_{Sc}$$

for the capacitor problem, or

$$g(\boldsymbol{\rho}) = 4\pi\epsilon[\Phi_{Sp} - \Phi^i(\rho, \phi, 0)]$$

for the polarizability problem, we can write the integral equation of the electrostatic problem for the disk as

$$g(\rho, \phi) = \int_0^{2\pi} \int_0^a \frac{f(\rho', \phi')\rho' \, d\rho' d\phi'}{|\boldsymbol{\rho} - \boldsymbol{\rho}'|}; \quad (0 \le \rho < a; 0 \le \phi \le 2\pi) \qquad (7.123)$$

The kernel of this two-dimensional integral equation can be represented as an exponential Fourier series by setting $k = 0$ in (4.76), renaming the variable $\mathbf{r}_0$ as $\mathbf{r}'$ and finally setting $z = z' = 0$:

$$\frac{1}{|\boldsymbol{\rho} - \boldsymbol{\rho}'|} = \sum_{m=-\infty}^{\infty} e^{jm(\phi-\phi')} K_m(\rho, \rho') \qquad (7.124)$$

where (by (B.8))

$$K_m(\rho, \rho') = K_{-m}(\rho, \rho') = \int_0^{\infty} J_m(\lambda\rho) J_m(\lambda\rho') \, d\lambda \qquad (7.125)$$

and thus, if we introduce Fourier series representations of f and g:

$$f = \sum_{-\infty}^{\infty} e^{jm\phi} f_m(\rho)$$

$$g = \sum_{-\infty}^{\infty} e^{jm\phi} g_m(\rho) \qquad (7.126)$$

we obtain a set of decoupled one-dimensional integral equations for $f_m(\rho)$:

$$\frac{g_m(\rho)}{2\pi} = \int_0^a K_m(\rho, \rho') f_m(\rho')\rho' \, d\rho'; \quad (0 \le \rho < a) \qquad (7.127)$$

These equations have been solved by a wide variety of methods in the literature.[11] We will use a bilinear expansion of the kernel K_m into orthogonal functions, analogous to our treatment of the static problem for the strip.

---

[11] See:

D. S. Jones, *Theory of Electromagnetism*, Oxford: Pergamon Press, 1964, Section 9.19.

I. N. Sneddon, *Mixed Boundary Value Problems in Potential Theory*. Amsterdam: North-Holland, 1966, Chapter 3.

V. M. Aleksandrov, *PMM: J. Appl. Math. Mech.*, vol. 31, pp. 1122-1136, 1967.

R. P. Kanwal, *Linear Integral Equations*, San Diego: Academic Press, 1971, Chapter 10.

R. Friedberg, *Amer. J. Phys.*, vol. 61, pp. 1084-1096, 1993.

and references cited therein.

Similar to the strip problem, we introduce the changes of variable $\rho = a \sin u$ and $\rho' = a \sin u'$. Because K_m is symmetric in both u and u' about the point $\pi/2$, it has an expansion of the form (B.61):[12]

$$K_m(a \sin u, a \sin u') = \tag{7.128}$$

$$\frac{\pi}{a} \sum_{k=0}^{\infty} (2k + m + \frac{1}{2}) \frac{[(2m + 2k - 1)!!]^2}{2^{2k}(k!)^2} P_{2k+m}^{-m}(\cos u) P_{2k+m}^{-m}(\cos u')$$

$$= \frac{\pi}{a} (-1)^m \sum_{k=0}^{\infty} (2k + m + \frac{1}{2}) \frac{(2k + 2m)!}{(2k)!} \frac{[(2m + 2k - 1)!!]^2}{2^{2k}(k!)^2}$$

$$\cdot P_{2k+m}^{m}(\cos u) P_{2k+m}^{-m}(\cos u')$$

where the double factorial notation signifies

$$(2n - 1)!! = 1 \cdot 3 \cdot 5 \cdots (2n - 1) = \frac{(2n)!}{2^n n!} \tag{7.129}$$

and P_n^m is the associated Legendre function. Now, as noted in Appendix B, the Legendre functions appearing in this bilinear expansion are all symmetric functions about $u = \pi/2$. Thus, the expansion theorem (B.61) applies, and can be expressed for this case as:

$$f_m(u) = \sum_{k=0}^{\infty} F_{2k+m}^{m} P_{2k+m}^{m}(\cos u) \tag{7.130}$$

$$F_{2k+m}^{m} = (-1)^m (4k + 2m + 1) \int_0^{\pi/2} P_{2k+m}^{-m}(\cos u) f_m(u) \sin u \, du$$

and

$$\int_0^{\pi/2} P_{2k+m}^{m}(\cos u) P_{2k'+m}^{-m}(\cos u) \sin u \, du = \frac{(-1)^m}{4k + 2m + 1} \delta_{kk'} \tag{7.131}$$

The integral equation in the new variables becomes

$$\frac{G_m(u)}{2\pi} = a \int_0^{a} K_m(a \sin u, a \sin u') S_m(u') \cos u' \, du'; \quad (0 \le u < \frac{\pi}{2}) \tag{7.132}$$

where

$$S_m(u) = f_m(a \sin u) \cos u \tag{7.133}$$

---

[12]See

C. J. Bouwkamp, *Indag. Math.*, vol. 12, pp. 208-215, 1950.
G. Ya. Popov, *J. Appl. Math. Mech.*, vol. 26, pp. 207-225, 1962.
G. Ya. Popov, *J. Appl. Math. Mech.*, vol. 27, pp. 1255-1271, 1963.
G. Ya. Popov, *Izv. VUZ Matem.*, no. 4 (53), pp. 77-85, 1966.
P. Wolfe, *J. Math. Phys.*, vol. 12, pp. 1215-1218, 1971.

for details.

and

$$G_m(u) = g_m(a \sin u) \tag{7.134}$$

When the bilinear series (7.128) is introduced into (7.132), we obtain a Legendre expansion of $G_m(u)$ of the form (7.130), whose coefficients can be expressed in terms of the expansion coefficients for $S_m(u)$. If we put

$$G_m(u) = \sum_{k=0}^{\infty} G_{2k+m}^m P_{2k+m}^m(\cos u) \tag{7.135}$$

and

$$S_m(u) = \sum_{k=0}^{\infty} S_{2k+m}^m P_{2k+m}^m(\cos u) \tag{7.136}$$

then (7.132) and (7.128) give

$$S_{2k+m}^m = \frac{1}{\pi^2} G_{2k+m}^m \frac{(2k)!}{(2k+2m)!} \frac{2^{2k}(k!)^2}{[(2k+2m-1)!!]^2} \tag{7.137}$$

Typical source functions $g_m(\rho)$ are $g_0(\rho) \equiv G_0^0 = \text{const}$ and $g_1(\rho) \equiv G_1^1 \rho/a$, which lead to the solutions

$$f_0(\rho) = \frac{a}{\pi^2} \frac{G_0^0}{\sqrt{a^2 - \rho^2}} \tag{7.138}$$

and

$$f_1(\rho) = \frac{\rho}{2\pi^2} \frac{G_1^1}{\sqrt{a^2 - \rho^2}} \tag{7.139}$$

respectively. These results will be used in the solution of the capacitance and polarizability problems for the circular disk below.

For the capacitance problem, we have $g_0(\rho) = 4\pi\epsilon\Phi_{Sc}$ and all other g_m's are zero. Then $f = f_0$ from (7.138):

$$f(\rho, \phi) = \frac{4\epsilon\Phi_{Sc}}{\pi} \frac{1}{\sqrt{a^2 - \rho^2}} \tag{7.140}$$

and the total charge on the disk is thus

$$Q = \frac{4\epsilon\Phi_{Sc}}{\pi} 2\pi \int_0^a \frac{\rho \, d\rho}{\sqrt{a^2 - \rho^2}} = 8\epsilon a \Phi_{Sc}$$

so that

$$C = \frac{Q_l}{\Phi_{Sc}} = 8\epsilon a \tag{7.141}$$

is the exact capacitance for an isolated circular disk.

For the polarizability problem, we see immediately as in the case of the strip that a z-directed incident field produces no polarization of the charge on the disk. By symmetry, the cases of x- and y-directed incident fields are equivalent, so we will consider only the case of an incident potential corresponding to $\mathbf{E}^i = \mathbf{u}_x E_0$ whence

$$\Phi_{Sp} - \Phi^i(x, y, 0) = x E_0$$

Then

$$g(\rho, \phi) = 4\pi\epsilon E_0 \rho \cos\phi = g_1(\rho)e^{j\phi} + g_{-1}(\rho)e^{-j\phi} \tag{7.142}$$

where $g_1(\rho) = g_{-1}(\rho) = 2\pi\epsilon E_0\rho$. By (7.139), the solutions $F_{\pm 1}$ are

$$f_{\pm 1}(\rho) = \frac{4\epsilon E_0}{\pi} \frac{\rho}{\sqrt{a^2 - \rho^2}} \tag{7.143}$$

so that

$$f(\rho, \phi) = f_1 e^{j\phi} + f_{-1}e^{-j\phi} = \frac{8\epsilon E_0}{\pi} \frac{x}{\sqrt{a^2 - \rho^2}} \tag{7.144}$$

which clearly integrates to a net charge on the disk of zero, as required for the potential problem. The dipole moment is

$$\begin{aligned}
\mathbf{p} &= \int_0^{2\pi} d\phi \int_0^a (x\mathbf{u}_x + y\mathbf{u}_y)\frac{8\epsilon E_0}{\pi} \frac{x}{\sqrt{a^2 - \rho^2}} \rho\, d\rho \\
&= 8\epsilon E_0 \mathbf{u}_x \int_0^a \frac{\rho^3\, d\rho}{\sqrt{a^2 - \rho^2}} \\
&= \frac{16\epsilon a^3}{3} E_0 \mathbf{u}_x
\end{aligned} \tag{7.145}$$

Since a similar result holds for a y-directed incident field, we have

$$\overset{\leftrightarrow}{\boldsymbol{\alpha}}_E = \frac{16 a^3}{3}(\mathbf{u}_x\mathbf{u}_x + \mathbf{u}_y\mathbf{u}_y) \tag{7.146}$$

as the exact polarizability dyadic for the circular disk.

7.7 INTEGRAL EQUATIONS FOR SCATTERING BY AN APERTURE IN A PLANE

Consider two half-spaces V_1 and V_2 separated by the plane $z = 0$. All of this plane is occupied by an infinitesimal perfectly conducting screen (this part is called S) with the exception of an aperture A through which fields can penetrate from one side of the screen to the other (Figure 7.9). Some sources $\mathbf{J}, \mathbf{M}$ (generally present on both sides of the screen) produce what we will call the short-circuit field $\mathbf{E}^{sc}, \mathbf{H}^{sc}$ *in the absence of the aperture*. In other words, $\mathbf{E}^{sc}, \mathbf{H}^{sc}$ would result if A were covered over by metal like the rest of the plane $z = 0$ (the aperture is then said to be "metallized"). We can readily find these fields, for example, by image theory. In particular, there corresponds to the

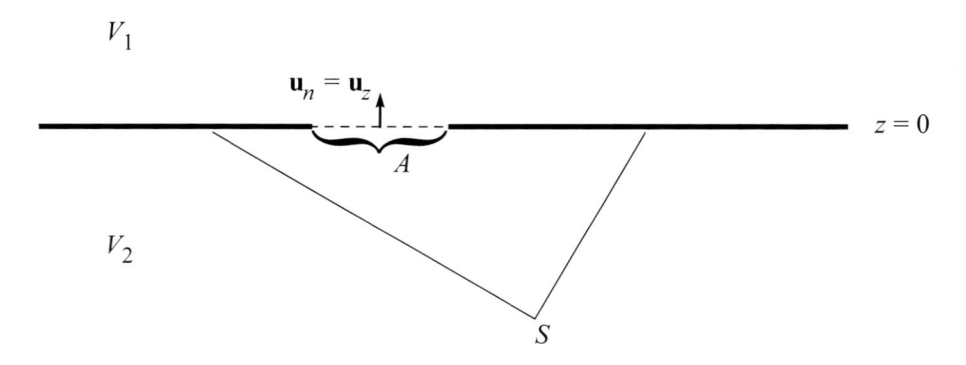

Figure 7.9   Aperture in a perfectly conducting plane.

short-circuit field a short-circuit surface current and charge density in the plane $z = 0$:

$$\rho_S^{\text{sc}}(x, y) = \epsilon\, E_z^{\text{sc}}(x, y, z)\big|_{z=0^-}^{0^+} \tag{7.147}$$

$$\mathbf{J}_S^{\text{sc}}(x, y) = \mathbf{u}_z \times \mathbf{H}^{\text{sc}}(x, y, z)\big|_{z=0^-}^{0^+} \tag{7.148}$$

The presence of the aperture produces an additional scattered field $\mathbf{E}^s$, $\mathbf{H}^s$, which must be such that the appropriate boundary conditions are met by the total field at S and at A: First, the tangential components of $\mathbf{E}^s$ are continuous everywhere in the plane $z = 0$. Next,

$$\mathbf{u}_z \times \mathbf{E}^s\big|_S = 0 \tag{7.149}$$

since $\mathbf{E}^{\text{sc}}\big|_S = 0$. Third,

$$\mathbf{u}_z \times \mathbf{H}^s\big|_{z=0^-}^{0^+} = -\mathbf{J}_S^{\text{sc}} \quad \text{in } A \tag{7.150}$$

since no surface current exists in A, and finally,

$$\epsilon E_z^s\big|_{z=0^-}^{0^+} = -\rho_S^{\text{sc}} \quad \text{in } A \tag{7.151}$$

since no surface charge exists in A. We have assumed that the medium parameters ϵ and μ are the same on either side of $z = 0$.

To form an integral equation for the yet unknown tangential electric field $\mathbf{E}_t$ in the aperture (in terms of which we can find the entire field anywhere—see Section 4.4), we first specify scalar Green's functions G_1 and G_2 defined for $\mathbf{r}, \mathbf{r}' \in V_1$, or V_2, respectively, and satisfying

$$(\nabla^2 + k^2)G_{1,2} = -\delta(\mathbf{r} - \mathbf{r}') \quad \text{in } V_{1,2} \tag{7.152}$$

$$\frac{\partial G_{1,2}}{\partial z'}\bigg|_{z'=0} = 0 \tag{7.153}$$

and the radiation condition as $r \to \infty$. These are also easily obtained from image theory using the free-space Green's function (7.13). We find that

$$G_1(\mathbf{r}, \mathbf{r}') = G_2(\mathbf{r}, \mathbf{r}') = \frac{1}{4\pi} \left\{ \frac{e^{-jk|\mathbf{r}-\mathbf{r}'|}}{|\mathbf{r}-\mathbf{r}'|} + \frac{e^{-jk|\mathbf{r}-\mathbf{r}'_i|}}{|\mathbf{r}-\mathbf{r}'_i|} \right\} \qquad (7.154)$$

where $\mathbf{r}'_i = x'\mathbf{u}_x + y'\mathbf{u}_y - z'\mathbf{u}_z$ is the image source point.

Let us now apply (7.15) with $\mathbf{P} = \mathbf{H}$ and $\mathbf{Q} = \mathbf{G}^c$ as given by (7.18). Using Maxwell's equations, we have

$$\mathbf{u}_c \cdot \mathbf{H}(\mathbf{r}) = \mathbf{u}_c \cdot \mathbf{H}^{sc}(\mathbf{r}) + j\omega\epsilon \oint_{S_T} \mathbf{G}^c \cdot \mathbf{E}(\mathbf{r}') \times \mathbf{u}'_n \, dS' - \oint_{S_T} (\nabla' \times \mathbf{G}^c) \cdot (\mathbf{u}'_n \times \mathbf{H}(\mathbf{r}')) \, dS' \qquad (7.155)$$

where S_T is boundary of the volume V to which Green's theorem is applied, to be specified below. If G in (7.18) is replaced by G_1 or G_2 above, and $\mathbf{u}_c$ is chosen to be either $\mathbf{u}_x$ or $\mathbf{u}_y$, then

$$(\nabla' \times \mathbf{G}^c) \cdot (\mathbf{u}'_n \times \mathbf{H}(\mathbf{r}'))|_{z'=0} = 0$$

by (7.153). We now apply (7.155) to the volumes $V_1(z' > 0)$ and $V_2(z' < 0)$ in turn, noting that the boundary surface S_T will in each case consist of the aperture A, the screen S, and an appropriate hemisphere that recedes to infinity. The integrals over these hemispheres will vanish by the radiation condition, and using (7.18) we obtain after some manipulation that

$$-j\omega\mu\mathbf{H}_t(\mathbf{r}_1) = -j\omega\mu\mathbf{H}_t^{sc}(\mathbf{r}_1) - (\nabla_t\nabla_t \cdot + k^2) \int_A \mathbf{M}_S(\boldsymbol{\rho}')G_1 \, dS' \qquad (7.156)$$

$$-j\omega\mu\mathbf{H}_t(\mathbf{r}_2) = -j\omega\mu\mathbf{H}_t^{sc}(\mathbf{r}_2) + (\nabla_t\nabla_t \cdot + k^2) \int_A \mathbf{M}_S(\boldsymbol{\rho}')G_2 \, dS' \qquad (7.157)$$

where

$$\mathbf{M}_S(\boldsymbol{\rho}') = \mathbf{E}_t(\mathbf{r}') \times \mathbf{u}_z|_{A(z'=0)} \qquad (7.158)$$

is the equivalent magnetic surface current in A (as seen from the *upper* half space). We obtain an integro-differential equation for $\mathbf{M}_S$ at A by requiring $\mathbf{H}_t$ to be continuous across A (Equation (7.150)):

$$(\nabla_t\nabla_t \cdot + k^2) \int_A \mathbf{M}_S(\boldsymbol{\rho}')K(\boldsymbol{\rho}, \boldsymbol{\rho}') \, dS' = j\omega\mu\mathbf{u}_z \times \mathbf{J}_S^{sc}(\boldsymbol{\rho}) \quad (\boldsymbol{\rho} \in A) \qquad (7.159)$$

where the kernel K is given by

$$K(\boldsymbol{\rho}, \boldsymbol{\rho}') = [G_1 + G_2]_{z=z'=0} = \frac{e^{-jk|\boldsymbol{\rho}-\boldsymbol{\rho}'|}}{\pi|\boldsymbol{\rho}-\boldsymbol{\rho}'|} \qquad (7.160)$$

This MFIE for the aperture problem is completely analogous to the EFIE (7.45) for the scattering from a conducting body. Most of the discussion of the latter can be carried over to the present case as well. Indeed, the relationship between the MFIE of the aperture scattering problem and the EFIE of the scattering problem for the complementary structure of a thin conducting *disk* occupying A, with the original screen removed, is another way of expressing Babinet's principle (Section 3.5).

In particular, analogously to (7.58), Equation (7.159) can be written in the form:

$$\frac{1}{\mu} \nabla_t \int_A \rho_{Sm}(\boldsymbol{\rho}') K(\boldsymbol{\rho}, \boldsymbol{\rho}') \, dS' + j\omega\epsilon \int_A \mathbf{M}_S(\boldsymbol{\rho}') K(\boldsymbol{\rho}, \boldsymbol{\rho}') \, dS' \;\; = \;\; -\mathbf{u}_z \times \mathbf{J}_S^{\mathrm{sc}}$$
$$(\boldsymbol{\rho} \in A) \qquad (7.161)$$

where ρ_{Sm} is the equivalent aperture magnetic charge density seen from the upper half space:

$$\rho_{Sm}(\boldsymbol{\rho}) = -\frac{\nabla_t \cdot \mathbf{M}_S}{j\omega} = \mu H_z(\boldsymbol{\rho}) \qquad (7.162)$$

A consequence of (7.161) obtained by taking its tangential curl is

$$\epsilon \mathbf{u}_z \cdot \nabla_t \times \int_A \mathbf{M}_S(\boldsymbol{\rho}') K(\boldsymbol{\rho}, \boldsymbol{\rho}') \, dS' = \rho_S^{\mathrm{sc}}(\boldsymbol{\rho}) \quad (\boldsymbol{\rho} \in A) \qquad (7.163)$$

where ρ_S^{sc} is defined in (7.147), or from (7.147) and (7.158):

$$\nabla_t \cdot \int_A \mathbf{E}_t(\boldsymbol{\rho}') K(\boldsymbol{\rho}, \boldsymbol{\rho}') \, dS' = E_z^{\mathrm{sc}}(\boldsymbol{\rho})\big|_{z=0^-}^{0^+} \qquad (7.164)$$

7.8 STATIC APERTURE PROBLEMS

7.8.1 ELECTROSTATIC APERTURE SCATTERING

The static forms of the integral equations for aperture scattering are derived in an analogous fashion. For the case of a short-circuit electrostatic potential Φ^i, we use a static Dirichlet Green's function for either V_1 or V_2, which obeys the boundary condition $G_d(\mathbf{r}, \mathbf{r}') = 0$ at $z' = 0$:

$$G_d(\mathbf{r}, \mathbf{r}') = \frac{1}{4\pi} \left\{ \frac{1}{|\mathbf{r} - \mathbf{r}'|} - \frac{1}{|\mathbf{r} - \mathbf{r}_i'|} \right\} \qquad (7.165)$$

and vanishes at infinity. If the point at infinity is taken to be at zero potential, then the same must be true of the short-circuit potential at $z = 0$ and of the scattered potential Φ^s at the screen S. By the usual Green's function methods, then, we obtain the following representations for the scattered potential at a point $\mathbf{r}_1 \in V_1$ or a point $\mathbf{r}_2 \in V_2$:

$$\Phi^s(\mathbf{r}_{1,2}) = \pm \int_A \Phi^s(\mathbf{r}') \frac{\partial G_d}{\partial z'} \, dS' = \mp 2 \frac{\partial}{\partial z} \int_A \Phi(\mathbf{r}') G_0(\mathbf{r}_{1,2}, \mathbf{r}') \, dS' \qquad (7.166)$$

where G_0 represents the infinite-space static Green's function

$$G_0 = \frac{1}{4\pi |\mathbf{r} - \mathbf{r}'|} \qquad (7.167)$$

The z-components of scattered electric field corresponding to these are then

$$E_z^s(\mathbf{r}_{1,2}) = \pm 2\frac{\partial^2}{\partial z^2}\int_A \Phi(\mathbf{r}')G_0(\mathbf{r}_{1,2},\mathbf{r}')\,dS' = \mp 2\nabla_t^2\int_A \Phi(\mathbf{r}')G_0(\mathbf{r}_{1,2},\mathbf{r}')\,dS'$$
(7.168)

by virtue of $\nabla^2 G_0 = 0$ for points $\mathbf{r}$ not in the aperture A.

To obtain the integral equation for the potential Φ in the aperture A, we enforce (7.151) there and use (7.168) to get

$$\nabla_t^2\int_A \Phi(\boldsymbol{\rho}')K_0(\boldsymbol{\rho},\boldsymbol{\rho}')\,dS' = \rho_S^{\mathrm{sc}}(\boldsymbol{\rho})/\epsilon$$
(7.169)

which is an integro-differential equation for $\boldsymbol{\rho} \in A$. The kernel K_0 is given by

$$K_0(\boldsymbol{\rho},\boldsymbol{\rho}') = \frac{1}{\pi|\boldsymbol{\rho}-\boldsymbol{\rho}'|}$$
(7.170)

This type of equation (as well as the EFIE for scattering from a conductor studied previously) is of the type first studied by Pocklington for the case of wire antennas,[13] so we will call it a *Pocklington equation*. Its derivation implicitly contains a boundary condition

$$\Phi(\boldsymbol{\rho})|_{C_A} = 0$$
(7.171)

for the potential at the edge C_A of the aperture.

In some contexts, the presence of the derivative in (7.169) can be inconvenient, and one might wish to have a pure integral equation to deal with, without having to introduce principal value integrals or other special ways of computing the integrals. In these cases, we can accomplish a conversion to a pure integral equation by using a *Hallén-Grinberg*[14] form of the equation (Hallén first used this method for the wire antenna problem, while Grinberg later generalized it for more general scattering objects). The idea is to treat the integral

$$f(\boldsymbol{\rho}) = \int_A \Phi(\boldsymbol{\rho}')K_0(\boldsymbol{\rho},\boldsymbol{\rho}')\,dS'$$
(7.172)

as an unknown function satisfying

$$\nabla_t^2 f(\boldsymbol{\rho}) = \rho_S^{\mathrm{sc}}(\boldsymbol{\rho})/\epsilon$$
(7.173)

for $\boldsymbol{\rho} \in A$. The only boundary condition on f is an indirect one resulting from condition (7.171) on Φ. A solution to (7.173) can be obtained by methods of

---

[13]Cf. D. S. Jones, *The Theory of Electromagnetism*. Oxford: Pergamon Press, 1964, Sections 3.14-3.19.

[14]See

D. S. Jones, *op. cit.*

G. A. Grinberg, *Sov. Phys. Tech. Phys.*, vol. 3, pp. 509-520, 1958.

two-dimensional Green's functions and Green's theorem (see Problem 7-1). We have a particular solution

$$f_p(\boldsymbol{\rho}) = -\frac{1}{2\pi\epsilon} \int_A \rho_S^{\mathrm{sc}}(\boldsymbol{\rho}') \ln \frac{\rho_0}{|\boldsymbol{\rho} - \boldsymbol{\rho}'|} \, dS' \tag{7.174}$$

where ρ_0 is an arbitrary constant. To (7.174) must be added an arbitrary solution f_h of the homogeneous equation

$$\nabla_t^2 f_h(\boldsymbol{\rho}) = 0 \tag{7.175}$$

which will generally contain an infinite number of arbitrary constants. Eventually, these constants must be determined by enforcing the condition (7.171) on Φ, and hence indirectly on f. The resulting Hallén-Grinberg equation is thus

$$\int_A \Phi(\boldsymbol{\rho}') K_0(\boldsymbol{\rho}, \boldsymbol{\rho}') \, dS' = f_p(\boldsymbol{\rho}) + f_h(\boldsymbol{\rho}) \tag{7.176}$$

together with the edge condition (7.171).

Regardless of whether the Pocklington or Hallén-Grinberg form of the equation is solved, once the aperture potential Φ is obtained, we can calculate the far-field value of Φ^s in V_1 or V_2 using (7.166) and techniques now familiar (see (7.71)). We have

$$\Phi^s(\mathbf{r}_{1,2}) \simeq \pm \frac{\mathbf{r} \cdot \mathbf{p}}{2\pi\epsilon r^3} \tag{7.177}$$

where the induced electric dipole moment is

$$\mathbf{p} = \mathbf{u}_z \epsilon \int_A \Phi \, dS \tag{7.178}$$

The factor of 2 appearing in (7.177) as compared to (7.77) is due to the fact that an infinite conducting plane screen is present at $z = 0$ in the case of the aperture problem. If the incident field is constant and has a discontinuity of

$$E_z^i(\boldsymbol{\rho}) \Big|_{z=0^-}^{0^+} = E_0 = -\rho_S^{\mathrm{sc}}/\epsilon \tag{7.179}$$

across the eventual location of the aperture, we define the electric polarizability dyadic of the aperture to be $\overset{\leftrightarrow}{\boldsymbol{\alpha}}_E = \mathbf{u}_z \mathbf{u}_z \alpha_{E,zz}$, where

$$p_{ez} = \epsilon \alpha_{E,zz} E_0 = -\rho_S^{\mathrm{sc}} \alpha_{E,zz} \tag{7.180}$$

7.8.2   MAGNETOSTATIC APERTURE SCATTERING

For the magnetostatic aperture problem, we make use of the Neumann static Green's function

$$G_n(\mathbf{r}, \mathbf{r}') = \frac{1}{4\pi} \left\{ \frac{1}{|\mathbf{r} - \mathbf{r}'|} + \frac{1}{|\mathbf{r} - \mathbf{r}_i'|} \right\} \tag{7.181}$$

which obeys the boundary condition (7.153) at the plane $z' = 0$. Applying Green's theorem to this Green's function and the scalar magnetic potential Ψ^s for the scattered magnetic field, we get

$$\Psi^s(\mathbf{r}_{1,2}) = \pm 2 \int_A G_0(\mathbf{r}, \mathbf{r}') H_z(\mathbf{r}') \, dS' \qquad (7.182)$$

Letting $\mathbf{r}_{1,2}$ approach A and enforcing continuity of total Ψ (which follows from the absence of surface current in A) we get the integral equation

$$\int_A H_z(\boldsymbol{\rho}') K_0(\boldsymbol{\rho}, \boldsymbol{\rho}') \, dS' = -\Psi_0(\boldsymbol{\rho}) \qquad (7.183)$$

for $\boldsymbol{\rho} \in A$, where $\Psi_0(\boldsymbol{\rho}) = \Psi^{\mathrm{sc}}|_{z=0^-}^{0^+}$ is the discontinuity of the short-circuit potential across the screen at A. Notice that this integral equation is formally identical to the integral equation (7.72) for the electrostatic charge distribution on a thin conducting disk of the same shape as A if we make the replacements

$$-\frac{1}{4}\Psi_0 \to \Phi_S - \Phi^{\mathrm{sc}}; \qquad H_z \to \frac{\rho_S^{\mathrm{tot}}}{\epsilon}$$

Solution techniques for the electrostatic disk problem are thus directly applicable to the magnetostatic aperture problem with no substantial change.

The far field of Ψ^s is evaluated as

$$\Psi^s(\mathbf{r}_{1,2}) \simeq \pm \frac{1}{2\pi} \left[\frac{\int_A H_z(\boldsymbol{\rho}') \, dS'}{r} + \frac{\mathbf{r} \cdot \int_A \boldsymbol{\rho}' H_z(\boldsymbol{\rho}') \, dS'}{r^3} + \cdots \right] \qquad (7.184)$$

as $r \to \infty$. But $\int_A H_z \, dS = \oint_S H_n \, dS = 0$ (where S is a closed surface—including a hemisphere at infinity—bounding V_2) by Gauss' law for the magnetic field, so (7.184) becomes

$$\Psi^s(\mathbf{r}_{1,2}) \simeq \pm \frac{1}{2\pi} \frac{\mathbf{r} \cdot \mathbf{m}}{r^3} \qquad (7.185)$$

where

$$\mathbf{m} = \int_A \boldsymbol{\rho}' H_z(\boldsymbol{\rho}') \, dS' \qquad (7.186)$$

In the polarizability problem where a uniform short-circuit field $\mathbf{H}^i$ is applied, define

$$\mathbf{H}_0 = \mathbf{H}_t^{\mathrm{sc}}|_{z=0^-}^{0^+} \qquad (7.187)$$

and thence the magnetic polarizability

$$\mathbf{m} = \overset{\leftrightarrow}{\boldsymbol{\alpha}}_M \cdot \mathbf{H}_0 \qquad (7.188)$$

By the equivalence noted above between the magnetostatic aperture problem and the electrostatic disk problem, we see immediately that

$$\overset{\leftrightarrow}{\boldsymbol{\alpha}}_M \bigg|_{\text{aperture } A} = \frac{1}{4} \overset{\leftrightarrow}{\boldsymbol{\alpha}}_E \bigg|_{\text{disk } A} \qquad (7.189)$$

Thus, for example, the magnetic polarizability of a circular aperture of radius a is, from (7.146),

$$\overset{\leftrightarrow}{\boldsymbol{\alpha}}_M = \frac{4a^3}{3}(\mathbf{u}_x\mathbf{u}_x + \mathbf{u}_y\mathbf{u}_y) \tag{7.190}$$

The far-field response of any aperture to a short-circuit static field is thus completely described by the electric and magnetic polarizabilities of the aperture. They also play a part in the approximate description of aperture scattering for the low-frequency time-harmonic problem.

7.8.3 EXAMPLE: ELECTRIC POLARIZABILITY OF A CIRCULAR APERTURE

As an example of the solution of the Hallén-Grinberg equation (7.176), consider once again the circular hole of radius a. By (7.174), and techniques by now familiar,

$$f_p(\boldsymbol{\rho}) = \frac{a^2 E_0}{2}\ln\frac{\rho_0}{a} + \frac{E_0}{4}(a^2 - \rho^2) \tag{7.191}$$

and is thus independent of ϕ. The solution of (7.175) can be written as

$$f_h(\boldsymbol{\rho}) = \sum_{m=-\infty}^{\infty} e^{jm\phi} A_m \rho^{|m|} \tag{7.192}$$

We do not include negative powers of ρ in this solution, because they do not lead to solutions of (7.175), but rather generate delta functions on the right hand side concentrated at $\boldsymbol{\rho} = 0$.

We now have in (7.176)

$$\int_0^{2\pi}\int_0^a \frac{\Phi(\boldsymbol{\rho}')\rho'\,d\rho'd\phi'}{|\boldsymbol{\rho} - \boldsymbol{\rho}'|} = \frac{\pi E_0}{4}(a^2 - \rho^2) + \sum_{m=-\infty}^{\infty} e^{jm\phi} A_m \rho^{|m|} \tag{7.193}$$

where we have absorbed a constant term into the yet undetermined A_0. We split (7.193) into an infinite set of equations (7.127) as in Section 7.6, where

$$f_m(\rho) = \Phi_m(\rho)$$

is the coefficient in the expansion

$$\Phi(\rho, \phi) = \sum_{-\infty}^{\infty} e^{jm\phi}\Phi_m(\rho) \tag{7.194}$$

and

$$\begin{aligned} g_0(\rho) &= \frac{\pi E_0}{4}(a^2 - \rho^2) + A_0 \\ g_m(\rho) &= A_m \rho^{|m|} \quad (m \neq 0) \end{aligned} \tag{7.195}$$

The solution of (7.127) for this case is found from the results of Section 7.6. We find, for example, that

$$\Phi_0(\rho) = \frac{E_0}{\pi} \sqrt{a^2 - \rho^2} + \left(\frac{A_0}{\pi^2} - \frac{E_0 a^2}{4\pi} \right) \frac{1}{\sqrt{a^2 - \rho^2}} \tag{7.196}$$

while

$$\Phi_1(\rho) = \frac{2A_1}{\pi^2} \frac{\rho}{\sqrt{a^2 - \rho^2}}$$

and so on. For all $m \neq 0$, it turns out that $\Phi_m(\rho) \to \infty$ as $\rho \to a$, unless $A_m = 0$. But by (7.171) we must have $\Phi_m(a) = 0$ for all m, and thus $A_m = 0$ for all $m \neq 0$. From (7.196), on the other hand, we see that

$$A_0 = \frac{\pi E_0 a^2}{4} \tag{7.197}$$

in order for $\Phi_0(a)$ to be zero. Thus finally we get

$$\Phi_0(\rho) = \frac{E_0}{\pi} \sqrt{a^2 - \rho^2} \tag{7.198}$$

From (7.180), then, we obtain the induced dipole moment p_z as:

$$p_z = \frac{\epsilon E_0}{\pi} \int_0^{2\pi} \int_0^a \rho \sqrt{a^2 - \rho^2}\, d\rho d\phi = \frac{2\epsilon E_0 a^3}{3}$$

and thus

$$\alpha_{E,zz} = \frac{2a^3}{3} \tag{7.199}$$

7.9   PROBLEMS

7–1 Denote by ∇_t^2 the transverse Laplacian $\partial^2/\partial x^2 + \partial^2/\partial y^2$ and let $f(\boldsymbol{\rho})$ be a solution of the two-dimensional static potential equation (Poisson's equation)

$$\nabla_t^2 f = -s_f$$

Show that

(a) $G(\boldsymbol{\rho}, \boldsymbol{\rho}') = \frac{1}{2\pi} \ln \frac{\rho_0}{|\boldsymbol{\rho}-\boldsymbol{\rho}'|}$ for some constant ρ_0 is a Green's function for this equation:

$$\nabla_t^2 G = -\delta(\boldsymbol{\rho} - \boldsymbol{\rho}')$$

(b) and that

$$f(\boldsymbol{\rho}) = \frac{1}{2\pi} \int \ln \frac{\rho_0}{|\boldsymbol{\rho} - \boldsymbol{\rho}'|} s_f(\boldsymbol{\rho}') \, dS' + f_h(\boldsymbol{\rho})$$

where $f_h(\boldsymbol{\rho})$ is a solution of the homogeneous equation

$$\nabla_t^2 f_h = 0$$

and the integral is carried out over all points where s_f is nonzero.

7–2 Find a general expression for the scalar electrostatic Green's function G for the grounded dielectric slab shown below.

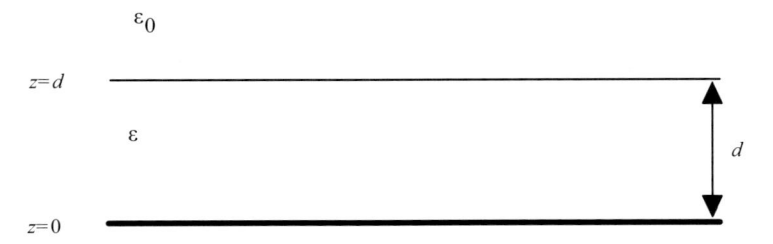

Your result should be in the form of a double Fourier integral or be a Sommerfeld integral of the form (4.75). The Green's function should satisfy

$$\nabla^2 G = -\delta(\mathbf{r} - \mathbf{r}')$$

in $0 < z < d$ and $z > d$;

$$G = 0 \qquad \text{at} \quad z = 0$$

G should be continuous at $z = d$;

$$\epsilon_0 \frac{\partial G}{\partial z}\bigg|_{z=d+} = \epsilon \frac{\partial G}{\partial z}\bigg|_{z=d-}$$

and G should go to 0 as $z \to \infty$.

When $z = z' = d$, show that your expression can be expressed in the form

$$G = \frac{1}{2\pi(\epsilon_r + 1)} \int_0^\infty J_0(\lambda|\boldsymbol{\rho} - \boldsymbol{\rho}'|) \frac{1 - e^{-2\lambda d}}{1 + \delta_\epsilon e^{-2\lambda d}} \, d\lambda$$

and that this can also be expressed in the alternate form

$$G = \frac{1}{2\pi(\epsilon_r + 1)} \left[\frac{1}{|\boldsymbol{\rho} - \boldsymbol{\rho}'|} + \frac{2\epsilon_r}{\epsilon_r - 1} \sum_{m=1}^\infty \frac{(-\delta_\epsilon)^m}{\sqrt{|\boldsymbol{\rho} - \boldsymbol{\rho}'|^2 + (2md)^2}} \right]$$

where $\epsilon_r = \epsilon/\epsilon_0$, $\boldsymbol{\rho}$ and $\boldsymbol{\rho}'$ are the components of $\mathbf{r} = \boldsymbol{\rho} + d\mathbf{u}_z$ and $\mathbf{r}' = \boldsymbol{\rho}' + d\mathbf{u}_z$, respectively, which are perpendicular to the z-direction, and

$$\delta_\epsilon = \frac{\epsilon_r - 1}{\epsilon_r + 1}$$

7–3 Derive (2.22)-(2.23) and (7.34) from (7.28) and (7.33).

7–4 Define

$$\Phi = \frac{1}{\epsilon} \int [\rho_V(\mathbf{r}') + \rho_{eq}(\mathbf{r}')] G(\mathbf{r}, \mathbf{r}') \, dV'$$

where $\rho_V = -\nabla \cdot \mathbf{J}_V/j\omega$ and $\rho_{eq} = -\epsilon \mathbf{u}_n \cdot \mathbf{E} \delta_S(n)$. Show that $\Phi = -\nabla \cdot \mathbf{A}/j\omega\mu\epsilon$, where $\mathbf{A}$ is given by (7.34).

7–5 Show that the first two terms on the right side of (7.55) approach zero in the limit as $S_0 \to 0$, regardless of the shape of S_0. [Hint: Try it first for the case when S_0 is a circle.]

7–6 Consider the source-free MFIE

$$\frac{1}{2}\mathbf{J}_S(\mathbf{r}) \times \mathbf{u}_n + \left\{ \oint_S \nabla'G \times \mathbf{J}_S(\mathbf{r}') \, dS' \right\}_{\text{tan}} = 0; \quad (\mathbf{r} \in S)$$

(compare (7.57)). Show that nontrivial solutions $\mathbf{J}_S$ to this equation can only exist at frequencies corresponding to resonant PMC cavity modes of the region interior to S (in other words, at frequencies where there are nontrivial solutions to Maxwell's equations inside S that obey the boundary condition $\mathbf{H}_{\text{tan}} = 0$ on S). [Hint: Use of equivalence and/or duality principles can be helpful here.]

7–7 Suppose that a material body occupying a volume V in a background medium of constant permittivity ϵ_b and permeability μ_b is characterized not only by the permittivity $\epsilon(\mathbf{r})$ in V, but also by the permeability $\mu(\mathbf{r})$ there. Show that the total fields $\mathbf{E}$ and $\mathbf{H}$ within V obey the system of volume integral equations

$$\begin{aligned}
\mathbf{E}(\mathbf{r}) = \; & \mathbf{E}^i(\mathbf{r}) + \left(\nabla\nabla \cdot + k_b^2\right) \int_V \frac{\epsilon(\mathbf{r}') - \epsilon_b}{\epsilon_b} \mathbf{E}(\mathbf{r}') G \, dV' \\
& - j\omega\mu_b \nabla \times \int_V \frac{\mu(\mathbf{r}') - \mu_b}{\mu_b} \mathbf{H}(\mathbf{r}') G \, dV'
\end{aligned}$$

and

$$\mathbf{H}(\mathbf{r}) = \mathbf{H}^i(\mathbf{r}) + \left(\nabla\nabla \cdot + k_b^2\right) \int_V \frac{\mu(\mathbf{r}') - \mu_b}{\mu_b} \mathbf{H}(\mathbf{r}') G\, dV'$$

$$+ j\omega\epsilon_b \nabla \times \int_V \frac{\epsilon(\mathbf{r}') - \epsilon_b}{\epsilon_b} \mathbf{E}(\mathbf{r}') G\, dV'$$

for $\mathbf{r} \in V$.

7–8 Show that the induced surface charge ρ_S on a conducting surface S placed in an incident static field $\mathbf{E}^i(\mathbf{r})$ satisfies the following integral equation of the second kind:

$$\rho_S(\mathbf{r}) = 2\epsilon E_n^i(\mathbf{r}) + \frac{1}{2\pi} \oint_S \frac{\rho_S(\mathbf{r}')}{|\mathbf{r} - \mathbf{r}'|^3} \mathbf{u}_n \cdot (\mathbf{r} - \mathbf{r}')\, dS'$$

for $\mathbf{r}$ on S, and that if S is smooth ($\mathbf{u}_n$ is continuous on it) there is no need to regard the integral as a principal value. [Hint: Take the gradient of (7.70)].

7–9 If S is the surface of a simply-connected perfect conductor that is immersed in an incident magnetostatic field corresponding to the scalar potential Ψ^i, show that the total scalar magnetic potential Ψ satisfies the integral equation

$$\Psi(\mathbf{r}) = 2\Psi^i(\mathbf{r}) + 2 \oint_S \Psi(\mathbf{r}') \frac{\partial G}{\partial n'}\, dS'$$

for $\mathbf{r} \in S$ [compare I. Lucas, *J. Appl. Phys.*, vol. 47, pp. 1645-1652, 1976].

7–10 The thin strip in Section 7.5 is grounded (i.e., held at zero potential), but immersed in the incident potential produced by a line source directly above it at height h:

$$\Phi^i(x,0) = \frac{\rho_l}{2\pi\epsilon} \ln \frac{R}{\sqrt{h^2 + x^2}}$$

where ρ_l is the charge per unit length of the line source.

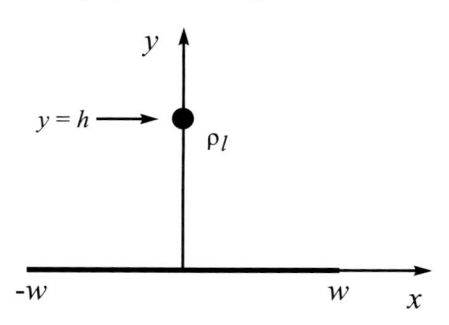

Find the solution for the total charge density $f(x)$ on the strip in $|x| < w$. Here R is a normalizing constant giving the distance from the line source at which the incident potential is zero. How does this solution compare with that when the total charge on the strip is required to be zero? In this latter case, the incident field and the resulting induced charge density on the strip will not depend on the constants R and ρ_0.

7–11 A slot of width $2a$ in the x-direction, which is infinitely long in the y-direction, exists in an otherwise perfectly conducting plane in $z = 0$ as shown.

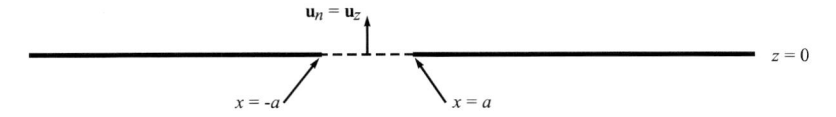

A time-harmonic incident wave $\mathbf{E}^i$, $\mathbf{H}^i$ (which are independent of y) exists in the absence of the slot. Show that an integral equation for the aperture field $\mathbf{E}_t(x)$ in $z = 0$ is:

$$\left(\frac{d^2}{dx^2} + k^2 \right) \int_{-a}^{a} E_y(x') H_0^{(2)}(k|x - x'|) \, dx' = \omega \mu J_{Sy}^i(x)$$

in $-a < x < a$ for the y-component of $\mathbf{E}_t$, and

$$k^2 \int_{-a}^{a} E_x(x') H_0^{(2)}(k|x - x'|) \, dx' = \omega \mu J_{Sx}^i(x)$$

in the same interval for the x-component of $\mathbf{E}_t$.

7–12 Let a simply-connected conducting disk of arbitrary shape A lie in the plane $z = 0$. Show that the "total" magnetic scalar potential

$$\Psi^{tot} = \Psi^{top} + \Psi^{bottom}$$

on both sides of the disk satisfies the (static) integro-differential equation

$$H_z^i \big|_{z=0^-} + \nabla_t^2 \int_A \Psi^{tot}(\mathbf{r}') \frac{dS'}{4\pi |\mathbf{r} - \mathbf{r}'|} = 0$$

for all $\mathbf{r} \in A$. Find a general relationship between the magnetic polarizability $\alpha_{M,zz}$ of this disk and the electric polarizability $\alpha_{E,zz}$ of an aperture of the same shape A in a conducting plane. Give the value of $\alpha_{M,zz}$ for a circular disk of radius a.

8 Approximation Methods

Most electromagnetic boundary problems, whether formulated as differential equations or as integral equations, cannot be solved in closed form. We therefore have to find approximate solutions, in either analytical or numerical form. In this chapter, we will present several types of analytical approximation methods that are particularly appropriate for problems formulated as integral equations (although they have much more general applicability), and apply them to some practical problems.

8.1 RECURSIVE PERTURBATION APPROXIMATION

Often, a good deal of insight into the solution of a problem can be obtained by studying some approximation of it. In this section, we examine the approximate solution of integral equations of the first kind based upon approximation of their kernels. For example, consider the equation

$$\int_{-w}^{w} f(x') K(x, x') \, dx' = g(x); \quad |x| < w \tag{8.1}$$

of the type that arose in Section 7.5 in connection with static strip problems. Let

$$K(x, x') = K_0(x, x') + K_1(x, x') \tag{8.2}$$

where K_0 is a kernel for which (8.1) can be solved exactly, while K_1 is, in some sense, "small," perhaps on the order of some small parameter δ. Then we may expect that the solution will take the form

$$f(x) = f_0(x) + f_1(x) \tag{8.3}$$

where f_0 is the solution of (8.1) with K_0 as kernel:

$$\int_{-w}^{w} f_0(x') K_0(x, x') \, dx' = g(x); \quad |x| < w \tag{8.4}$$

while f_1 is reckoned to be small of the same order δ as K_1. Substituting (8.3) and (8.2) into (8.1) and using (8.4) now gives

$$\int_{-w}^{w} f_0(x') K_1(x, x') \, dx' + \int_{-w}^{w} f_1(x') K_0(x, x') \, dx' + \int_{-w}^{w} f_1(x') K_1(x, x') \, dx' = 0 \tag{8.5}$$

for $|x| < w$. If both f_1 and K_1 are small of order δ, then the last term in (8.5) should be much smaller yet (of order δ^2), and we will, as a first approximation, neglect it. Then (8.5) becomes approximately

$$\int_{-w}^{w} f_1(x') K_0(x, x') \, dx' = -\int_{-w}^{w} f_0(x') K_1(x, x') \, dx' \tag{8.6}$$

for $|x| < w$. Since we have assumed that the kernel K_0 gives rise to an integral equation that can be solved exactly for any right-hand side, we can proceed in a recursive manner, solving (8.4) for f_0, inserting this solution into the right side of (8.6), which now becomes known, and finally solving (8.6) for (an approximation to) f_1.

This method is known as a *perturbation method*, and can be made more systematic and extended to higher-order corrections as follows. Suppose that the kernel K has an expansion of the form

$$K = K_0 + K_1 + K_2 + \cdots \tag{8.7}$$

where K_1 is of order δ, K_2 is of order δ^2, etc. Correspondingly, instead of (8.3), we now write

$$f(x) = f_0(x) + f_1(x) + f_2(x) + \cdots \tag{8.8}$$

where f_1 is supposed to be of order δ, f_2 of order δ^2, and so on. Thus (for instance) f_1^2, f_2, and $f_1 K_1$ are all of the same order: δ^2. Substituting (8.7) and (8.8) into (8.1), we now gather together in separate groups all terms of the same order δ^m, and demand that they be separately equal. Equations (8.4) and (8.6) again arise (the latter now as an exact equation), and form the first two in a sequence of recurrent equations whose general form is

$$\int_{-w}^{w} f_N(x') K_0(x, x')\, dx' = -\sum_{n=0}^{N-1} \int_{-w}^{w} f_n(x') K_{N-n}(x, x')\, dx' \tag{8.9}$$

Each of these can be solved once the previous one in the sequence has been solved. In practice, the analytical forms of the right sides of these equations become rapidly more complicated as N increases, and it is usually not practical to find more than the first few correction terms.

8.1.1 EXAMPLE: STRIP OVER A GROUND PLANE

As an example, consider a thin conducting strip of width $2w$ located at a height h above an infinite, perfectly conducting ground plane as shown in Figure 8.1. Using image theory, it is not difficult to come up with the integral equation for the total charge distribution $\rho_S(x)$ on the strip as

$$\Phi_{Sc} = \frac{1}{2\pi\epsilon} \int_{-w}^{w} \rho_S(x') \ln \frac{\sqrt{(x - x')^2 + 4h^2}}{|x - x'|}\, dx' \tag{8.10}$$

on $|x| < w$, where Φ_{Sc} is the potential of the strip, whereas the ground plane is at zero potential. Now, if $w \ll h$, then $|x - x'| \ll 2h$ for any x and x' on the strip, and the kernel of the integral equation (8.10) can be approximated by

$$\ln \frac{\sqrt{(x - x')^2 + 4h^2}}{|x - x'|} = \ln(2h/|x - x'|) + \frac{(x - x')^2}{8h^2} + \cdots \tag{8.11}$$

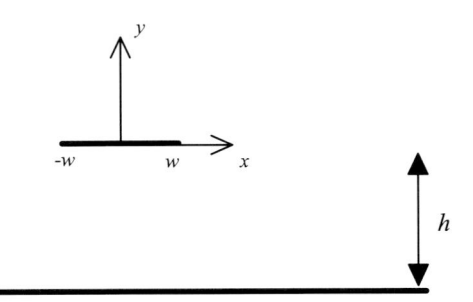

Figure 8.1　Strip above a ground plane.

In this case, provided we choose $\rho_0 = 2h$, (8.10) is precisely (7.100), which we have already solved exactly. From that solution, the zeroth-order approximation $\rho_{S0}(x)$ is obtained as

$$\rho_{S0}(x) = \frac{2\epsilon\Phi_{Sc}}{\ln(4h/w)}\frac{1}{\sqrt{w^2 - x^2}} \tag{8.12}$$

To this order, the capacitance per unit length of this system is

$$C_l \simeq \frac{2\pi\epsilon}{\ln(4h/w)} \tag{8.13}$$

which is obtained by integrating (8.12) over the strip.

The first-order correction $\rho_{S1}(x)$ to this approximation is found by solving (8.6):

$$\int_{-w}^{w} \rho_{S1}(x') \ln\frac{2h}{|x - x'|}\, dx' = -\int_{-w}^{w} \rho_{S0}(x')\frac{(x - x')^2}{8h^2}\, dx'$$
$$= -\frac{w^2}{8h^2}\frac{\pi\epsilon\Phi_{Sc}}{\ln(4h/w)}\left(\frac{2x^2}{w^2} + 1\right) \tag{8.14}$$

Using the results of Section 7.5, we now put $f(x) = \rho_{S1}(x)$, and observe from (8.14) that the nonzero Fourier coefficients of $G(u)$ are

$$G_0 = -\frac{\pi w^2}{4h^2}\frac{\epsilon\Phi_{Sc}}{\ln(4h/w)} \tag{8.15}$$

and

$$G_2 = -\frac{\pi w^2}{8h^2}\frac{\epsilon\Phi_{Sc}}{\ln(4h/w)} \tag{8.16}$$

The solution to (8.14) is thus, in terms of the variable u:

$$F(u)\sin u = \frac{1}{\pi w}\left[\frac{G_0}{\ln(4h/w)} + 2G_2\cos 2u\right] \tag{8.17}$$

and so:

$$\rho_{S1}(x) = -\frac{1}{\sqrt{w^2 - x^2}} \left(\frac{\epsilon \Phi_{Sc}}{\ln(4h/w)} \frac{w^2}{4h^2} \right) \left[\frac{1}{\ln(4h/w)} + 2\frac{x^2}{w^2} - 1 \right] \tag{8.18}$$

But

$$\int_{-w}^{w} \rho_{S1}(x') \, dx' = \frac{G_0}{\ln(4h/w)} \tag{8.19}$$

so that from (8.15) (or (8.18)) we have

$$\int_{-w}^{w} \rho_{S1}(x') \, dx' = -\frac{\pi w^2}{4h^2} \frac{\epsilon \Phi_{Sc}}{\ln^2(4h/w)} \tag{8.20}$$

The total charge per unit length to this new approximation is then

$$Q_l = \int_{-w}^{w} \rho_{S0}(x') \, dx' + \int_{-w}^{w} \rho_{S1}(x') \, dx' = \frac{2\pi\epsilon \Phi_{Sc}}{\ln(4h/w)} \left[1 - \frac{w^2}{8h^2 \ln(4h/w)} \right] \tag{8.21}$$

Thus a more refined approximation for C_l is

$$C_l \simeq \frac{2\pi\epsilon}{\ln(4h/w)} \left[1 - \frac{w^2}{8h^2 \ln(4h/w)} \right] \tag{8.22}$$

If $2w/h \leq 1$, the correction term in the square brackets is $\leq 1.5\%$. Further improvement is possible by going to even higher approximations; this is left as an exercise. However, we will see below that there are more accurate approximations possible that can be obtained with much less work.

8.1.2 PHYSICAL OPTICS APPROXIMATION

For the problem of scattering by a perfectly conducting object, consider the MFIE (7.57), which we rewrite here as

$$\left\{ \int \mathbf{J}_S(\mathbf{r}') \times [\mathbf{u}'_n \delta(\mathbf{r} - \mathbf{r}') - 2\nabla' G \delta_S(n')] \, dV' \right\}_{\text{tan}} = 2\mathbf{H}^i_{\text{tan}}(\mathbf{r}); \quad (\mathbf{r} \in S) \tag{8.23}$$

where $\delta_S(n)$ is the Dirac delta function concentrated on the surface of the scatterer. We identify the terms

$$\mathbf{K}_0(\mathbf{r}, \mathbf{r}') = \mathbf{u}'_n \delta(\mathbf{r} - \mathbf{r}') \tag{8.24}$$

$$\mathbf{K}_1(\mathbf{r}, \mathbf{r}') = -2\nabla' G \delta_S(n') \tag{8.25}$$

so that (8.23) has the same form as (8.1)-(8.2), but generalized to three-dimensional integrals. The zero-order approximation corresponding to the solution of (8.4) is straightforward:

$$\left\{ \int \mathbf{J}_{S0}(\mathbf{r}') \times [\mathbf{u}'_n \delta(\mathbf{r} - \mathbf{r}')] \, dV' \right\}_{\text{tan}} = 2\mathbf{H}^i_{\text{tan}}(\mathbf{r}); \quad (\mathbf{r} \in S) \tag{8.26}$$

or

$$\mathbf{J}_{S0}(\mathbf{r}) = 2\mathbf{u}_n \times \mathbf{H}_{\tan}^i(\mathbf{r}) \tag{8.27}$$

because of the sifting property of the delta function. This is *exactly* the result that would be obtained if the surface S were an infinite perfectly conducting plane (at $z = 0$, say), as can readily be shown by image theory.

So why are we entitled to claim that $\mathbf{K}_1$ is "small" in some sense compared to K_0? In the case when S is the plane $z = 0$, the normal derivative $\partial G/\partial z' = 0$ when $z' = 0$, while $\nabla_t'G$ will contribute nothing to the tangential part of the left side of (8.23), because $\mathbf{J}_S$ is also tangential to S. We can expect that the contribution of $\mathbf{K}_1$ will be small if S is locally "almost" a plane, which in practice will mean that the radii of curvature of S are large compared to a wavelength. The next order of approximation corresponding to (8.6) can be found in a similar way:

$$\mathbf{J}_{S1}(\mathbf{r}) = 4\mathbf{u}_n \times \left[\oint_S \mathbf{H}_{\tan}^i(\mathbf{r}') \frac{\partial G(\mathbf{r}, \mathbf{r}')}{\partial n'} \, dS' \right]_{\tan} \tag{8.28}$$

In applications, this correction turns out to be small on the "illuminated" portion of the scatterer surface (the part that is not in the geometrical shadow region), but more significant on the shadowed side. In fact, on the shadowed side, this correction tends to cancel out the contribution from $\mathbf{J}_{S0}$. The approximation of $\mathbf{J}_S$ by $\mathbf{J}_{S0}$ on the illuminated portion of S and by zero on the shadowed portion is called the *Physical Optics Approximation*, and is used in some situations as an alternative to GTD or other geometrical theories of diffraction.

8.2 OPERATOR FORMALISM FOR APPROXIMATION METHODS

The complexity of the equations that arise in the recursive perturbation technique can obscure the essential features of the method. In this subsection, we will introduce an operator/inner-product formalism to help clean up the mathematics, and which will allow us to obtain formulas of higher accuracy more readily for certain quantities that depend on the solution to our integral equation.[1]

We begin by rewriting (8.1) as an operator equation:

$$Lf = g \tag{8.29}$$

where L is an *operator*, an entity that takes one function $f(x)$ and produces from it another function of x. The rule by which the new function is created

---

[1] A good introduction to these ideas can be found in

I. Stakgold, *Green's Functions and Boundary Value Problems*. New York: Wiley, 1979, Chapters 4 and 5, and Section 8.4.

defines the operator L. In the present case, L is defined by:

$$Lf = \int_{-w}^{w} f(x')K(x,x')\,dx' \tag{8.30}$$

Most operators cannot be applied to any arbitrary function f; only for a certain set of functions will the definition of the operator make sense. This set is known as the *domain* of L, and is denoted $\mathcal{D}(L)$. We say that

$$f \in \mathcal{D}(L)$$

if f is in the domain of L.

Note that what we will do in this section depends very little on the specific form (8.30) of the operator L. The operator might be a matrix that multiplies a column vector f to produce another column vector, or a differential operator that takes a function f and calculates some combination of its derivatives, etc. Only a few basic properties of L are essential: here we will ask that it be *linear*, so that

$$L(af) = aLf \tag{8.31}$$

for a constant a, and

$$L(f_1 + f_2) = Lf_1 + Lf_2 \tag{8.32}$$

for any f_1 and f_2 in the domain of L. Some examples of linear operators besides (8.30) are differential operators such as

$$Lf(x) = \frac{df}{dx}, \qquad Lf(x) = \frac{d^2 f}{dx^2}, \qquad Lf(\mathbf{r}) = \nabla f, \qquad \text{etc.} \tag{8.33}$$

and multiplication operators:

$$Lf(x) = g(x)f(x) \tag{8.34}$$

for some given (generalized) function $g(x)$, to name only a few. Operators may be characterized as *local* (when $Lf(x)$ depends only on the behavior of f near the point x—examples include differential operators like (8.33) and multiplication operators like (8.34)) or *nonlocal* (when $Lf(x)$ depends on the behavior of $f(x)$ at all points x, as for instance in the case of an integral operator like (8.30)).

As we saw in the example of the previous subsection, it is often not the solution f of the operator equation that interests us most as engineers, but instead some *functional* of that solution. A *linear functional* of f is defined as a rule that assigns a number $F = F[f]$ to any given function f, such that $F[af] = aF[f]$ and $F[f+g] = F[f]+F[g]$ for any constant a and any functions f and g. For the case of the operator (8.30), we choose

$$F = \int_{-w}^{w} f(x)h(x)\,dx \tag{8.35}$$

for some given function h.

In a more compact abstract notation, we say

$$F = (f, h) \tag{8.36}$$

where $(\cdot, \cdot)$ is what we will call a *scalar product* (more precisely, an *indefinite scalar product*), defined in the present case of real-valued functions f and h as

$$(f, h) = \int_{-w}^{w} f(x)h(x)\,dx \tag{8.37}$$

The specific definition of the scalar product is not important for what we are about to do; only a few basic properties are needed. These are:

$$(h, f) = (f, h) \tag{8.38}$$

for any f and h,

$$(af, h) = a(f, h) \tag{8.39}$$

for any constant a,

$$(f + g, h) = (f, h) + (g, h) \tag{8.40}$$

for any (real) f, g and h.

The example that originally motivated this discussion involved only functions f, g, h, etc. that are real, and such that the operator L always produces real functions from real functions. It is possible to generalize these ideas to the case when the functions and operators are complex. Properties (8.38)-(8.40) remain true, but it is often useful to consider an *(indefinite) inner product* $\langle \cdot, \cdot \rangle$ which obeys

$$\langle h, f \rangle = \langle f, h \rangle^* \tag{8.41}$$

rather than the scalar product version (8.38). Clearly, if $(\cdot, \cdot)$ is a scalar product, then $\langle \cdot, \cdot \rangle$ is an inner product. Moreover, if we restrict ourselves to real functions, the concepts of scalar and inner products are equivalent. We shall henceforth use only the scalar product in our formulas, although they will remain true if the scalar product is replaced by an inner product (with only minor, if any, changes needed).

If it is additionally true that $(f, f) \geq 0$ or $\langle f, f \rangle \geq 0$, with equality holding only if $f \equiv 0$, then the scalar or inner product is said to be *definite*, and we denote the quantity

$$\|f\| \equiv \sqrt{(f, f)} \geq 0 \tag{8.42}$$

(if it exists) as the *norm* of f. The magnitude of the inner product can be bounded above using the *Schwarz inequality*:

$$|(f, g)| \leq \|f\|\|g\| \tag{8.43}$$

Because our original problem of capacitance deals only with real functions, (8.37) is indeed an inner product, and so can be associated with a norm.

However, a norm is not necessary for much of what we will do below, and is mostly required when estimates of error in approximate solutions are needed.

An operator related to L is the *adjoint* operator $L^\dagger$, defined by

$$(p, Lf) = (L^\dagger p, f) \tag{8.44}$$

for any $f \in \mathcal{D}(L)$ and $p \in \mathcal{D}(L^\dagger)$ (the existence of the inner products in (8.44) is used to define the domain $\mathcal{D}(L^\dagger)$ of the adjoint operator). The adjoint of an operator is analogous to the Hermitian transpose of a matrix. An operator is said to be *self-adjoint*, with respect to the inner product used in (8.44), if $L = L^\dagger$, i.e., if

$$(p, Lf) = (Lp, f) \tag{8.45}$$

for any f and p in the domain of L. The operator defined in (8.30) can be shown to be self-adjoint since $K(x, x') = K(x', x)$. The operator of multiplication by a given function is obviously self-adjoint, while the operator $L = d/dx$ with the inner product (8.37) is not. In fact,

$$\left(p, \frac{df}{dx}\right) = \int_{-w}^{w} p(x) \frac{df(x)}{dx}\, dx = [p(x)f(x)]_{x=-w}^{w}$$
$$- \int_{-w}^{w} \frac{dp(x)}{dx} f(x)\, dx = [p(x)f(x)]_{x=-w}^{w} - \left(\frac{dp}{dx}, f\right) \tag{8.46}$$

If the domain of $L = d/dx$ is restricted to functions that obey certain boundary conditions [$f(-w) = f(w) = 0$, say], then we see that $L^\dagger = -L$ for this operator. On the other hand, the operator $L = d^2/dx^2$ subject to the same boundary conditions can be shown to be self-adjoint. These examples emphasize that the boundary conditions are an essential part of the definition of a differential operator; the specification of the domain of the operator is as important as the "formal" definition in terms of derivatives.

We now define an operator equation adjoint to the original equation (8.29):

$$L^\dagger p = h \tag{8.47}$$

The solution p of this equation has the property that

$$F = (h, f) = (L^\dagger p, f) = (p, Lf) = (p, g) \tag{8.48}$$

are all equivalent ways of expressing the functional F if f is the solution of (8.29). In essence, this property is the analog of Lorentz reciprocity. Indeed, Lorentz reciprocity is (8.48) when L is chosen to be the Maxwell differential operator. To see this, write Maxwell's curl equations in the matrix-operator form

$$Lf \equiv Df - \omega Mf = g \tag{8.49}$$

where

$$f = \begin{bmatrix} \mathbf{E} \\ \mathbf{H} \end{bmatrix} \tag{8.50}$$

$$D = j \begin{bmatrix} 0 & -\nabla \times \\ \nabla \times & 0 \end{bmatrix} \tag{8.51}$$

$$M = \begin{bmatrix} \epsilon & 0 \\ 0 & \mu \end{bmatrix} \tag{8.52}$$

$$g = -j \begin{bmatrix} \mathbf{J} \\ \mathbf{M} \end{bmatrix} \tag{8.53}$$

Define the indefinite scalar product

$$(f, g) = \int_V (\mathbf{E} \cdot \mathbf{J} - \mathbf{H} \cdot \mathbf{M}) \, dV \tag{8.54}$$

where the integral is carried out over a volume V bounded by a surface S. Denoting

$$f = \begin{bmatrix} \mathbf{E}^a \\ \mathbf{H}^a \end{bmatrix}; \qquad p = \begin{bmatrix} \mathbf{E}^b \\ \mathbf{H}^b \end{bmatrix}$$

it can be shown that

$$(p, Lf) = (Lp, f) - j \int_S \left(\mathbf{E}^a \times \mathbf{H}^b - \mathbf{E}^b \times \mathbf{H}^a \right) \cdot \mathbf{u}_n \, dS \tag{8.55}$$

Under appropriate boundary conditions on S (such as those used to prove the radiation condition in Section 3.2.3, for example), the surface integral can be shown to vanish, and under such conditions the operator L is self-adjoint. Equation (8.48) is thus seen to be equivalent to (3.58).

A significant property of operator equations is the principle known as the *Fredholm alternative*. It is important to know whether an equation such as (8.29) has a unique solution or not. The answer to this question depends on whether the homogeneous version of the equation,

$$L f_e = 0 \tag{8.56}$$

has a nontrivial solution or not. If it does, we say that 0 is an eigenvalue of the operator L, corresponding to an eigenfunction $f_e \neq 0$. In such a case, if a solution f of (8.29) exists, then for any constant A, the function $f + A f_e$ is also a solution, and the solution is clearly not unique. A similar statement holds for (8.47). The Fredholm alternative states that *either*:

The operator equations

$$L f = 0 \tag{8.57}$$

and

$$L^\dagger p = 0 \tag{8.58}$$

have only the trivial solutions $f = 0$ and $p = 0$, in which case equations (8.29) and (8.47) have solutions, which are unique;

or:

Equations (8.57) and (8.58) have nontrivial solutions f_e and p_e, in which case (8.29) has a solution only if

$$(g, p_e) = (Lf, p_e) = (f, L^\dagger p_e) = 0 \qquad (8.59)$$

for any p_e that satisfies (8.58), and likewise (8.47) has a solution only if

$$(h, f_e) = 0 \qquad (8.60)$$

for any f_e that satisfies (8.57). Equations (8.59)-(8.60) are known as solvability conditions for (8.29) and (8.47).

A perturbation method[2] can now be described for the approximate calculation of the functional F. Let the operator L have an expansion

$$L = L_0 + L_1 + L_2 + \cdots \qquad (8.61)$$

where each L_i is "smaller" in some sense than the previous one L_{i-1}. This expansion is analogous to (8.7); indeed, we can take L_i to be the integral operator with kernel K_i. We now assume that f has the expansion (8.8), and likewise

$$p(x) = p_0(x) + p_1(x) + p_2(x) + \cdots \qquad (8.62)$$

We substitute these expansions into (8.29), and group terms of equal smallness to get the operator analog of (8.9):

$$L_0 f_0 = g \qquad (8.63)$$

$$L_0 f_1 = -L_1 f_0 \qquad (8.64)$$

$$L_0 f_2 = -L_1 f_1 - L_2 f_0 \qquad (8.65)$$

and so on. Likewise,

$$L_0^\dagger p_0 = h \qquad (8.66)$$

$$L_0^\dagger p_1 = -L_1^\dagger p_0 \qquad (8.67)$$

$$L_0^\dagger p_2 = -L_1^\dagger p_1 - L_2^\dagger p_0 \qquad (8.68)$$

This perturbation procedure can work only if the appropriate solvability conditions from the Fredholm alternative are satisfied. For example, (8.64) has a

---

[2]B. Noble, in *Electromagnetic Waves*, R. E. Langer, ed. Madison, WI: University of Wisconsin Press, 1962, pp. 323-360.

solution only if

$$(L_1 f_0, p_e) = 0 \tag{8.69}$$

for any p_e such that $L^\dagger p_e = 0$.

The functional F then has the expansion

$$F = (h, f) = (h, f_0) + (h, f_1) + \cdots \tag{8.70}$$

The second order term in the expansion (8.70) can be expressed as

$$(h, f_1) = (L_0^\dagger p_0, f_1) = (p_0, L_0 f_1) = -(p_0, L_1 f_0) \tag{8.71}$$

so that only the solution f_0 of the lowest-order (zero-order) operator equation (8.63) (and the solution p_0 of its corresponding adjoint operator equation) are needed to get first-order accuracy for the functional F:

$$F \simeq (h, f_0) - (p_0, L_1 f_0) = (L_0^\dagger p_0, f_0) - (p_0, L_1 f_0) = (p_0, L_0 f_0 - L_1 f_0) \tag{8.72}$$

8.2.1 EXAMPLE: STRIP OVER A GROUND PLANE (REVISITED)

Consider again the thin conducting strip of Figure 8.1. We can identify the various operators and functions from the general perturbation approach given above as follows: $f_0 = \rho_{S0}$ is given by (8.12), $g = 2\pi\epsilon\Phi_{Sc}$, $h = 1$,

$$p_0 = \frac{f_0}{2\pi\epsilon\Phi_{Sc}} \tag{8.73}$$

$$L_0 f = \int_{-w}^{w} f(x') \ln \frac{2h}{|x - x'|} \, dx' \tag{8.74}$$

$$L_1 f = \frac{1}{8h^2} \int_{-w}^{w} f(x')(x - x')^2 \, dx' \tag{8.75}$$

and

$$F = \Phi_{Sc} C_l = (1, f) \tag{8.76}$$

so from (8.76) and (8.72),

$$C_l = 2\pi\epsilon \left[(p_0, L_0 p_0) - (p_0, L_1 p_0) \right] \tag{8.77}$$

The integrals in (8.77) can be readily evaluated as before using the changes of variables $x = w \cos u$ and $x' = w \cos u'$, and the result is exactly (8.22), without having had to solve for f_1 or p_1.

Note that (8.22) becomes infinite when $h = w/4$, a manifestation of the Fredholm alternative, because for this problem we have nontrivial solutions of $L_0 f = 0$ (see (7.113)), and as the reader may verify, the solvability condition (8.69) is not satisfied.

8.3   VARIATIONAL APPROXIMATION[3]

Consider again the expansions (8.8) and (8.61). Suppose that the lowest-order approximation f_0 is chosen to be "close" somehow to the exact solution f of (8.29), and that an approximation p_0 is similarly chosen for p, but that we do not explicitly know a corresponding approximation L_0 that is "close" to L. We suppose only that it exists, and that (8.63) is satisfied for our choice of f_0. We then take $L_1 = L - L_0$, so that L_2, L_3 and all higher terms in the expansion (8.61) are zero. Since our approximate solutions f_0 and p_0 are no longer the first in a sequence of recursively determined functions, we will rename them as f_t and p_t, and refer to them hereafter as "trial functions," hopefully close to the exact solutions we seek.

The first-order accurate approximation (8.72) to F is now

$$
\begin{aligned}
F &\simeq (L_0^\dagger p_t, f_t) + (p_t, L_0 f_t) - (p_t, L f_t) \\
&= (h, f_t) + (p_t, g) - (p_t, L f_t) \\
&\equiv F_0
\end{aligned}
\tag{8.78}
$$

This is different in general from (8.72) because now L_1 is the *entire* correction to L_0, and not just the first term in an expansion. Thus, even when no expansion (8.61) of our operator is explicitly available (i.e., we have no suitable L_0, etc.), if good approximations f_t and p_t to the solutions of (8.29) and (8.47) are available in some way we can still get the benefits of a first-order accurate approximation for F as we did in the perturbation method. This is true despite the fact that the trial functions are only *zeroth-order* approximations. Put another way, the error $F_0 - F$ is of second order, while the errors $f_t - f$ and $p_t - p$ are of first order (much larger). Explicitly (the reader is invited to fill in the algebra),

$$
F_0 = F - (p_t - p, L(f_t - f))
\tag{8.79}
$$

The approximation (8.78) is called a *variational* approximation, and is the basis of many numerical methods for the computation of electromagnetic parameters.

The variational expression (8.78) has the disadvantage that it is dependent on the amplitudes of the trial functions. The variational property of (8.78) can be used to optimize the variational expression by automatically finding the best amplitudes for a given set of trial functions. We proceed as follows. Suppose that f_b and p_b are some given *fixed*-amplitude basis functions, and that we insert the trial functions $f_t = A f_b$ and $p_t = B p_b$ into (8.78), with A and B being adjustable constants. Then (8.78) becomes

$$
F_0 = A(h, f_b) + B(p_b, g) - AB(p_b, L f_b)
\tag{8.80}
$$

---

[3]This section provides only a brief introduction to variational methods. For more detail, see E. F. Kuester and D. C. Chang, *Sci. Rept. No. 80*, Electromagnetics Laboratory, Univ. of Colorado, Boulder, August 1986, and references therein.

Suppose for a moment that the two basis functions f_b and p_b are identical, and that A and B are likewise the same. Then the error expression (8.79) is a quadratic function of A, and with suitable properties of L, will attain a minimum at the value of A for which $\partial F_0/\partial A = 0$. In the most general case, we cannot expect this minimum or extremal property to hold, but because (8.80) is variational, its value tends to "flatten out" near the optimum values of A and B as these amplitudes are varied. We formalize this property by stating that the optimum values of A and B occur when

$$\frac{\partial F_0}{\partial A} = \frac{\partial F_0}{\partial B} = 0 \tag{8.81}$$

(we say that (8.80) has been made *stationary*), from which we find

$$A_{\text{opt}} = \frac{(p_b, g)}{(p_b, Lf_b)}; \qquad B_{\text{opt}} = \frac{(h, f_b)}{(p_b, Lf_b)} \tag{8.82}$$

When these are inserted back into (8.80), we obtain the *amplitude-optimized* variational approximation

$$F_{0,\text{opt}} = \frac{(h, f_b)(p_b, g)}{(p_b, Lf_b)} = \frac{(h, f_t)(p_t, g)}{(p_t, Lf_t)} \tag{8.83}$$

that uses the same information as (8.78), but in the most accurate manner possible. Note that (8.83) is independent of the amplitudes of the trial functions, like (8.78) needs only lowest-order approximations to f and p and requires no approximation to the operator L.

8.3.1 THE GALERKIN-RITZ METHOD

A widely used numerical procedure called the Galerkin-Ritz (or Galerkin, or Galerkin-Petrov) method is a generalization of this procedure. Let f_{bn} and p_{bm}, $m, n = 1, 2, \ldots, N$ be given sets of basis functions. We write our trial functions as linear combinations of them:

$$f_t = \sum_{n=1}^{N} A_n f_{bn} \tag{8.84}$$

$$p_t = \sum_{m=1}^{N} B_m p_{bm} \tag{8.85}$$

where A_n and B_m are constants to be determined. The functional F_0 from (8.78) is now

$$F_0 = \sum_{n=1}^{N} A_n(h, f_{bn}) + \sum_{m=1}^{N} B_m(p_{bm}, g) - \sum_{n=1}^{N}\sum_{m=1}^{N} A_n B_m(p_{bm}, Lf_{bn}) \tag{8.86}$$

We next enforce the stationary conditions

$$\frac{\partial F_0}{\partial A_n} = \frac{\partial F_0}{\partial B_m} = 0 \qquad \text{for all } m, n = 1, 2, \ldots, N \qquad (8.87)$$

from which there result

$$0 = \frac{\partial F_0}{\partial A_n} = (h, f_{bn}) - \sum_{m=1}^{N} B_m (p_{bm}, L f_{bn}) \qquad \text{for all } n = 1, 2, \ldots, N \quad (8.88)$$

$$0 = \frac{\partial F_0}{\partial B_m} = (p_{bm}, g) - \sum_{n=1}^{N} A_n (p_{bm}, L f_{bn}) \qquad \text{for all } m = 1, 2, \ldots, N \quad (8.89)$$

These equations can be written in the compact matrix form

$$[C][A] = [G] \qquad \text{and} \qquad [C]^T[B] = [H] \qquad (8.90)$$

where the superscript ^{T} denotes the matrix transpose,

$$[A] = \begin{bmatrix} A_1 \\ A_2 \\ \vdots \\ A_N \end{bmatrix} \quad [B] = \begin{bmatrix} B_1 \\ B_2 \\ \vdots \\ B_N \end{bmatrix} \quad [G] = \begin{bmatrix} (p_{b1}, g) \\ (p_{b2}, g) \\ \vdots \\ (p_{bN}, g) \end{bmatrix} \quad [H] = \begin{bmatrix} (h, f_{b1}) \\ (h, f_{b2}) \\ \vdots \\ (h, f_{bN}) \end{bmatrix}$$

$$(8.91)$$

are $N \times 1$ column vectors, and

$$[C] = \begin{bmatrix} (p_{b1}, L f_{b1}) & (p_{b1}, L f_{b2}) & \cdots & (p_{b1}, L f_{bN}) \\ (p_{b2}, L f_{b1}) & (p_{b2}, L f_{b2}) & \cdots & (p_{b2}, L f_{bN}) \\ \vdots & \vdots & \ddots & \vdots \\ (p_{bN}, L f_{b1}) & (p_{bN}, L f_{b2}) & \cdots & (p_{bN}, L f_{bN}) \end{bmatrix} \qquad (8.92)$$

is a square $N \times N$ matrix.

Equations (8.90) are readily solved for $[A]$ and $[B]$ by matrix inversion or other suitable means if the matrix $[C]$ is nonsingular. This will be the case if the operator L is invertible (i.e., if (8.57)-(8.58) have only trivial solutions) and the set of basis functions f_{bn} and p_{bm} are chosen properly.[4] Since very efficient algorithms are available for the numerical solution of algebraic equations like (8.90), Galerkin-type methods have become very popular for the approximate solution of many problems in electromagnetics.[5] With the values of $[A]$ and $[B]$

---

[4]This question is addressed in S. G. Mikhlin and K. L. Smolitskiy, *Approximate Methods for Solution of Differential and Integral Equations.* New York: American Elsevier, 1967.

[5]See, for example, R. F. Harrington, *Field Computation by Moment Methods.* Malabar, FL: Krieger, 1968.

thus obtained, the corresponding value of the functional F_0 can be computed by substitution back into (8.86) and using (8.90):

$$F_0 = [H]^T [C]^{-1} [G] \tag{8.93}$$

This generalization of (8.83) is seen to be a special case of the Galerkin-Ritz approximation for $N = 1$.

8.3.2 EXAMPLE: STRIP OVER A GROUND PLANE (RE-REVISITED)

We address one more time the thin conducting strip problem of Figure 8.1. Following similar calculations to those in Section 8.2.1, we obtain from (8.78) the following (unoptimized) variational approximation to the capacitance per unit length:

$$C_l = \frac{2\pi\epsilon}{\ln(4h/w)} \left\{ 1 - \frac{1}{2\pi^2 \ln(4h/w)} \int_0^\pi \int_0^\pi \ln\left[1 + \frac{w^2}{4h^2}(\cos u - \cos u')^2\right] du' du \right\} \tag{8.94}$$

The corresponding Galerkin-Ritz variational expression that comes from (8.83) is:

$$C_l = \frac{2\pi\epsilon}{\ln(4h/w) + \frac{1}{2\pi^2} \int_0^\pi \int_0^\pi \ln\left[1 + \frac{w^2}{4h^2}(\cos u - \cos u')^2\right] du' du} \tag{8.95}$$

The double integral can be carried out by numerical integration, or by expansion of the integrand into a power series in $(w/2h)^2$ and integrating term by term.

In Figure 8.2 we compare the various approximations we have found for the capacitance of this problem. We see that our lowest-order perturbation approximation (8.13) is quite accurate even for fairly wide strips relative to h. With the first-order perturbation approximation (8.22), the error is only about 3% even for $2w/h = 2$. The unoptimized variational approximation (8.94) is a definite improvement over (8.22), but all three of these approximations are seen to break down for large enough values of w/h. In fact, they all have singularities for $w/h = 4$ (where (8.63) fails to have a solution according to the Fredholm alternative), and so must be expected to break down well before that value is reached. On the other hand, the proper adjustment of trial function amplitude that goes into the Galerkin-Ritz approximation (8.95) is seen to repair this defect, and actually retains dramatically improved accuracy far beyond the range of small w/h for which the trial functions were originally constructed. In fact, even for $w/h = 10$, the error in C_l is less than 10%. This is a characteristic feature of the optimized variational approximation— it retains accuracy far beyond what would reasonably be expected from the information that goes into it.

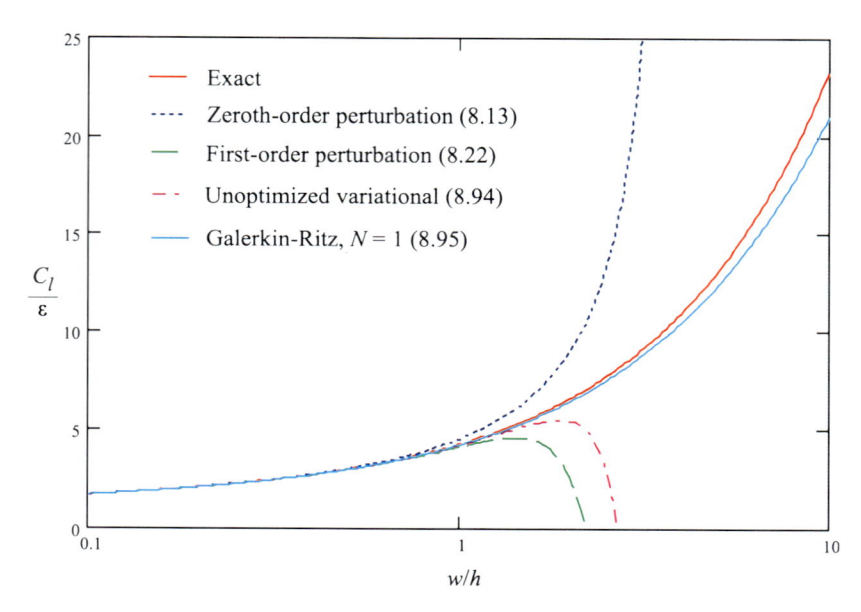

Figure 8.2 Comparison of exact values, perturbation approximations, and variational approximations for capacitance per unit length of strip over ground plane, normalized to ϵ.

8.4 PROBLEMS

8–1 (a) If $|\boldsymbol{\rho}-\boldsymbol{\rho}'| << 2d$, show that the Green's function of Problem 7-2 can be written approximately as

$$G \simeq \frac{1}{2\pi(\epsilon_r + 1)|\boldsymbol{\rho} - \boldsymbol{\rho}'|} - \frac{\epsilon_r}{2\pi d(\epsilon_r^2 - 1)} \ln\left(\frac{2\epsilon_r}{\epsilon_r + 1}\right)$$

(b) Suppose that a planar, perfectly conducting disk of surface S, when placed in free space perpendicular to the z-axis and charged to unit voltage with respect to infinity, has the total surface charge density $\rho_{S0}(\boldsymbol{\rho})$ and the resulting capacitance

$$C_0 = \int_S \rho_{S0}(\boldsymbol{\rho})\, dS$$

If this same disk is placed on the top of the dielectric slab of Problem 7-2, and if the "diameter" (maximum distance between any two points) of the disk is small compared to d, show that the total charge density $\rho_S(\boldsymbol{\rho})$ on the disk is approximately proportional to $\rho_{S0}(\boldsymbol{\rho})$, and show that the capacitance of the disk is now

$$C \simeq \frac{\epsilon_r + 1}{2} \frac{C_0}{1 - \dfrac{C_0}{4\pi\epsilon_0 df(\epsilon_r)}}$$

where $f(\epsilon_r) = (\epsilon_r - 1)/[\epsilon_r \ln(\frac{2\epsilon_r}{\epsilon_r+1})]$.

8–2 (a) Obtain the next term in the approximation of the kernel of (8.11) as:

$$K_2 = -\frac{1}{4}\left[\frac{(x - x')^4}{(2h)^4}\right]$$

(b) Find ρ_{S2}.

(c) Find C_l to second order in $(w/h)^2$ using the results of (a) and (b). Plot this vs. w/h, and compare it to the zero-order and first-order approximations (8.13) and (8.22).

8–3 Provide the details of the derivation of (8.28).

8–4 Prove the Schwarz inequality (8.43). Under what condition does the equal sign hold in (8.43)? [Hint: examine the norm of the function $f + Ag$ for any constant A, and find the value of A that minimizes that norm.]

8–5 Consider the inner product

$$(f, g) \equiv \int_a^b f(x)g^*(x)\, dx$$

where f and g are *complex* functions, and * denotes the complex conjugate. Prove the Schwarz inequality (8.43) for this inner product. Under what condition does the equal sign hold? [Hint: examine the norm of the function $f + Ag^*$ for any constant A, and find the value of A that minimizes that norm.]

8–6 Prove (8.55).

8–7 Provide the details of the derivations of (8.94) and (8.95).

8–8 Compute explicitly the Galerkin-Ritz approximation (8.93) for the functional F_0 in the case $N = 2$.

A Generalized Functions

A.1 INTRODUCTION

When solving certain electromagnetic radiation and scattering problems, it is common practice to simplify or idealize the problems by examining a limiting case of an actual source distribution. An example encountered early in the study of electromagnetic fields is the concept of a surface current—one that exists only on a surface (of infinitesimal thickness). In similar fashion, we may idealize the current on a wire into that of a line source—current on a wire of zero radius. In these and certain other cases, the volume current density must become infinite on a certain set of points, while the total current (the integral of this density) through a transverse surface remains finite.

Of interest is a way to mathematically express this type of source distribution. It cannot be done with ordinary functions defined in the context of classical analysis. It *can* be done using what are called *generalized functions*.

Generalized functions[1] can be defined in a variety of ways, the most physically intuitive of which is through the use of limits as we shall do here. Consider a sequence of ordinary functions $f_n(x)$ as suggested by Figure A.1. There may be no classical sense in which the sequence $f_n(x)$ approaches a well-defined limit function $f(x)$, but instead the limit

$$\lim_{n \to \infty} \int_{-\infty}^{\infty} f_n(x)g(x)\,dx$$

may exist for every "good" function[2] $g(x)$. We will say that the sequence f_n converges *weakly* to the generalized function $f(x)$ by *defining* the integral of

---

[1]See

M. J. Lighthill, *Introduction to Fourier Analysis and Generalised Functions*. Cambridge: Cambridge University Press, 1958.

M. Bouix, *Les Fonctions Généralisées ou Distributions*. Paris: Masson et Cie., 1964.

B. W. Roos, *Analytic Functions and Distributions in Physics and Engineering*. New York: Wiley, 1969.

V. S. Vladimirov, *Generalized Functions in Mathematical Physics*. Moscow: Mir, 1979.

D. S. Jones, *The Theory of Generalised Functions*. Cambridge: Cambridge University Press, 1982.

A. H. Zemanian, *Distribution Theory and Transform Analysis*. New York: Dover, 1987.

R. P. Kanwal, *Generalized Functions: Theory and Applications*. Boston: Birkhäuser, 2004.

[2]We call a function $g(x)$ "good" if it is infinitely differentiable, and together with all its derivatives decays faster than *any* algebraic power of $|x|$: $|x|^{-N}$ ($N > 0$) as $|x| \to \infty$.

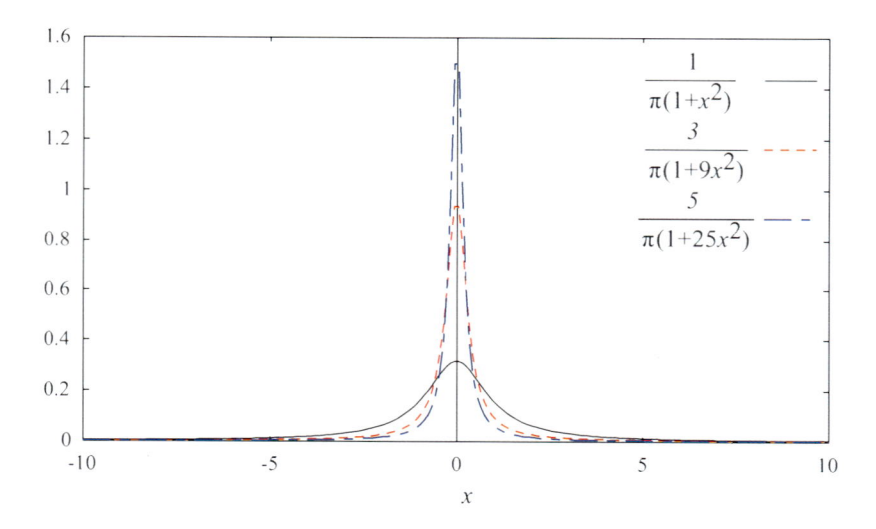

Figure A.1 Sequence of ordinary functions tending to a generalized function.

fg to be the limit:

$$\int_{-\infty}^{\infty} f(x)g(x)\, dx \equiv \lim_{n \to \infty} \int_{-\infty}^{\infty} f_n(x)g(x)\, dx \tag{A.1}$$

for any good function $g(x)$. In the classical theory of generalized functions (sometimes called *distribution theory*, as originated by Laurent Schwartz), this is the only meaning we can give to the generalized function f.

Note then that an ordinary function can also be a generalized function, but that we treat it differently in this context than when we regard it as a classical function. For example, $f(x) \equiv 1$ defines (pointwise) an ordinary function, which we could also understand as a generalized function defined by:

$$\int_{-\infty}^{\infty} f(x)g(x)\, dx = \int_{-\infty}^{\infty} g(x)\, dx \tag{A.2}$$

for all good $g(x)$. Consider now the function $f_1(x)$ defined to be 1 if x is not equal to an integer, and zero if x is equal to an integer. Clearly, $f(x) \neq f_1(x)$ in the sense of ordinary functions, but because of the properties of Riemann integrals, f and f_1 *are* (weakly) equal in the sense of generalized functions: We write $f \approx f_1$.

An example of such a function is $\exp[-(x - x_0)^2/a^2]$ for any fixed constants a and x_0. For some purposes, we need not have a function decay at infinity, but only need to have it grow no faster than algebraically. In this case we will say that the function is "fairly good."

Though this mathematical definition of a generalized function may seem at first to be rather artificial and nit-picking, it is actually closer in some ways to physical reality than is the classical pointwise definition of an ordinary function. When we consider, for example, the voltage $V(x)$ on a transmission line aligned along the x-direction, defined for every point x of the line, we are talking about a thing we will never be able to measure precisely. This is because any real measurement device must occupy a nonzero range of x values, and so will respond not only to the voltage at one particular point, but to the voltages within the range occupied by the measuring device (to say nothing of the fact that the device will itself disturb the very voltage we are trying to measure). The response of the measuring device is thus a weighted average of voltage values in the vicinity of the device; we can regard this weight function as a good function $M(x)$ that characterizes the device. The device response $R_M(V)$ then has the form

$$R_M(V) = \int_{-\infty}^{\infty} V(x)M(x)\,dx \tag{A.3}$$

Two voltages $V_1(x)$ and $V_2(x)$ are said to be weakly equal in the sense of generalized functions if their responses $R_M(V_1)$ and $R_M(V_2)$ with respect to *any* given measurement weight function $M(x)$ are the same. In this way, we allow for a voltage (or any other physical quantity) to have a more general behavior than that exhibited by classical functions.

One can prove that many properties of good functions also extend to generalized functions. In fact, we can define the derivative of a generalized function $f(x)$ either as the limit of the sequence of derivatives $f_n'(x)$, or by requiring that integration by parts formulas hold as if ordinary functions were involved. Thus, if $g(x)$ is a good function and $f(x)$ is a generalized function,

$$\int_{-\infty}^{\infty} f'(x)g(x)\,dx \equiv -\int_{-\infty}^{\infty} f(x)g'(x)\,dx \tag{A.4}$$

(for all good $g(x)$) defines the weak derivative $f'(x)$ of f.

Example 1 Let $f(x) = \vartheta(x)$, the Heaviside unit step function

$$\vartheta(x) = \begin{cases} 1 & (x > 0) \\ 0 & (x < 0) \end{cases} \tag{A.5}$$

(which could also be viewed as an ordinary function). As a generalized function, $\vartheta(x)$ needs no specification of its value at $x = 0$ (at least in terms of the weak concept of equality, which is the only one we have considered so far), since this would make no difference to the value of an integral involving $\vartheta(x)$ as defined by (A.1). By (A.4), we can define

$\vartheta'(x)$ by

$$
\begin{aligned}
\int_{-\infty}^{\infty} \vartheta'(x)g(x)\,dx &= -\int_{-\infty}^{\infty} \vartheta(x)g'(x)\,dx \\
&= -\int_{0}^{\infty} g'(x)\,dx \qquad\qquad\text{(A.6)} \\
&= g(0)
\end{aligned}
$$

That is, the result of multiplying any $g(x)$ by $\vartheta'(x)$ and integrating is to pick out the value of g at $x = 0$. This property characterizes the so-called *Dirac delta-function* (actually a generalized function):

$$
\int_{-\infty}^{\infty} \delta(x)g(x)\,dx = g(0) \qquad\qquad\text{(A.7)}
$$

for all good $g(x)$. Thus we have $\vartheta'(x) \approx \delta(x)$. In similar fashion, we can define derivatives of $\delta(x)$, e.g.,

$$
\int_{-\infty}^{\infty} \delta'(x)g(x)\,dx = -g'(0) \qquad\qquad\text{(A.8)}
$$

and so on.   □

Example 2 Consider an ordinary function $F(x)$ such that

$$
\int_{-\infty}^{\infty} F(x)\,dx = q \qquad\qquad\text{(A.9)}
$$

and put

$$
f_n(x) = nF(nx) \qquad\qquad\text{(A.10)}
$$

(see Figure A.1). Then

$$
\lim_{n\to\infty} f_n(x) \approx q\delta(x) \qquad\qquad\text{(A.11)}
$$

in the sense of generalized functions. To see this, put

$$
\begin{aligned}
G_n(x) &= \int_{-1}^{x} f_n(y)\,dy \\
&= \int_{-1}^{x} nF(ny)\,dy \\
&= \int_{-n}^{nx} F(u)\,du
\end{aligned}
$$

If $x < 0$, then the limit of G_n as $n \to \infty$ will be zero (if it wasn't, then q in (A.9) would have been infinite). On the other hand, if $x > 0$, this limit is equal to q. Hence,

$$
\lim_{n\to\infty} G_n(x) \approx q\vartheta(x) \qquad\qquad\text{(A.12)}
$$

and by the previous example, we can take derivatives and get

$$\lim_{n \to \infty} nF(nx) \approx q\delta(x) \tag{A.13}$$

in the sense of generalized functions. □

Although this definition of a generalized function does not strictly make it possible to evaluate one at a point, we can define a sort of "set" of its values on an open set of points.[3] For example, a generalized function $f(x)$ is said to equal 0 for $x > 0$ if

$$\int_{-\infty}^{\infty} f(x)g(x)\,dx = 0$$

for all good $g(x)$ that obey $g(x) \equiv 0$ for $x \leq 0$. Thus, $\delta(x) = 0$ for $x < 0$ and $x > 0$.

Changes of variable are usually straightforward. Thus, for example, $\delta(ax)$, where $a > 0$ is a constant, obeys

$$\int_{-\infty}^{\infty} g(x)\delta(ax)\,dx = \frac{1}{a}\int_{-\infty}^{\infty} g\left(\frac{u}{a}\right)\delta(u)\,du = \frac{1}{a}g(0)$$

for all good $g(x)$. Hence we conclude that

$$\delta(ax) = \frac{1}{a}\delta(x) \tag{A.14}$$

If $p(x)$ is a monotonically increasing real function of x with a single zero at $x = x_0$, then

$$\delta(p(x)) = \frac{1}{p'(x_0)}\delta(x - x_0) \tag{A.15}$$

If $p(x)$ were instead monotonically decreasing in x, then we have instead

$$\delta(p(x)) = -\frac{1}{p'(x_0)}\delta(x - x_0) \tag{A.16}$$

More generally, if $p(x)$ is a real function with two zeros (at x_{01} and x_{02}, say), then

$$\delta(p(x)) = \frac{1}{|p'(x_{01})|}\delta(x - x_{01}) + \frac{1}{|p'(x_{02})|}\delta(x - x_{02}) \tag{A.17}$$

with an obvious generalization to the case when p has more than two zeroes.

---

[3]At the expense of rather cumbersome mathematical technicalities, point values of generalized functions *can* be defined. Essentially (and very roughly), we do this by defining certain types of "infinity" as allowed values of the function. See

H. A. Biagioni, *A Nonlinear Theory of Generalized Functions*, Lecture Notes in Mathematics No. 1421. Berlin: Springer-Verlag, 1990.

Just as the Dirac delta function and its derivatives are approximated by a sequence of progressively narrower ordinary functions, an ordinary function that is in some sense concentrated near a single point can be approximated by a series of delta functions and their derivatives. Let $f(x)$ be a generalized function. Using the techniques already described, we can derive the Taylor series expansion

$$f(x - x') \approx \sum_{n=0}^{\infty} \frac{1}{n!} \left(-x' \frac{d}{dx} \right)^n f(x) \tag{A.18}$$

where x' is independent of x. Next, let $g(x)$ be a good function. Using the sifting property of the delta function, and applying (A.18) to $\delta(x - x')$, we obtain

$$g(x) = \int g(x')\delta(x - x') \, dx' = \sum_{n=0}^{\infty} (-1)^n \frac{g_n}{n!} \left(\frac{d}{dx} \right)^n \delta(x) \tag{A.19}$$

where

$$g_n = \int (x')^n g(x') \, dx'$$

are the nth order moments of the function $g(x)$, the integrations in each case being carried out over all x'. If $g(x)$ is concentrated in a narrow region about the point $x = 0$, the g_n should decrease rapidly with increasing n, and g is then accurately represented by the first few terms of the series (A.19). It should be noted that by considering a sequence of good functions converging to a generalized function, (A.19) can be given validity even when g is a generalized function, provided the moments g_n can be appropriately defined.

A.2   MULTIPLICATION OF GENERALIZED FUNCTIONS

The definition of the product of a generalized function with a good function is straightforward. If q is a good function and f a generalized function, then we put

$$\int_{-\infty}^{\infty} [f(x)q(x)] \, g(x) \, dx = \int_{-\infty}^{\infty} f(x) \, [q(x)g(x)] \, dx \tag{A.20}$$

for all good functions g. The integral on the right side of (A.20) is well defined because the product of the two good functions g and q is always a good function.

Example 3 We have

$$g(x)\delta(x) = g(0)\delta(x) \tag{A.21}$$

for any good function g.

Example 4 A "ramp function" $R(x)$ equal to x for $x > 0$ but vanishing for $x < 0$ can be defined as:

$$R(x) = x\vartheta(x) \qquad (A.22)$$

The derivative of the ramp function can then be calculated straightforwardly:

$$R'(x) = x\delta(x) + \vartheta(x) = \vartheta(x) \qquad (A.23)$$

as a consequence of (A.21).

Multiplication of two generalized functions has historically proved rather hard to define in general. It is not possible to define a multiplication rule for all generalized functions such that *all* properties of the product obeyed by ordinary functions are preserved, so long as the only notion of equality for two generalized functions is that of weak equality $\approx$ implied in (A.1). This is because only good functions appear multiplying the sequence f_n that converges to the generalized function f, while to define the product of any two generalized functions, we need at least to examine the limit of an integral involving the product of f_n with a generalized function.[4]

As an example, consider the two sequences illustrated in Figure A.2 that converge to Heaviside unit step functions $\vartheta_{(\alpha)}(x)$ and $\vartheta_{(\beta)}(x)$ in the sense of (A.1).[5] The subscripts $_{(\alpha)}$ and $_{(\beta)}$ are meant to carry information on the sequences themselves; that is, these subscripts help us "remember" how the sequences approached step functions "microscopically" near $x = 0$ as $n \to \infty$. As far as what has been discussed up to this point is concerned, these two step functions are the same. However, if we ask what is the product of each of these generalized functions with a delta-function $\delta(x)$ (an operation that is as yet undefined), the apparent answer is that $\vartheta_{(\alpha)}(x)\delta(x) = \frac{1}{2}\delta(x)$, while

---

[4]For the method we use here, see:

B. Fisher, *Quart. J. Math.* (2), vol. 22, pp. 291-298, 1971.

D. S. Jones, *Quart. J. Math.* (2), vol. 24, pp. 145-163, 1973.

D. S. Jones, *Proc. Roy. Soc. London*, vol. A371, pp. 479-508, 1980.

H. A. Biagioni, *loc. cit.*

J.-F. Colombeau, *J. Acoustique*, vol. 1, pp. 9-14, 1988.

Y. A. Barka, J.-F. Colombeau, and B. Perrot, *J. Acoustique*, vol. 2, pp. 333-346, 1989.

J.-F. Colombeau, *Bull. Amer. Math. Soc.*, vol. 23, pp. 251-268, 1990.

Yu. V. Egorov, *Russ. Math. Surveys*, vol. 45, no. 5, pp. 1-49, 1990.

M. Oberguggenberger, *Multiplication of Distributions and Applications to Partial Differential Equations*. Harlow, UK: Longman, 1992.

J. F. Colombeau, *Multiplication of Distributions*, Lecture Notes in Mathematics No. 1532. Berlin: Springer-Verlag, 1992.

A. Gsponer, *Eur. J. Phys.*, vol. 30, pp. 109-126, 2009.

[5]In principle, we should multiply the functions shown by some good function that is equal to 1 up to some large value of x, and then falls to zero as $x \to \infty$, in order that each of the functions in this sequence is truly a good function.

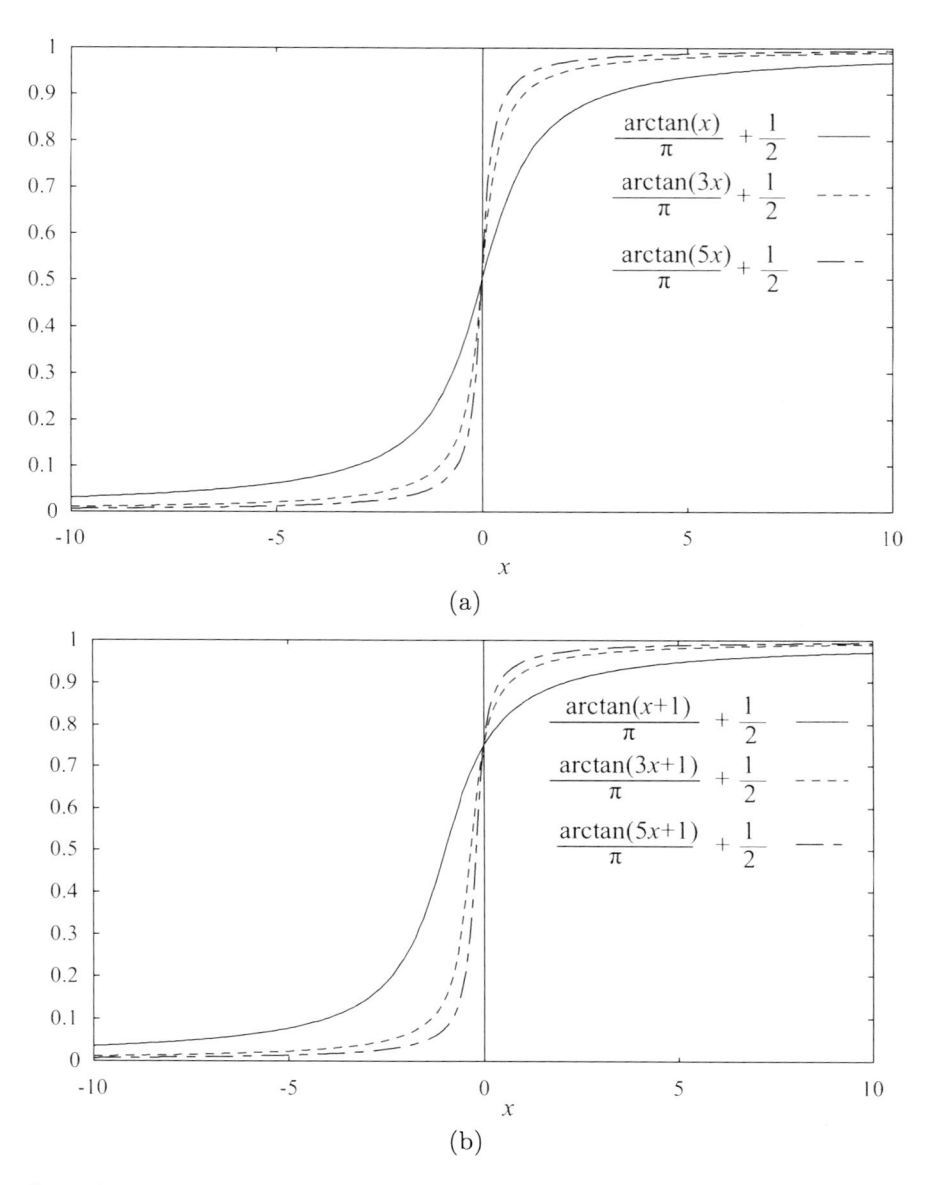

Figure A.2 Sequences of functions converging to two Heaviside unit step functions as $n \to \infty$: (a) $\vartheta_{(\alpha),n} \to \vartheta_{(\alpha)}$ and (b) $\vartheta_{(\beta),n} \to \vartheta_{(\beta)}$.

$\vartheta_{(\beta)}(x)\delta(x) = \frac{3}{4}\delta(x)$. We are led to the conclusion that, despite the fact that both step functions are the same in the sense of (A.1), they cannot be equal if we insist on being able to multiply them with other generalized functions, at least as we have presented the idea here.

One answer to this dilemma is to acknowledge that (A.1) is not a strong enough condition to define an equality that is up to this task. Instead, it only defines a weak equality condition that we will call "association" (not to be confused with the associative property of multiplication or addition). We have said previously that two generalized functions defined as limits of sequences of good functions $f_{(\alpha),n}$ and $f_{(\beta),n}$ are *associated with* (or *weakly equal to*) each other (denoted by $f_{(\alpha)} \approx f_{(\beta)}$) if for all good functions g, the limit of the left side of (A.1) is the same for $f_n = f_{(\alpha),n}$ as it is for $f_n = f_{(\beta),n}$. We will define true (*strong*) equality $f_{(\alpha)} = f_{(\beta)}$, on the other hand, by requiring that their defining sequences obey

$$f_{(\alpha),n}(x) = f_{(\beta),n}(x) \tag{A.24}$$

for all large enough values of n. What this means, roughly, is that it matters not only what the sequences converge to, but how they got there as well.

Having defined strong equality of two generalized functions, we may now define the product of *any* two generalized functions as follows. Let f_n and t_n be two sequences whose limits are the generalized functions f and t. The product $s = ft = tf$, if it exists,[6] is defined as the limit of the sequence $s_n = f_n t_n$:

$$s = \lim_{n \to \infty} f_n t_n \tag{A.25}$$

with equality understood in the sense of (A.24) above. We would say $s \approx ft$ if merely

$$s \approx \lim_{n \to \infty} f_n t_n \tag{A.26}$$

with weak equality understood in the sense of (A.1).

Note that strong equality works as we expect equality to do in ordinary algebra and calculus. However, we must be careful when manipulating equations of association. For example, $f_{(\alpha)} \approx f_{(\beta)}$ does *not* imply that $gf_{(\alpha)} \approx gf_{(\beta)}$ for any generalized function g. Indeed, there are generalized functions $g \approx 0$ for which g^2 is not weakly equal to zero. However, if g is a good function, then it *is* true that $gf_{(\alpha)} \approx gf_{(\beta)}$. Moreover, if $f_{(\alpha)} \approx f_{(\beta)}$, then $f'_{(\alpha)} \approx f'_{(\beta)}$; that is, differentiation works as expected for weak equality as well as for strong equality.

We are led to the conclusion that, if multiplication of arbitrary generalized functions is to be permitted, then there may exist for example not merely *one* unit step function or *one* delta function, but associated classes of them, all behaving identically as far as (A.1) is concerned, but that do not satisfy (A.24) among different members of the same class. Thus, different members of the same class cannot be expected to behave the same when multiplied by other generalized functions. In some applications, there will be no need to

---

[6]If the limit is infinite, as is the case when f and t are both delta functions, the product can still be defined as a sort of "generalized number," in a sense made more precise in the work of Colombeau (*loc. cit.*). We do not require this refinement of the theory for our applications.

distinguish between strong and weak equality of generalized functions. We will simply refer to "the" step function or "the" delta function when no ambiguity can arise. However, in those cases where the distinction is important, it is well to remember that the central issue is in how the defining sequence f_n for the generalized function approaches its limit. Often, this is connected with some physical insight into the details of the process that is being idealized by modeling it with a generalized function. Let us try to clarify these points through some examples.

Example 5 An important product arising in applications is $\vartheta(x)\delta(x)$. As pointed out above, this cannot be done without specifying "which" step function and delta function are involved.

Let $G_{(\alpha),n}$ be a sequence converging to some unit step function $\vartheta_{(\alpha)}$. Then the derivatives $G'_{(\alpha),n}$ of this sequence converge to a corresponding delta function $\delta_{(\alpha)}$. Now

$$s_n \equiv G_{(\alpha),n} G'_{(\alpha),n} = \frac{1}{2} \left[(G_{(\alpha),n})^2 \right]'$$

and $(G_{(\alpha),n})^2$ converges to *some* unit step function $\vartheta_{(\beta)}$ for $x \neq 0$; that is, it approaches the association class of unit step functions weakly. We conclude that

$$\vartheta_{(\alpha)}(x)\delta_{(\alpha)}(x) = \frac{1}{2}\delta_{(\beta)}(x) \tag{A.27}$$

We note that (A.27) could have been obtained by taking the derivative of

$$[\vartheta_{(\alpha)}(x)]^2 = \vartheta_{(\beta)}(x)$$

and using the product rule for differentiation. However, the statement $\vartheta^2(x) = \vartheta(x)$ sometimes quoted is not correct, and we have merely

$$\vartheta^2(x) \approx \vartheta(x)$$

It is because of (A.27) that we sometimes see the value of $\frac{1}{2}$ assigned to $\vartheta(0)$. Although this has no direct meaning in the sense of weak equality of generalized functions implied by (A.1), it may be given one in the sense of strong equality as in (A.24). Note, however, that if $\alpha \neq \beta$, then no matter what γ is,

$$\vartheta_{(\alpha)}(x)\delta_{(\beta)}(x) \neq \frac{1}{2}\delta_{(\gamma)}(x)$$

in general, and that the actual value of this product, $A\delta_{(\gamma)}(x)$ for some constant A, will depend on how the sequences defining these generalized functions approach their limits. Note too that we have not given any indication of *which* delta function $\delta_{(\beta)}(x)$ in the association class (A.27) holds true for. Usually it does not matter, since weak equality is all that is needed once a product of generalized functions has been evaluated. To find $\delta_{(\beta)}$ would require more specific knowledge of the sequence $G_{(\alpha),n}$.

We next examine how products of the form $\vartheta_{(\alpha)}(x)\delta_{(\beta)}(x)$ can take on different values depending on how the sequences representing the two functions approach their limits. Consider two ordinary good functions F_1 and F_2 such that

$$\int_{-\infty}^{\infty} F_1(x)\,dx = \int_{-\infty}^{\infty} F_2(x)\,dx = 1 \tag{A.28}$$

Then (see Example 2), we have

$$f_n^{(1)}(x) \equiv nF_1(nx) \to \delta_{(1)}(x) \quad \text{and} \quad f_n^{(2)}(x) \equiv nF_2(nx) \to \delta_{(2)}(x) \tag{A.29}$$

as $n \to \infty$. If further we define

$$G_n^{(1,2)}(x) \equiv \int_{-\infty}^{x} f_n^{(1,2)}(y)\,dy = \int_{-\infty}^{nx} F_{1,2}(u)\,du \tag{A.30}$$

then

$$\lim_{n\to\infty} G_n^{(1,2)}(x) = \vartheta_{(1,2)}(x)\,du \tag{A.31}$$

are the step functions corresponding to (A.29).

Now let us compute the limit of

$$I = \int_{-\infty}^{\infty} g(x)G_n^{(1)}(x)F_m^{(2)}(x)\,dx \tag{A.32}$$

for any good function $g(x)$ as m and n approach ∞, in order to impart a meaning to the product $\vartheta_{(1)}(x)\delta_{(2)}(x)$. Using the definitions of $G_n^{(1)}$ and $F_m^{(2)}$ above, we have

$$I = \int_{-\infty}^{\infty} g(x)\left[\int_{-\infty}^{x} nF_1(ny)\,dy\right] mF_2(mx)\,dx \tag{A.33}$$

Applying the change of variables $u = mx$ and $v = ny$ gives

$$I = \int_{-\infty}^{\infty} g\left(\frac{u}{m}\right) F_2(u) \left[\int_{-\infty}^{\frac{n}{m}u} F_1(v)\,dv\right] du \tag{A.34}$$

Let us now examine various ways of taking the limit $(m,n) \to \infty$.

1. If $n \to \infty$ with m fixed,

$$I \to \int_{-\infty}^{\infty} g\left(\frac{u}{m}\right) F_2(u)\,du$$

Now letting $m \to \infty$, we get $I \to g(0)$ and hence

$$\vartheta_{(1)}(x)\delta_{(2)}(x) \approx \delta(x) \tag{A.35}$$

for some delta function $\delta(x)$.

2. If $m \to \infty$ with n fixed,

$$I \to \int_{-\infty}^{\infty} g\left(0\right) F_2(u) \left[\int_{-\infty}^{0} F_1(v)\, dv\right] du = Ag(0)$$

where

$$A = \int_{-\infty}^{0} F_1(v)\, dv$$

In this case,

$$\vartheta_{(1)}(x)\delta_{(2)}(x) \approx A\delta(x) \tag{A.36}$$

3. If we maintain $\frac{n}{m} = b$ as $(m, n) \to \infty$, where b is a positive constant, then

$$I = \int_{-\infty}^{\infty} g\left(\frac{u}{m}\right) F_2(u) \left[\int_{-\infty}^{bu} F_1(v)\, dv\right] du$$

As $m \to \infty$,

$$I \to \int_{-\infty}^{\infty} g\left(0\right) F_2(u) \left[\int_{-\infty}^{bu} F_1(v)\, dv\right] du = A_b g(0)$$

where now

$$A_b = \int_{-\infty}^{\infty} F_2(u) \left[\int_{-\infty}^{bu} F_1(v)\, dv\right] du$$

Thus we now have

$$\vartheta_{(1)}(x)\delta_{(2)}(x) \approx A_b \delta(x) \tag{A.37}$$

which evidently contains 1 and 2 above as special cases.
We see how the "memory" of the limiting process of the sequences approaching the step and delta functions influences the value of their product.

□

Example 6 For any constants a and b, $a\vartheta_{(\alpha)}(x) + b\vartheta_{(\beta)}(x) = (a + b)\vartheta_{(\gamma)}(x)$ for some γ; likewise $a\delta_{(\alpha)}(x) + b\delta_{(\beta)}(x) = (a+b)\delta_{(\gamma)}(x)$. If only weak equality is needed, then $a\vartheta_{(\alpha)}(x) + b\vartheta_{(\beta)}(x) \approx (a + b)\vartheta(x)$ and $a\delta_{(\alpha)}(x) + b\delta_{(\beta)}(x) \approx (a + b)\delta(x)$.   □

Products of three or more generalized functions become very cumbersome to calculate. Likewise, double products such as $\delta^2(x)$ are also more troublesome, and are so rarely needed that we will not pursue them here. The interested reader is referred to the previously cited literature for their evaluation.

Solution of equations involving multiplication is also a tricky process. For example, in ordinary functions, the equation

$$xf(x) = 1 \tag{A.38}$$

has only one solution: $f(x) = 1/x$. If we seek generalized function solutions of the weak version of (A.38), however:

$$xf(x) \approx 1 \tag{A.39}$$

we find infinitely many, since its homogeneous counterpart

$$xf(x) \approx 0 \tag{A.40}$$

has the general solution $f(x) \approx C\delta(x)$ for an arbitrary constant C. The situation is analogous to that which arises in the solution of differential equations, where a general solution consists of both homogeneous and particular terms.

A particular solution of (A.39) is

$$f(x) \approx \mathrm{PV}\left(\frac{1}{x}\right) \tag{A.41}$$

where $\mathrm{PV}(\frac{1}{x})$ is the generalized function defined by the ordinary function $1/x$ and the requirement that integrals involving it be carried out in the Cauchy principal value sense:

$$\int_{-\infty}^{\infty} \mathrm{PV}\left(\frac{1}{x}\right) g(x)\,dx \equiv \lim_{\epsilon \to 0^+} \left\{ \int_{-\infty}^{-\epsilon} \frac{g(x)}{x}\,dx + \int_{\epsilon}^{\infty} \frac{g(x)}{x}\,dx \right\}$$

(recall that generalized functions are only defined in terms of their integrals, and that $1/x$ would give an undefined improper integral without further specification). The general solution of (A.39) is thus

$$f(x) \approx \mathrm{PV}\left(\frac{1}{x}\right) + C\delta(x) \tag{A.42}$$

and the constant C cannot be determined without additional constraints or other information about $f(x)$, just as the solution of a differential equation is not fixed uniquely until the boundary conditions are enforced. On the other hand, if we seek the solution of (A.38) with *strong equality* in terms of generalized functions, the definition (A.25) leads to the conclusion that the only solution is $f(x) = \mathrm{PV}(\frac{1}{x})$.

Example 7 In the sense of generalized functions,

$$[\ln|x|]' \approx \mathrm{PV}\left(\frac{1}{x}\right) \tag{A.43}$$

To show this, let $g(x)$ be any good function. Then, in the sense of generalized functions,

$$\int_{-\infty}^{\infty} [\ln|x|]' g(x)\, dx = -\int_{-\infty}^{\infty} \ln|x| g'(x)\, dx \tag{A.44}$$

by (A.4). But the right side of (A.44) is now an ordinary integral and can be evaluated as such. Thus,

$$-\int_{-\infty}^{\infty} \ln|x| g'(x)\, dx = \lim_{\epsilon \to 0}\left\{ -\int_{-\infty}^{-\epsilon} \ln|x| g'(x)\, dx - \int_{\epsilon}^{\infty} \ln|x| g'(x)\, dx\right\}$$

$$= \lim_{\epsilon \to 0}\left\{ \int_{-\infty}^{-\epsilon} \frac{g(x)}{x}\, dx + \int_{\epsilon}^{\infty} \frac{g(x)}{x}\, dx\right\} \tag{A.45}$$

where we have used ordinary integration by parts. But this is simply the definition of the Cauchy principal value,

$$\text{PV} \int_{-\infty}^{\infty} \frac{g(x)}{x}\, dx$$

and so (A.43) is proven. $\qquad\square$

Example 8 The general solution of

$$x f(x) \approx \vartheta(x) \tag{A.46}$$

is

$$f(x) \approx x_+^{-1} + C\delta(x) \tag{A.47}$$

where x_+^{-1} is the generalized function defined by

$$x_+^{-1} \equiv [\vartheta(x)\ln|x|]' \tag{A.48}$$

We prove this as in Example 7. Let $g(x)$ be any good function. Then we have

$$\int_{-\infty}^{\infty} x x_+^{-1} g(x)\, dx = -\int_{-\infty}^{\infty} (xg)' \vartheta(x)\ln|x|\, dx \tag{A.49}$$

and since the right side is now an ordinary integral,

$$-\int_{-\infty}^{\infty} (xg)' \vartheta(x)\ln|x|\, dx = -\int_{0}^{\infty} (xg)' \ln|x|\, dx \tag{A.50}$$

$$= \int_{0}^{\infty} g\, dx = \int_{-\infty}^{\infty} g(x)\vartheta(x)\, dx$$

which proves (A.47). An alternate way of characterizing (A.48) is as follows:

$$
\begin{aligned}
\int_{-\infty}^{\infty} g(x) x_+^{-1}\, dx &= -\int_{-\infty}^{\infty} g'(x)\vartheta(x)\ln|x|\, dx \\
&= -\int_{0}^{\infty} g'(x)\ln|x|\, dx \quad\quad\quad\text{(A.51)} \\
&= \lim_{\epsilon\to 0+} \left\{ g(\epsilon)\ln\epsilon + \int_{\epsilon}^{\infty} \frac{g(x)}{x}\, dx \right\} \\
&= \lim_{\epsilon\to 0+} \left\{ g(0)\ln\epsilon + \int_{\epsilon}^{\infty} \frac{g(x)}{x}\, dx \right\}
\end{aligned}
$$

$\square$

Example 9 Consider the equality

$$
a\delta(x) + b\delta'(x) = 0 \quad\quad\quad\text{(A.52)}
$$

for some constants a and b. If we multiply both sides of (A.52) by an arbitrary good function g and integrate, we have

$$
ag(0) - bg'(0) = 0 \quad\quad\quad\text{(A.53)}
$$

But since $g(0)$ and $g'(0)$ are arbitrary, we conclude that we must have $a = b = 0$.

Example 10 Let us solve the strong equality

$$
\left[a + b\vartheta_{(\beta)}(x)\right] f(x) = \delta_{(\alpha)}(x) \quad\quad\quad\text{(A.54)}
$$

for $f(x)$, with a given step function $\vartheta_{(\beta)}(x)$, a given delta function $\delta_{(\alpha)}(x)$, and constants a and b such that $a \neq 0$ and $a + b \neq 0$. The conditions on a and b are necessary to prevent the factor multiplying f on the left side of (A.54) from vanishing in either $x < 0$ or $x > 0$, which would result in the possibility of arbitrary homogeneous solutions of the type $\vartheta(x)$ or $\vartheta(-x)$. Formally, we could divide both sides of (A.54) by the factor in square brackets to solve the equation. Note that

$$
\frac{1}{a + b\vartheta_{(\beta)}(x)} = \frac{1}{a} + \left(\frac{1}{a+b} - \frac{1}{a}\right)\vartheta_{(\theta)}(x) \quad\quad\quad\text{(A.55)}
$$

where $\vartheta_{(\theta)}$ is the step function defined by:

$$
\vartheta_{(\theta)}(x) = \frac{(a+b)\vartheta_{(\beta)}(x)}{a + b\vartheta_{(\beta)}(x)} \quad\quad\quad\text{(A.56)}
$$

(in the sense that its defining sequence is that given by the right side of (A.56) with $\vartheta_{(\beta)}$ replaced by its defining sequence). Then f is given by

$$f(x) = \frac{1}{a}\delta_{(\alpha)}(x) + A\left(\frac{1}{a+b} - \frac{1}{a}\right)\delta_{(\gamma)}(x) \qquad (A.57)$$

where A and $\delta_{(\gamma)}$ are such that

$$\vartheta_{(\theta)}(x)\delta_{(\alpha)}(x) = A\delta_{(\gamma)}(x) \qquad (A.58)$$

If the association class of f is all that is needed, we have:

$$f(x) \approx \left(\frac{1-A}{a} + \frac{A}{a+b}\right)\delta(x) \qquad (A.59)$$

$\square$

A.3 FOURIER TRANSFORMS AND FOURIER SERIES OF GENERALIZED FUNCTIONS

We may also define the Fourier transform of a generalized function. Let us first summarize the properties of the Fourier transform for classical (good) functions.

For a good function $g(x)$, the classical Fourier transform pair is defined by

$$G(\alpha) = \frac{1}{2\pi}\int_{-\infty}^{\infty} e^{j\alpha x} g(x)\, dx$$

$$g(x) = \int_{-\infty}^{\infty} e^{-j\alpha x} G(\alpha)\, d\alpha \qquad (A.60)$$

and it can be shown that $G(\alpha)$ is also a good function. If $f(x)$ and $F(\alpha)$ are another such pair, we have the *convolution theorem*:

$$\frac{1}{2\pi}\int_{-\infty}^{\infty} f(x')g(x-x')\, dx' = \int_{-\infty}^{\infty} F(\alpha)G(\alpha)e^{-j\alpha x}\, d\alpha \qquad (A.61)$$

The *uncertainty principle*

$$\int_{-\infty}^{\infty} x^2 |f(x)|^2\, dx \int_{-\infty}^{\infty} \alpha^2 |F(\alpha)|^2\, d\alpha \geq \frac{\pi}{2}E^2 \qquad (A.62)$$

where

$$E = \int_{-\infty}^{\infty} |F(\alpha)|^2\, d\alpha = \frac{1}{2\pi}\int_{-\infty}^{\infty} |f(x)|^2\, dx \qquad (A.63)$$

is true for the class of functions (or generalized functions) for which the indicated integrals exist.[7] Note that Dirac delta functions are not included in this class.

---

[7] A. Papoulis, *Systems and Transforms with Applications in Optics*. New York: McGraw-Hill, 1968, pp. 193-199.

No.	$f(x)$	$F(\alpha)$
A.1.1	$f'(x)$	$-j\alpha F(\alpha)$
A.1.2	$f(-x)$	$F(-\alpha)$
A.1.3	$F(x)$	$f(-\alpha)/2\pi$
A.1.4	even function of x	even function of α
A.1.5	odd function of x	odd function of α
A.1.6	$f(x)e^{-j\alpha x}$	$F(\alpha - a)$
A.1.7	$f(x - x_0)$	$F(\alpha)e^{j\alpha x_0}$
A.1.8	$\delta(x)$	$1/2\pi$
A.1.9	$\vartheta(x)$	$-\frac{1}{2\pi j}\mathrm{PV}(\frac{1}{\alpha}) + \frac{1}{2}\delta(\alpha)$
A.1.10	$\mathrm{PV}(\frac{1}{x})$	$\frac{j}{2}\mathrm{sgn}(\alpha)$
A.1.11	$\mathrm{sgn}(x)$	$-\frac{1}{\pi j}\mathrm{PV}(\frac{1}{\alpha})$

Table A.1
Some Fourier transform pairs and rules for generalized functions.

If $f(x)$ is instead a generalized function, we cannot use (A.60) to define its Fourier transform (because $e^{j\alpha x}$ is not a good function); strictly speaking, neither of the pair (A.60) is defined at all. However, the integrals on both sides of (A.61) are well-defined if g and G are good functions, and so we may use it to define $F(\alpha)$ (weakly) for any generalized function f, and thereby attach a *formal* meaning to (A.60):

$$\int_{-\infty}^{\infty} F(\alpha)G(\alpha)\, d\alpha \equiv \frac{1}{2\pi} \int_{-\infty}^{\infty} f(x)g(-x)\, dx \qquad (\text{A.64})$$

for all good $g(x)$ with transform $G(\alpha)$.

All rules that apply to the Fourier transform of ordinary functions and do not involve the multiplication of two or more functions can also be shown to hold for generalized functions. In some cases, properties involving defined products of generalized functions may also be valid, but these must be used with care. A brief list of rules that do hold and of some elementary transform pairs is given in Table A.1.

Example 11 The Fourier transform of $\delta(x)$ is $1/2\pi$. This does *not* follow from (A.60) (although a *formal* application of the properties of the delta function does give the right result[8]), but must be proved from

---

[8]Generally, such abuses of formal manipulations, when they give results that make sense at all, do give the right answers. It should be kept in mind, however, that the formal manipulations do not constitute a proof that the results are correct.

(A.64). We have

$$
\begin{aligned}
\int_{-\infty}^{\infty} F(\alpha)G(\alpha)\,d\alpha &= \frac{1}{2\pi}\int_{-\infty}^{\infty}\delta(x)g(-x)\,dx \\
&= \frac{g(0)}{2\pi} \\
&= \frac{1}{2\pi}\int_{-\infty}^{\infty}G(\alpha)\,d\alpha
\end{aligned}
$$

for all good $G(\alpha)$ if $f(x) = \delta(x)$. Hence, $F(\alpha) \approx 1/2\pi$.   □

Example 12 To find the Fourier transform of $\vartheta(x)$, which we will denote $h(\alpha)$, we note that its derivative has Fourier transform $1/2\pi$, and thus

$$
\frac{1}{2\pi} \approx -j\alpha h(\alpha)
$$

Hence by (A.42),

$$
h(\alpha) \approx -\frac{1}{2\pi j}\mathrm{PV}(\frac{1}{\alpha}) + C\delta(\alpha)
$$

for some constant C. But the Fourier transform of $\vartheta(-x)$ is $h(-\alpha)$ and $\vartheta(x) + \vartheta(-x) \approx 1$. Therefore,

$$
h(\alpha) + h(-\alpha) \approx \delta(\alpha)
$$

and so $C = \frac{1}{2}$. Thus,

$$
h(\alpha) \approx -\frac{1}{2\pi j}\mathrm{PV}(\frac{1}{\alpha}) + \frac{1}{2}\delta(\alpha) \tag{A.65}
$$

as in entry A.1.9 of the table.   □

Example 13 The function $\vartheta(x)e^{-jax}$ has a Fourier transform that can be found using entries A.1.6 and A.1.9 of the table:

$$
-\frac{1}{2\pi j}\mathrm{PV}(\frac{1}{\alpha - a}) + \frac{1}{2}\delta(\alpha - a) \tag{A.66}
$$

This is straightforward to interpret and use if a is real, but care must be used if $\mathrm{Im}(a) \neq 0$. If $\mathrm{Im}(a) > 0$, the Fourier transform does not exist even in the sense of generalized functions as employed in the Appendix, because of the exponential growth of the function as $x \to +\infty$. Fortunately, this case does not usually arise in practical applications since such exponential growth of fields is precluded on physical grounds. On the other hand, if $\mathrm{Im}(a) < 0$, the Fourier transform exists even in the sense of ordinary functions:

$$
-\frac{1}{2\pi j}\frac{1}{\alpha - a} \qquad [\mathrm{Im}(a) < 0] \tag{A.67}
$$

We can regard (A.67) as a special case of (A.66) if the need for the principal value is discarded, and the delta function is zero for all real α. In a similar way, we find that the Fourier transform of $\vartheta(-x)e^{+jax}$ is

$$\frac{1}{2\pi j}\frac{1}{\alpha + a} \qquad [\text{Im}(a) < 0] \qquad (A.68)$$

This interpretation is restricted to values of a with negative imaginary part, which is not obvious without the context of the original functions whose Fourier transforms we desired. $\qquad \square$

Example 14 The *signum* function $\text{sgn}(x)$ is defined to be the algebraic sign of x, that is, $\text{sgn}(x) = 1$ if $x > 0$ and $\text{sgn}(x) = -1$ if $x < 0$. Clearly $\text{sgn}(x) \approx 2\vartheta(x) - 1$, so by the previous examples, and entry A.1.3 of the table, we have its Fourier transform as:

$$-\frac{1}{\pi j}\text{PV}(\frac{1}{\alpha})$$

per entry A.1.11 of the table. $\qquad \square$

Example 15 Formally taking the imaginary part of entry A.1.10 of the table for $\alpha = 1$ yields the known integral

$$\int_{-\infty}^{\infty} \frac{\sin x}{x} \, dx = \pi$$

If this result is used together with (A.9)-(A.11), we obtain

$$\lim_{n\to\infty} \frac{\sin nx}{x} \approx \pi\delta(x) \qquad (A.69)$$

$\qquad \square$

The Fourier series can also be defined for generalized functions in an analogous way. If $f(\phi)$ is a good function, periodic in ϕ with period 2π, we introduce the transform pair

$$
\begin{aligned}
f_m &= \frac{1}{2\pi}\int_0^{2\pi} f(\phi)e^{-jm\phi} \, d\phi \\
f(\phi) &= \sum_{m=-\infty}^{\infty} f_m e^{jm\phi}
\end{aligned}
\qquad (A.70)
$$

The convolution theorem for this transform pair is:

$$\frac{1}{2\pi}\int_0^{2\pi} f(\phi')g(\phi - \phi') \, d\phi' = \sum_{m=-\infty}^{\infty} f_m g_m e^{jm\phi} \qquad (A.71)$$

where f and g are both good, periodic functions. If f is a generalized function, the summation in the second of Equations (A.70) will generally not converge in a classical sense, but meaning can be given to the Fourier series by insisting that for all good functions g, f_m are chosen such that (A.71) always holds (the summation now being valid, classically speaking).

An easy consequence of the complex exponential Fourier series (A.70) is the trigonometric Fourier series for a periodic function $f(\phi)$:

$$
\begin{aligned}
a_m &= \frac{1}{\pi} \int_0^{2\pi} f(\phi) \cos m\phi \, d\phi \\
b_m &= \frac{1}{\pi} \int_0^{2\pi} f(\phi) \sin m\phi \, d\phi \\
f(\phi) &= \frac{a_0}{2} + \sum_{m=1}^{\infty} (a_m \cos m\phi + b_m \sin m\phi)
\end{aligned}
\tag{A.72}
$$

If $f(\phi)$ is an even function, then $b_m \equiv 0$, while if f is odd, then $a_m \equiv 0$.

Example 16 The delta function itself is not periodic, but a periodic function can be constructed from it:

$$
\delta_{2\pi}(\phi) \equiv \sum_{n=-\infty}^{\infty} \delta(\phi + 2n\pi)
\tag{A.73}
$$

Then a Fourier series transform pair is found to be

$$
\begin{aligned}
\frac{1}{2\pi} e^{-jm\phi_0} &= \frac{1}{2\pi} \int_0^{2\pi} \delta_{2\pi}(\phi - \phi_0) e^{-jm\phi} \, d\phi \\
\delta_{2\pi}(\phi - \phi_0) &= \sum_{m=-\infty}^{\infty} \frac{1}{2\pi} e^{jm(\phi - \phi_0)}
\end{aligned}
\tag{A.74}
$$

for some constant ϕ_0, the validity of which is purely formal, being understood in the sense of generalized functions. The corresponding Fourier cosine series for the delta function is

$$
\delta_{2\pi}(\phi - \phi_0) = \frac{1}{2\pi} + \sum_{m=1}^{\infty} \frac{1}{\pi} \cos m(\phi - \phi_0)
\tag{A.75}
$$

$\square$

Example 17 The result of Example 16 can be used to prove the *Poisson summation formula.* Let $g(x)$ be a good function, and let K be a positive constant. Then

$$
\begin{aligned}
\sum_{m=-\infty}^{\infty} g(Km) &= \sum_{m=-\infty}^{\infty} \int_{-\infty}^{\infty} e^{-j\alpha Km} G(\alpha)\, d\alpha \\
&= \int_{-\infty}^{\infty} G(\alpha) \left(\sum_{m=-\infty}^{\infty} e^{-j\alpha Km} \right) d\alpha \\
&= 2\pi \int_{-\infty}^{\infty} G(\alpha) \delta_{2\pi}(-\alpha K)\, d\alpha \\
&= \frac{2\pi}{K} \sum_{n=-\infty}^{\infty} G\left(\frac{2n\pi}{K}\right)
\end{aligned}
\tag{A.76}
$$

where (A.15), (A.16), and (A.74) have been used. Formula (A.76) can be extended to generalized functions in the usual way. $\qquad\square$

A.4   MULTIDIMENSIONAL GENERALIZED FUNCTIONS

For applications involving more than one space dimension and/or the time coordinate t, we encounter generalized functions of more than one variable. For example, the "point source" delta function in three dimensions is expressed as $\delta(\mathbf{r}-\mathbf{r}_0) \equiv \delta(x-x_0)\delta(y-y_0)\delta(z-z_0)$, because for any good function $f(\mathbf{r})$, we have:

$$
\int f(\mathbf{r}) \delta(\mathbf{r}-\mathbf{r}_0)\, dV = f(\mathbf{r}_0)
$$

where $dV = dx\, dy\, dz$ and the volume integral is over all space. Because the three delta functions appearing in this product involve three different variables, this is called a *direct product* rather than the ordinary product discussed earlier in this appendix, and encounters no special problems.

The Fourier transform can be extended (formally) to functions of the three Cartesian space variables as a composition of single-variable transforms:

$$
\begin{aligned}
F(\boldsymbol{\alpha}) = F(\alpha,\beta,\gamma) &= \frac{1}{(2\pi)^3} \int_{-\infty}^{\infty}\int_{-\infty}^{\infty}\int_{-\infty}^{\infty} e^{j\alpha x + j\beta y + j\gamma z} f(x,y,z)\, dx\, dy\, dz \\
&= \frac{1}{(2\pi)^3} \int e^{j\boldsymbol{\alpha}\cdot\mathbf{r}} f(\mathbf{r})\, dV \\
f(\mathbf{r}) = f(x,y,z) &= \int_{-\infty}^{\infty}\int_{-\infty}^{\infty}\int_{-\infty}^{\infty} e^{-j\alpha x - j\beta y - j\gamma z} F(\alpha,\beta,\gamma)\, d\alpha\, d\beta\, d\gamma \\
&= \int e^{-j\boldsymbol{\alpha}\cdot\mathbf{r}} F(\boldsymbol{\alpha})\, d\boldsymbol{\alpha}
\end{aligned}
\tag{A.77}
$$

The more compact forms are introduced via the notations $\mathbf{r} = x\mathbf{u}_x + y\mathbf{u}_y + z\mathbf{u}_z$ for the position vector and $\boldsymbol{\alpha} = \alpha\mathbf{u}_x + \beta\mathbf{u}_y + \gamma\mathbf{u}_z$ for the vector transform variable, while combining the triple integrals into a single integral sign and suppressing the limits where no confusion can arise.

When the time coordinate t is involved, tradition dictates a slightly different treatment of t and its transform variable ω:

$$
\begin{aligned}
F(\boldsymbol{\alpha}, \omega) &= \frac{1}{(2\pi)^3} \int_{-\infty}^{\infty} \int e^{j\boldsymbol{\alpha}\cdot\mathbf{r} - j\omega t} f(\mathbf{r}, t) \, dV \, dt \\
f(\mathbf{r}, t) &= \frac{1}{2\pi} \int_{-\infty}^{\infty} \int e^{j\omega t - j\boldsymbol{\alpha}\cdot\mathbf{r}} F(\boldsymbol{\alpha}, \omega) \, d\boldsymbol{\alpha} \, d\omega
\end{aligned}
\qquad (\text{A.78})
$$

Example 18 A point source $\delta(\mathbf{r} - \mathbf{r}_0)$ can be expressed formally via its Fourier transform:

$$
\delta(\mathbf{r} - \mathbf{r}_0) = \frac{1}{(2\pi)^3} \int e^{-j\boldsymbol{\alpha}\cdot(\mathbf{r} - \mathbf{r}_0)} \, d\boldsymbol{\alpha}
$$

$\square$

Example 19 The generalized function $f(\mathbf{r}, t) = [\delta(t - r/c) - \delta(t + r/c)]/(4\pi r)$ arises in the calculation of fields radiated by point sources. Here $r = \sqrt{x^2 + y^2 + z^2}$ is the radial coordinate in the spherical coordinate system (r, θ, ϕ). A *formal* manipulation using (A.77) gives its spatial Fourier transform as:

$$
\begin{aligned}
F(\boldsymbol{\alpha}, t) &= \frac{1}{(2\pi)^3} \int e^{j\boldsymbol{\alpha}\cdot\mathbf{r}} \frac{\delta(t - r/c) - \delta(t + r/c)}{4\pi r} \, dx \, dy \, dz \\
&= \frac{1}{32\pi^4} \int_0^{\infty} \int_0^{\pi} \int_0^{2\pi} e^{j|\boldsymbol{\alpha}|r\cos\theta} \frac{\delta(t - r/c) - \delta(t + r/c)}{r} r^2 \\
&\quad \times \sin\theta \, d\phi \, d\theta \, dr
\end{aligned}
$$

by converting to a spherical coordinate system where the direction of $\boldsymbol{\alpha}$ is regarded as the "z" axis (that is, the direction in which $\theta = 0$). Carrying out the θ and ϕ integrations by ordinary calculus, and formally carrying out the r integration, we obtain:

$$
F(\boldsymbol{\alpha}, t) = \frac{c}{(2\pi)^3} \frac{\sin(|\boldsymbol{\alpha}|ct)}{|\boldsymbol{\alpha}|}
\qquad (\text{A.79})
$$

where $|\boldsymbol{\alpha}| = \sqrt{\alpha^2 + \beta^2 + \gamma^2}$. This result must of course be verified by the more careful methods of generalized function theory in order to be regarded as genuine. $\square$

Example 20 The function $1/r$ and its derivatives appear in many electromagnetic field calculations. Although it possesses a singularity at the origin, no special treatment is needed in order to use it as a generalized function: the integral over all space

$$
\int_{\mathbb{R}^3} g(\mathbf{r}) \frac{1}{r} \, dV
$$

is defined for all good g, since the differential volume element $dV = r^2 \sin\theta dr d\theta d\phi$ contains the factor r^2 sufficient to render the integral unambiguously convergent. Likewise, none of the first derivatives of $1/r$ with respect to the space coordinates $x_1 \equiv x$, $x_2 \equiv y$ or $x_3 \equiv z$ requires special treatment, because the $1/r^2$ behavior they contain is compensated by the factor of r^2 from dV.

This nice behavior is gone, however, when we examine second derivatives of the form

$$\frac{\partial^2}{\partial x_i \partial x_j}\left(\frac{1}{r}\right)$$

because the singularity now has strength $1/r^3$ and must be treated with care.[9] So we proceed in a similar fashion to Example 7. By the properties of the derivatives of generalized functions, we have that

$$\int_{\mathbb{R}^3} g(\mathbf{r})\left[\frac{\partial^2}{\partial x_i \partial x_j}\left(\frac{1}{r}\right)\right] dV \equiv -\int_{\mathbb{R}^3} \frac{\partial g(\mathbf{r})}{\partial x_i}\left[\frac{\partial}{\partial x_j}\left(\frac{1}{r}\right)\right] dV$$

$$= -\int_{\mathbb{R}^3 - V_0} \frac{\partial g(\mathbf{r})}{\partial x_i}\left[\frac{\partial}{\partial x_j}\left(\frac{1}{r}\right)\right] dV - \int_{V_0} \frac{\partial g(\mathbf{r})}{\partial x_i}\left[\frac{\partial}{\partial x_j}\left(\frac{1}{r}\right)\right] dV$$

$$\text{(A.80)}$$

where V_0 (the so-called *principal volume*) is any region of space containing the origin. Denoting the unit vectors $\mathbf{u}_x \equiv \mathbf{u}_1$ and so on, we may integrate the first integral on the right side of (A.80) by parts as follows:

$$-\int_{\mathbb{R}^3 - V_0} \frac{\partial g(\mathbf{r})}{\partial x_i}\left[\frac{\partial}{\partial x_j}\left(\frac{1}{r}\right)\right] dV$$

$$= \oint_{S_0} (\mathbf{u}_n \cdot \mathbf{u}_i) g(\mathbf{r})\left[\frac{\partial}{\partial x_j}\left(\frac{1}{r}\right)\right] dS + \int_{\mathbb{R}^3 - V_0} g(\mathbf{r})\left[\frac{\partial^2}{\partial x_i \partial x_j}\left(\frac{1}{r}\right)\right] dV$$

$$\text{(A.81)}$$

where S_0 is the closed surface bounding V_0, and $\mathbf{u}_n$ is the unit normal vector to S_0 pointing away from V_0.

We now let the principal volume shrink to zero, always containing the origin as it does so. If the *shape* of the principal volume is maintained,

---

[9]See:

J. Van Bladel, *IRE Trans. Ant. Prop.*, vol. 9, pp. 563-566, 1961.

A. D. Yaghjian, *Proc. IEEE*, vol. 68, pp. 248-263, 1980.

S. W. Lee et al., *IEEE Trans. Ant. Prop.*, vol. 28, pp. 311-317, 1980.

A. D. Yaghjian, *Electromagnetics*, vol. 2, pp. 161-167, 1982.

J. S. Asvestas, *IEEE Trans. Ant. Prop.*, vol. 31, pp. 174-177, 1983.

J. Van Bladel, *Singular Electromagnetic Fields and Sources.* New York: IEEE, 1995, Chapter 3.

as well as its relative position with respect to the origin as $V_0 \rightarrow 0$, it can be seen that the surface integral in (A.81) is independent of the *size* of V_0. The second integral on the right side of (A.80) vanishes in this limit, since g is a good function. From that result and (A.81) we then obtain:

$$\int_{\mathbb{R}^3} g(\mathbf{r}) \left[\frac{\partial^2}{\partial x_i \partial x_j} \left(\frac{1}{r} \right) \right] dV$$
$$= \mathrm{PV}_{V_0} \int g(\mathbf{r}) \left[\frac{\partial^2}{\partial x_i \partial x_j} \left(\frac{1}{r} \right) \right] dV - 4\pi \mathbf{u}_j \cdot \overset{\leftrightarrow}{\mathbf{L}}_{V_0} \cdot \mathbf{u}_i g(0) \quad \text{(A.82)}$$

where the principal value integral (with respect to V_0) is defined to be:

$$\mathrm{PV}_{V_0} \int f(\mathbf{r}) \, dV \equiv \lim_{V_0 \to 0} \int_{\mathbb{R}^3 - V_0} f(\mathbf{r}) \, dV \quad \text{(A.83)}$$

and $\overset{\leftrightarrow}{\mathbf{L}}_{V_0}$ is the so-called *depolarization dyadic* defined as:

$$\overset{\leftrightarrow}{\mathbf{L}}_{V_0} \equiv \oint_{S_0} \frac{\mathbf{r}\mathbf{u}_n}{4\pi r^3} \, dS \quad \text{(A.84)}$$

Note that both (A.83) and (A.84) depend on the shape of V_0 and its position relative to the origin as $V_0 \rightarrow 0$, in such a way that the overall value of (A.82) is independent of the principal volume and how it shrinks to zero. More compactly, we can write:

$$\frac{\partial^2}{\partial x_i \partial x_j} \left(\frac{1}{r} \right) = \mathrm{PV}_{V_0} \frac{\partial^2}{\partial x_i \partial x_j} \left(\frac{1}{r} \right) - 4\pi \mathbf{u}_j \cdot \overset{\leftrightarrow}{\mathbf{L}}_{V_0} \cdot \mathbf{u}_i \delta(\mathbf{r}) \quad \text{(A.85)}$$

The depolarization dyadic can be evaluated for several simple shapes. For a sphere or a cube centered at the origin,

$$\overset{\leftrightarrow}{\mathbf{L}}_{\text{sphere}} = \overset{\leftrightarrow}{\mathbf{L}}_{\text{cube}} = \frac{1}{3} \overset{\leftrightarrow}{\mathbf{I}} \quad \text{(A.86)}$$

For a very thin needle of circular or square cross-section oriented along the z-axis,

$$\overset{\leftrightarrow}{\mathbf{L}}_{\text{needle}} = \frac{1}{2} \left(\mathbf{u}_x \mathbf{u}_x + \mathbf{u}_y \mathbf{u}_y \right) \quad \text{(A.87)}$$

while for a very flat cylindrical box (a "pillbox") of arbitrary cross-section oriented perpendicular to the z-axis,

$$\overset{\leftrightarrow}{\mathbf{L}}_{\text{pillbox}} = \mathbf{u}_z \mathbf{u}_z \quad \text{(A.88)}$$

In general, it can be shown that the depolarization dyadic is symmetrical:

$$\mathbf{u}_i \cdot \overset{\leftrightarrow}{\mathbf{L}}_{V_0} \cdot \mathbf{u}_j = \mathbf{u}_j \cdot \overset{\leftrightarrow}{\mathbf{L}}_{V_0} \cdot \mathbf{u}_i \quad \text{(A.89)}$$

Also for arbitrary V_0, it can be shown that the trace of the polarization dyadic is 1:

$$\sum_{i=1}^{3} \mathbf{u}_i \cdot \overleftrightarrow{\mathbf{L}}_{V_0} \cdot \mathbf{u}_i = 1 \tag{A.90}$$

The demonstrations of these properties are left as exercises. □

Example 21 It is straightforward to show that

$$\mathrm{PV}_{V_0} \nabla^2 \left(\frac{1}{r} \right) = 0 \tag{A.91}$$

by direct calculation before the limit $V_0 \to 0$ is taken. Using (A.85) and (A.90), we obtain the important result

$$\nabla^2 \left(\frac{1}{r} \right) = -4\pi \delta(\mathbf{r}) \tag{A.92}$$

An alternative derivation of this result is given by the calculation of the potential of a point charge in Section 4.1.

A.5 PROBLEMS

A–1 Show that $x\delta'(x) \approx -\delta(x)$.

A–2 Show that $g(x)\delta'(x) \approx g(0)\delta'(x) - g'(0)\delta(x)$ for any good function $g(x)$. If we take the limit of this result for a sequence of good functions that approaches a generalized function, the result also holds true in the strong sense if g is a generalized function.

A–3 In Example 10, if $\vartheta_{(\theta)}(x)$ is given by (A.56) and obeys (A.58), evaluate $\vartheta_{(\beta)}(x)\delta_{(\alpha)}(x)$ in terms of the constant A.

A–4 (a) Show that the Fourier transform of $\ln|x|$ is the generalized function

$$\frac{1}{2}\mathrm{PV}(\frac{1}{\alpha}) - \alpha_+^{-1} + C\delta(\alpha)$$

where C is a constant.

(b) Show that C from part (a) is equal to $-\gamma_E$, where γ_E is Euler's constant:

$$\gamma_E = -\lim_{\epsilon\to 0^+}\left\{\ln\epsilon + \int_\epsilon^\infty \frac{e^{-u}}{u}\,du\right\}$$

$$= .5772\ldots$$

A–5 Prove that the Fourier transform of $f'(x)$, where $f(x)$ is a generalized function, is $-j\alpha F(\alpha)$, where $F(\alpha)$ is the Fourier transform of $f(x)$. [Hint: this property is true for good functions.]

A–6 If $F(\alpha)$ is the Fourier transform of a generalized function $f(x)$, then prove that $F(-\alpha)$ is the Fourier transform of $f(-x)$. [Hint: this property is true for good functions.]

A–7 Prove the uncertainty principle (A.62), and show that equality holds only for Gaussian functions

$$f(x) = Ae^{-b^2x^2}$$

where A and b are real constants.

A–8 Show that $\delta(\mathbf{r} - \mathbf{r}_0) \equiv \delta(x - x_0)\delta(y - y_0)\delta(z - z_0)$ is expressible as

(a)

$$\delta(\mathbf{r} - \mathbf{r}_0) = \frac{\delta(r - r_0)\delta(\theta - \theta_0)\delta_{2\pi}(\phi - \phi_0)}{r^2\sin\theta}$$

in spherical coordinates, if $r_0 \neq 0$ and $0 < \theta_0 < \pi$;

(b)

$$\delta(\mathbf{r} - \mathbf{r}_0) = \frac{\delta(\rho - \rho_0)\delta_{2\pi}(\phi - \phi_0)\delta(z - z_0)}{\rho}$$

in cylindrical coordinates, if $\rho_0 \neq 0$.

A–9 Carry out the detailed calculations to obtain (A.79). Verify these formal calculations by appealing to the convolution theorem definition of the Fourier transform.

A–10 Derive (A.89) and (A.90).

A–11 In the xy-plane, with $\boldsymbol{\rho} = x\mathbf{u}_x + y\mathbf{u}_y$ and $\rho = |\boldsymbol{\rho}|$, show that

$$\lim_{a \to 0^+} \frac{a}{(a^2 + \rho^2)^{3/2}} = 2\pi\delta(\boldsymbol{\rho})$$

B Special Functions

In this appendix, we will summarize some useful properties of special functions that are of use in applications. We make no attempt at completeness, nor even coherence. For complete collections of such properties, the reader should consult the literature.[1]

B.1 GAMMA FUNCTION

The Gamma function $\Gamma(z)$ is defined for any z with positive real part by the integral

$$\Gamma(z) = \int_0^\infty t^{z-1} e^{-t}\, dt \tag{B.1}$$

When $\nu = m$, an integer, we can evaluate the integral explicitly to get

$$\Gamma(m+1) = m! \tag{B.2}$$

Various identities follow from its definition, notably

$$\Gamma(z+1) = z\Gamma(z) \tag{B.3}$$

$$\Gamma\left(\frac{1}{2}\right) = \sqrt{\pi} \tag{B.4}$$

$$\Gamma(z)\Gamma(1-z) = \frac{\pi}{\sin \pi z} \tag{B.5}$$

B.2 BESSEL FUNCTIONS

Bessel's differential equation

$$z^2 \frac{d^2 f}{dz^2} + z \frac{df}{dz} + (z^2 - \nu^2)f = 0 \tag{B.6}$$

is a linear, second-order ordinary differential equation, and therefore will have two linearly independent solutions. One of these can be obtained by power series methods, and is given by

$$
\begin{aligned}
f(z) &= J_\nu(z) \\
&= \sum_{m=0}^\infty \frac{(-1)^m (z/2)^{2m+\nu}}{m!\,\Gamma(m+\nu+1)}
\end{aligned}
\tag{B.7}
$$

---

[1]See, for example,

M. Abramowitz and I. A. Stegun, *Handbook of Mathematical Functions*. Washington, DC: US Government Printing House, 1964.

where Γ is the gamma function or factorial function (see Section B.1). The function $J_\nu(z)$ is known as the Bessel function of the *first* kind of *index* ν (or *order* ν) and argument z.

Clearly $J_{-\nu}(z)$ is also a solution of (B.6). However, when $\nu = n$ is an integer, it turns out that $J_{-n}(z)$ is not an independent solution; in fact

$$J_{-n}(z) = (-1)^n J_n(z) \tag{B.8}$$

So instead of using $J_{-\nu}(z)$ as our second solution, we define the Bessel function of the second kind $Y_\nu(z)$ as:

$$Y_\nu(z) = \frac{J_\nu(z)\cos\nu\pi - J_{-\nu}(z)}{\sin\nu\pi} \tag{B.9}$$

Even though this expression becomes indeterminate $(0/0)$ when $\nu \to n$, the limit exists and is still independent of $J_n(z)$. From (B.9) we find that

$$Y_{-\nu}(z) = Y_\nu(z)\cos\nu\pi + J_\nu(z)\sin\nu\pi \tag{B.10}$$

and in particular

$$Y_{-n}(z) = (-1)^n Y_n(z) \tag{B.11}$$

Plots of J_0, J_1, Y_0, and Y_1 are given in Figure B.1.

It is sometimes useful to define solutions of Bessel's equation that are analogous to complex exponential functions (i.e., $e^{\pm jz}$). These are the *Hankel functions* of the first and second kind:

$$H_\nu^{(1)}(z) \equiv J_\nu(z) + jY_\nu(z) \tag{B.12}$$

$$H_\nu^{(2)}(z) \equiv J_\nu(z) - jY_\nu(z) \tag{B.13}$$

The asymptotic large argument behavior of the Hankel functions is

$$H_\nu^{(1)}(z) \quad \sim \quad \sqrt{\frac{2}{\pi z}} e^{j(z-\pi/4-\nu\pi/2)}$$

$$H_\nu^{(2)}(z) \quad \sim \quad \sqrt{\frac{2}{\pi z}} e^{-j(z-\pi/4-\nu\pi/2)} \tag{B.14}$$

The small argument behavior of the Bessel functions is as follows:

$$J_\nu(z) \sim \frac{1}{\Gamma(\nu+1)}\left(\frac{z}{2}\right)^\nu \quad (z \to 0) \quad (\nu \geq 0) \tag{B.15}$$

$$Y_0(z) \sim \frac{2}{\pi}(\ln\frac{z}{2} + \gamma_E) \tag{B.16}$$

$$Y_\nu(z) \sim -\frac{\Gamma(\nu)}{\pi}\left(\frac{2}{z}\right)^\nu \quad (\nu > 0) \tag{B.17}$$

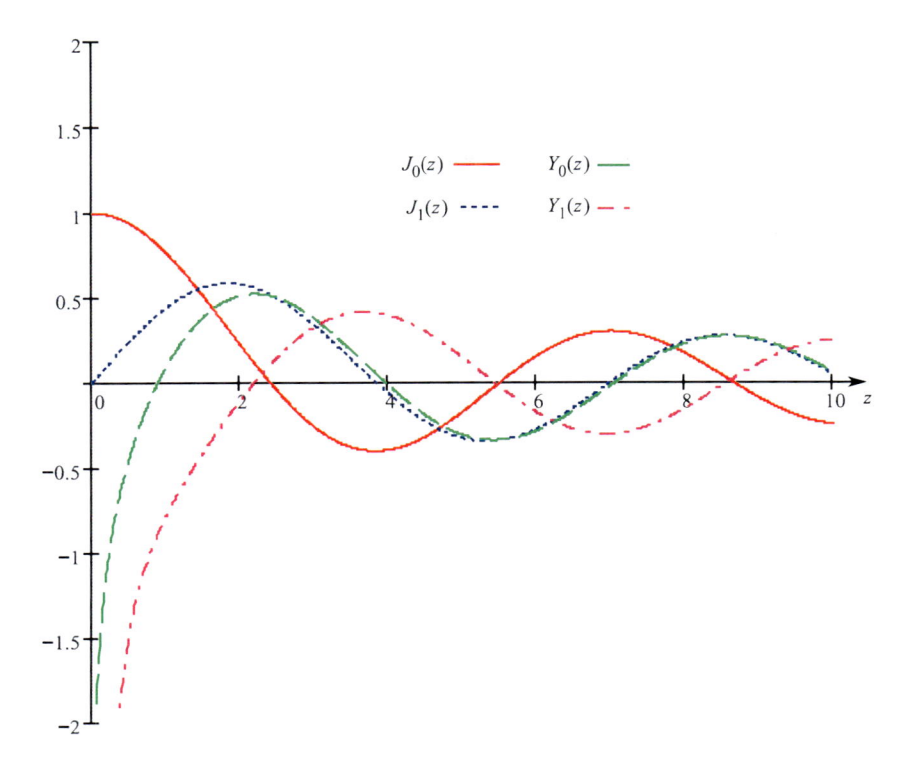

Figure B.1 Graphs of $J_0(z)$, $Y_0(z)$, $J_1(z)$, and $Y_1(z)$.

For the Hankel function of the second kind and zeroth order we have in particular

$$H_0^{(2)}(z) \sim 1 - \frac{2j}{\pi} \left[\ln \frac{z}{2} + \gamma_E \right] \qquad (z \to 0) \qquad (B.18)$$

Here $\gamma_E = 0.5772...$ is Euler's constant.[2]

The derivatives of Bessel functions can be expressed in terms of Bessel functions of different orders. Thus, for example,

$$\mathcal{Z}'_\nu(z) = -\mathcal{Z}_{\nu+1}(z) + \frac{\nu}{z} \mathcal{Z}_\nu(z) \qquad (B.19)$$

$$\mathcal{Z}'_\nu(z) = \mathcal{Z}_{\nu-1}(z) - \frac{\nu}{z} \mathcal{Z}_\nu(z) \qquad (B.20)$$

where $\mathcal{Z}$ represents any of the Bessel functions J, Y, $H^{(1)}$, or $H^{(2)}$, and the superscript $'$ denotes differentiation with respect to the argument z. The

---

[2]Often, this constant is referred to simply as γ, or in some other notation. Since γ is used for a rather different purpose in this book, we have concocted a distinct notation for this number.

Wronskian relation between the Bessel function of the first kind and the Bessel function of the second kind is

$$J_\nu(z)Y_\nu'(z) - J_\nu'(z)Y_\nu(z) = \frac{2}{\pi z} \tag{B.21}$$

Bessel functions possess many forms of integral representation, of which we present a few here:

$$J_m(z) = \frac{e^{-jm\pi/2}}{2\pi} \int_{-\pi}^{\pi} e^{jz\cos\phi}e^{jm\phi}\,d\phi \tag{B.22}$$

holds when the order m is an integer. When the order of the Bessel function J_ν is not an integer, we have the following generalization of (B.22):

$$J_\nu(z) = \frac{e^{-j\nu\pi/2}}{2\pi} \left\{ \int_{-\pi}^{\pi} e^{jz\cos\chi + j\nu\chi}\,d\chi - 2\sin\nu\pi \int_0^\infty e^{-\nu t - jz\cosh t}\,dt \right\} \tag{B.23}$$

Finally, we present some miscellaneous results for Bessel functions that arise in applications. The indefinite integral

$$\int xJ_1(bx)J_1(cx)\,dx = \frac{cxJ_1(bx)J_0(cx) - bxJ_0(bx)J_1(cx)}{b^2 - c^2} \tag{B.24}$$

is valid if b and c are constants. The indefinite integral

$$\int x^{\nu+1} \mathcal{Z}_\nu(x)\,dx = x^{\nu+1} \mathcal{Z}_{\nu+1}(x) \tag{B.25}$$

is valid for any ν.

B.3 SPHERICAL BESSEL FUNCTIONS

The equation

$$\frac{d}{dz}\left(z^2\frac{df}{dz}\right) + [z^2 - n(n+1)]f = 0 \tag{B.26}$$

occurs as a result of separating variables in the Helmholtz equations for spherical coordinates. The change of variable $f = z^{-1/2}g$ results in Bessel's differential equation as encountered previously, with order $n + \frac{1}{2}$. Its general solution is thus

$$f(z) = Bj_n(z) + Cy_n(z) \tag{B.27}$$

where the *spherical Bessel functions* are defined as

$$j_n(z) = \left(\frac{\pi}{2z}\right)^{1/2} J_{n+\frac{1}{2}}(z)$$

$$y_n(z) = \left(\frac{\pi}{2z}\right)^{1/2} Y_{n+\frac{1}{2}}(z) \tag{B.28}$$

and B and C are constants. We also define the spherical Hankel functions in analogy with ordinary Hankel functions to be:

$$h_n^{(1)}(z) = j_n(z) + jy_n(z) \quad ; \quad h_n^{(2)}(z) = j_n(z) - jy_n(z) \tag{B.29}$$

Note that

$$j_0(z) = \frac{\sin z}{z} \quad ; \quad y_0(z) = -\frac{\cos z}{z} \tag{B.30}$$

so that the zeroth-order spherical Hankel function

$$h_0^{(2)}(z) \equiv j_0(z) - jy_0(z) = j\frac{e^{-jz}}{z} \tag{B.31}$$

is nothing other than an outgoing spherical wave (cf. (4.22)). Further examples are:

$$j_1(z) = \frac{\sin z}{z^2} - \frac{\cos z}{z}; \quad y_1(z) = -\frac{\cos z}{z^2} - \frac{\sin z}{z} \tag{B.32}$$

$$j_2(z) = \left(\frac{3}{z^3} - \frac{1}{z}\right)\sin z - \frac{3\cos z}{z^2}; \quad y_2(z) = \left(-\frac{3}{z^3} + \frac{1}{z}\right)\cos z - \frac{3\sin z}{z^2} \tag{B.33}$$

and

$$h_1^{(2)}(z) = \left(\frac{j}{z^2} - \frac{1}{z}\right)e^{-jz}; \quad h_2^{(2)}(z) = \left(\frac{3j}{z^3} - \frac{3}{z^2} - \frac{j}{z}\right)e^{-jz} \tag{B.34}$$

The Wronskian relation for spherical Bessel functions is:

$$j_n(z)y_n'(z) - j_n'(z)y_n(z) = \frac{1}{z^2} \tag{B.35}$$

It is occasionally more convenient to utilize *Riccati-Bessel functions*, defined by

$$\psi_n(z) \equiv zj_n(z) \quad ; \quad \chi_n(z) \equiv -zy_n(z) \quad ; \quad \zeta_n(z) \equiv zh_n^{(2)}(z) = \psi_n(z) + j\chi_n(z) \tag{B.36}$$

The Riccati-Bessel functions obey the differential equation

$$z^2\frac{d^2 f}{dz^2} + [z^2 - n(n+1)]f = 0 \tag{B.37}$$

B.4 FRESNEL INTEGRALS

The Fresnel integral can be derived in many ways. We will start with a consideration of a function $F_1(b)$ defined by the definite integral:

$$F_1(b) = \int_0^\infty e^{-bx^2}\, dx \tag{B.38}$$

This integral converges for all b such that $\text{Re}(b) > 0$, and its values for b imaginary can be found by taking the limit as $\text{Re}(b) \to 0^+$. It can be evaluated in an elementary way by the following trick. Express

$$[F_1(b)]^2 = \int_0^\infty e^{-bx^2}\, dx \int_0^\infty e^{-by^2}\, dy = \int_0^{\pi/2}\int_0^\infty e^{-b\rho^2}\rho\, d\rho d\phi$$

where we have made the polar-coordinate change of variables $x = \rho\cos\phi$ and $y = \rho\sin\phi$. Performing the ϕ-integration, and making the further change of variable $u = \rho^2$ leads to

$$[F_1(b)]^2 = \frac{\pi}{4b}$$

or, since $F_1(b) > 0$ if b is real and positive,

$$F_1(b) = \frac{1}{2}\sqrt{\frac{\pi}{b}} \tag{B.39}$$

The square root of b in (B.39) is taken to have its argument between $-\pi/4$ and $\pi/4$.

Now let a be a real number. Consider the integral

$$I(a) = \int_0^\infty e^{-ja^2(u^2+1)}\frac{du}{u^2+1} \tag{B.40}$$

Differentiating this integral with respect to a gives

$$I'(a) = -2ja\int_0^\infty e^{-ja^2(u^2+1)}\, du = -\sqrt{\pi}\text{sgn}(a)e^{-ja^2+j\pi/4} \tag{B.41}$$

where we have used the result (B.39) for $b = ja^2$, $\text{sgn}(a) = 2\vartheta(a) - 1 \ (= +1$ if $a > 0$ and $= -1$ if $a < 0$) is the *signum* (algebraic sign) function, and $\vartheta(a)$ is the unit step function. Upon integrating this result with respect to a and using the initial condition

$$I(0) = \int_0^\infty \frac{du}{u^2+1} = \frac{\pi}{2}$$

we have that

$$I(a) = \frac{\pi}{2} - \sqrt{\pi}\text{sgn}(a)e^{j\pi/4}\int_0^a e^{-jt^2}\, dt \tag{B.42}$$

The Fresnel integral function $F(z)$ is defined as:

$$F(z) = e^{jz^2}\int_{-\infty}^z e^{-jt^2}\, dt \tag{B.43}$$

and thus (B.42) can be expressed in terms of it:

$$I(a) = \pi\vartheta(a) - \sqrt{\pi}\text{sgn}(a)e^{-ja^2+j\pi/4}F(a) \tag{B.44}$$

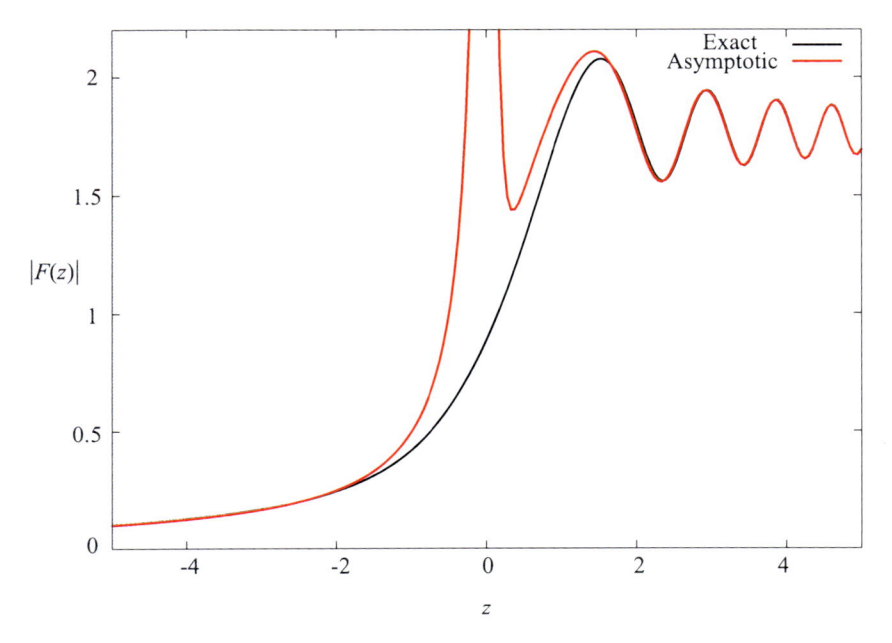

Figure B.2 Magnitude of the Fresnel integral and its large argument approximation versus its argument.

The Fresnel integral has the large argument expansion

$$F(z) \sim e^{jz^2 - j\pi/4} \sqrt{\pi} \vartheta(z) - \frac{1}{2jz} \tag{B.45}$$

as $|z| \to \infty$, while on the other hand

$$F(0) = \frac{\sqrt{\pi}}{2} e^{-j\pi/4} \tag{B.46}$$

The magnitude of the Fresnel integral is plotted along with its asymptotic approximation (B.45) in Figure B.2. From this we see that (B.45) is moderately accurate if $|z| > 1$, and very accurate if $|z| > 2$.

Other functions related to $F(z)$, and also given the name Fresnel integrals, are often encountered. We note in particular the functions

$$C(z) = \int_0^z \cos \frac{\pi t^2}{2} \, dt \tag{B.47}$$

and

$$S(z) = \int_0^z \sin \frac{\pi t^2}{2} \, dt \tag{B.48}$$

These functions are both real when their argument is real, and are related to F by

$$F(z) = \sqrt{\frac{\pi}{2}} e^{jz^2} \left[C\left(z\sqrt{\frac{2}{\pi}}\right) - jS\left(z\sqrt{\frac{2}{\pi}}\right) + \frac{1-j}{2} \right] \qquad (B.49)$$

B.5   LEGENDRE FUNCTIONS

The differential equation satisfied by Legendre functions is

$$\frac{1}{\sin\theta}\frac{d}{d\theta}\left(\sin\theta\frac{d\Theta}{d\theta}\right) + \left[n(n+1) - \frac{m^2}{\sin^2\theta}\right]\Theta = 0 \qquad (B.50)$$

Solutions of (B.50) are called *associated Legendre functions* $P_n^m(\cos\theta)$ and $Q_n^m(\cos\theta)$. When $n \geq 0$ is an integer, P_n^m is a well-behaved function of θ whereas Q_n^m has singularities at points where $\cos\theta = \pm 1$ ($\theta = 0$ or π). In problems where the entire range of θ from 0 to π are involved, we must thus eliminate Q_n^m from our solution. Some particular instances of the P_n^m for integer values of m and n are:

$$
\begin{aligned}
P_0^0(\cos\theta) &= 1 \\
P_1^0(\cos\theta) &= \cos\theta \\
P_1^1(\cos\theta) &= \sin\theta \\
P_2^0(\cos\theta) &= \frac{1}{2}(3\cos^2\theta - 1) \\
P_2^1(\cos\theta) &= 3\cos\theta\sin\theta \\
P_2^2(\cos\theta) &= 3\sin^2\theta
\end{aligned}
$$

In general, for nonnegative integers m, and any integers n,

$$P_n^m(x) = (-1)^m (1-x^2)^{m/2} \left(\frac{d}{dx}\right)^m P_n(x) \qquad (B.51)$$

where

$$P_n(x) \equiv P_n^0(x) \qquad (B.52)$$

is called simply the Legendre function of the first kind. Since $P_n(\cos\theta)$ is an even function about $\theta = \pi/2$ if n is even, while it is an odd function if n is odd, (B.51) shows that $P_n^m(\cos\theta)$ is even or odd accordingly as $m + n$ is. Moreover, (B.51) also shows that

$$P_n^m(\cos\theta) \equiv 0 \qquad \text{if } |m| > n \qquad (B.53)$$

Additionally, P_n^m is always real for real θ. From (B.51), we can prove the following derivative formula for $x = \cos\theta$:

$$\frac{dP_n^m(\cos\theta)}{d\theta} = m\cot\theta P_n^m(\cos\theta) + P_n^{m+1}(\cos\theta) \qquad (B.54)$$

Certain special values of the argument are of interest, notably:

$$P_n^{-m}(1) = 0 \qquad \text{if } m \neq 0 \tag{B.55}$$

and

$$P_n(1) = 1 \quad ; \qquad P_n(-1) = (-1)^n \tag{B.56}$$

An expansion theorem analogous to the Fourier series exists for the P_n^m functions. We have

$$f_m(\theta) = \sum_{n=0}^{\infty} A_n^m P_n^m(\cos\theta)$$

$$A_n^m = (-1)^m \left(n + \frac{1}{2}\right) \int_0^\pi P_n^{-m}(\cos\theta) f_m(\theta) \sin\theta \, d\theta \tag{B.57}$$

where P_n^{-m} is related to P_n^m by:

$$P_n^{-m}(\cos\theta) = (-1)^m \frac{(n-m)!}{(n+m)!} P_n^m(\cos\theta) \tag{B.58}$$

The expansion theorem (B.57) implies the orthogonality relation

$$\int_0^\pi P_n^m(\cos\theta) P_{n'}^{-m}(\cos\theta) \sin\theta \, d\theta = \frac{2(-1)^m}{2n+1} \delta_{nn'} \quad (|m| \leq n) \tag{B.59}$$

where

$$\delta_{mm'} = 0 \qquad \text{if } m \neq m'$$
$$= 1 \qquad \text{if } m = m' \tag{B.60}$$

is called the Kronecker delta function.

If the function $f_m(\theta)$ is even about $\theta = \pi/2$, then (B.57) reduces to

$$f_m(\theta) = \sum_{\substack{n=0 \\ (n-m=\text{even})}}^{\infty} A_n^m P_n^m(\cos\theta)$$

$$\tag{B.61}$$

$$A_n^m = (-1)^m (2n+1) \int_0^{\pi/2} P_n^{-m}(\cos\theta) f_m(\theta) \sin\theta \, d\theta \qquad [(m-n) = \text{even}]$$

$$= 0 \qquad\qquad\qquad\qquad\qquad\qquad\qquad\qquad\qquad [(m-n) = \text{odd}]$$

If the function $f_m(\theta)$ is odd about $\theta = \pi/2$, then (B.57) becomes

$$f_m(\theta) = \sum_{\substack{n=0 \\ (n-m=\text{odd})}}^{\infty} A_n^m P_n^m(\cos\theta)$$

(B.62)

$$A_n^m = (-1)^m (2n+1) \int_0^{\pi/2} P_n^{-m}(\cos\theta) f_m(\theta) \sin\theta \, d\theta \qquad [(m-n) = \text{odd}]$$

$$= 0 \qquad\qquad\qquad\qquad\qquad\qquad\qquad\qquad [(m-n) = \text{even}]$$

Certain integrals are important in practical applications. Notable is the following integral representation of the spherical Bessel function j_n:

$$j_n(z) = \frac{j^n}{2} \int_{-1}^{1} e^{-jzu} P_n(u) \, du = \frac{j^n}{2} \int_0^{\pi} e^{-jz\cos\theta} P_n(\cos\theta) \sin\theta \, d\theta \quad \text{(B.63)}$$

When this result is combined with (B.57) we have the spherical harmonic expansion of the exponential function

$$e^{-jz\cos\theta} = \sum_{n=0}^{\infty} (-j)^n (2n+1) j_n(z) P_n(\cos\theta)$$

(B.64)

B.6 CHEBYSHEV POLYNOMIALS

The Chebyshev polynomial of the first kind is defined as

$$T_m(z) \equiv \cos(m \cos^{-1} z)$$

(B.65)

Virtually all its properties can be obtained from corresponding trigonometric identities using a change of variable. Noteworthy is *Rodrigues's formula*:

$$A_m T_m(z) = (-1)^m \sqrt{1-z^2} \frac{d^m}{dz^m}[(1-z^2)^{m-1/2}]$$

(B.66)

where $A_0 = 1$, while for $m > 0$,

$$A_m = 1 \cdot 3 \cdot 5 \cdot \ldots \cdot (2m-1)$$

A recurrence relation for the T_m is

$$T_{m+1}(z) = 2z T_m(z) - T_{m-1}(z)$$

(B.67)

B.7 EXPONENTIAL INTEGRALS

The exponential integral $E_m(x)$ is defined as

$$E_m(x) = \int_x^{\infty} \frac{e^{-t}}{t^m} dt$$

(B.68)

The important special case of $m = 1$ merits attention; its small argument behavior is

$$E_1(x) \simeq -\gamma_E - \ln x + x + O(x^2) \tag{B.69}$$

as $x \to 0$, while as $x \to \infty$,

$$E_1(x) \sim \frac{e^{-x}}{x}\left[1 - \frac{1}{x} + O(x^{-2})\right] \tag{B.70}$$

B.8 POLYLOGARITHMS

The polylogarithm[3] of order n is defined as

$$\mathrm{Li}_n(z) = \sum_{m=1}^{\infty} \frac{z^m}{m^n} \tag{B.71}$$

for all $|z| < 1$, and by analytic continuation elsewhere. The dilogarithm Li_2 is of particularly frequent appearance, and can be approximated[4] to within about 0.1% accuracy between $-1 < z < 1$ by the Hermite-Padé approximant:

$$\mathrm{Li}_2(z) \simeq \frac{326z + 29z^2 + (206 + 6z)(1 - z)\ln(1 - z)}{96z + 120} \tag{B.72}$$

There are many related functions that also arise in applications. One is:

$$Q(z) = \sum_{m=1}^{\infty} z^m \ln\left(1 + \frac{1}{m}\right) \tag{B.73}$$

for which an approximation good to within about 0.1% on $-1 < z < 1$ is:

$$\begin{aligned}
Q(z) \simeq{}& -\ln(1 - z) - \frac{1}{2}\mathrm{Li}_2(z) + z\left(\ln 2 - \frac{1}{2}\right) + z^2\left(\ln\frac{3}{2} - \frac{3}{8}\right) \\
& + z^3\left(\ln\frac{4}{3} - \frac{5}{18}\right) + z^4\left(\ln\frac{5}{4} - \frac{7}{32}\right) + z^5\left(\ln\frac{6}{5} - \frac{9}{50}\right)
\end{aligned} \tag{B.74}$$

Plots of $\mathrm{Li}_2(z)$ and $Q(z)$ are shown in Figure B.3.

---

[3]L. Lewin, *Polylogarithms and Associated Functions*. New York: North Holland, 1981.

[4]See:

E. S. Ginsberg, *Commun. ACM*, vol. 18, pp. 200-202, 1975.
R. Morris, *Math. Comp.*, vol. 33, pp. 778-787, 1979.
S. Paszkowski, *Numer. Alg.*, vol. 10, pp. 337-361, 1995.

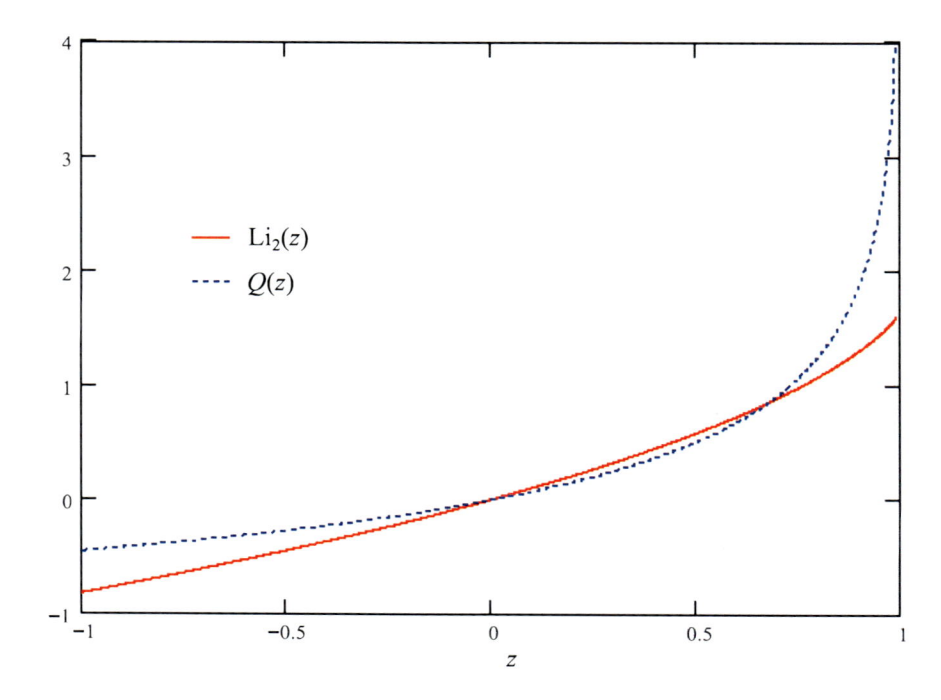

Figure B.3 Plots of $\mathrm{Li}_2(z)$ and $Q(z)$.

B.9 PROBLEMS

B–1 Show directly (without using (4.59)-(4.61)) that the function $f(\rho)$ defined by

$$f(\rho) = \frac{1}{2\pi} \int_0^{2\pi} e^{-jk\rho \cos \phi - jm\phi}\, d\phi$$

is a solution of Bessel's differential equation:

$$\left(\frac{1}{\rho} \frac{\partial}{\partial \rho} \rho \frac{\partial}{\partial \rho} + k^2 - \frac{m^2}{\rho^2} \right) f(\rho) = 0$$

when m is an integer. Show further that

$$f(\rho) \to \frac{e^{-jm\pi/2}}{m!} \left(\frac{k\rho}{2} \right)^m$$

as $k\rho \to 0$. (Hint: Use integration by parts.) This proves that

$$f(\rho) = e^{-jm\pi/2} J_m(k\rho)$$

and hence verifies (B.22).

B–2 Similarly to the previous problem, show that (B.23) obeys (B.6) for non-integer values of $\nu > 0$, and that as $k\rho \to 0$, it approaches

$$\frac{(z/2)^\nu}{\Gamma(\nu + 1)}$$

B–3 Prove (B.63) by (a) showing that the right side satisfies the differential equation (B.26), and (b) proving that both sides have the same behavior as $z \to 0$.

B–4 Prove (B.54) using (B.51).

C Rellich's Theorem

In this Appendix we will prove a result known as Rellich's theorem that is an important part of the demonstration of uniqueness for time-harmonic electromagnetic fields. Rellich's theorem is stated as follows. Let $\mathbf{E}$ and $\mathbf{H}$ be solutions of Maxwell's equations with no impressed sources for all r greater than some R_0 in a homogeneous medium whose material parameters are μ and ϵ. The field is not assumed to obey a radiation condition (Rellich's theorem is used in the proof of a uniqueness theorem that invokes the radiation condition, and we wish to avoid circular arguments). Then if

$$\lim_{R\to\infty} \frac{1}{2} \oint_{S_R} \left[\sqrt{\frac{\epsilon}{\mu}} |\mathbf{E}_{\text{tan}}|^2 + \sqrt{\frac{\mu}{\epsilon}} |\mathbf{H}_{\text{tan}}|^2 \right] dS = 0 \tag{C.1}$$

where S_R is a sphere of radius R centered at the origin, then $\mathbf{E}$ and $\mathbf{H}$ must vanish identically for all $r > R_0$. In particular $\mathbf{E}_{\text{tan}}$ and $\mathbf{H}_{\text{tan}}$ are zero as $r \to R_0^+$.

To prove this theorem, let v represent any of the Cartesian components of $\mathbf{E}$ or $\mathbf{H}$. We can write v as an expansion in spherical harmonics (5.99), with

$$v_n^m(r) = c_n^m h_n^{(1)}(kr) + d_n^m h_n^{(2)}(kr) \tag{C.2}$$

where c_n^m and d_n^m are complex constants. Note that we do *not* assume c_n^m to be zero as would be implied by the radiation condition. Using the asymptotic forms (B.14) to obtain the large argument forms of the spherical Bessel functions, we have

$$v_n^m(r) \sim \frac{2}{\pi kr} \left[c_n^m e^{j[kr-(n+1)\pi/2]} + d_n^m e^{-j[kr-(n+1)\pi/2]} + O(r^{-1}) \right] \tag{C.3}$$

as $r \to \infty$. Using (5.100) we can write

$$
\begin{aligned}
\oint_{S_R} |v|^2 \, dS &= 4\pi R^2 \sum_{n=0}^{\infty} \sum_{m=-n}^{n} |v_n^m(R)|^2 \frac{(n+m)!}{(2n+1)(n-m)!} \\
&= \frac{16}{\pi k^2} \left\{ \sum_{n=0}^{\infty} \sum_{m=-n}^{n} [|c_n^m|^2 + |d_n^m|^2 \right. \\
&\quad \left. + 2(-1)^{n+1}\text{Re}(c_n^m d_n^{m*} e^{2jkR})] \frac{(n+m)!}{(2n+1)(n-m)!} + O(R^{-1}) \right\} \\
&= A + \text{Re}[Be^{2jkR}] + O(R^{-1}) \tag{C.4}
\end{aligned}
$$

as $R \to \infty$, where " $*$ " denotes complex conjugate, and A and B are the constants

$$A = \frac{16}{\pi k^2} \sum_{n=0}^{\infty} \sum_{m=-n}^{n} (|c_n^m|^2 + |d_n^m|^2) \frac{(n+m)!}{(2n+1)(n-m)!}$$

$$B = \frac{32}{\pi k^2} \sum_{n=0}^{\infty} \sum_{m=-n}^{n} (-1)^{n+1} c_n^m d_n^{m*} \frac{(n+m)!}{(2n+1)(n-m)!}$$

A is real and nonnegative, while B is generally complex.

Now if

$$\lim_{R \to \infty} \oint_{S_R} |v|^2 \, dS = \lim_{R \to \infty} \oint_{S_R} \{A + \mathrm{Re}[Be^{2jkR}]\} \, dS = 0 \qquad (C.5)$$

then clearly both A and B must be zero. Only if all the c_n^m and d_n^m are zero can A be zero, so by (5.99) v itself (and hence all components of $\mathbf{E}$ and $\mathbf{H}$) vanishes identically throughout the volume over which (5.99)/(C.2) holds. This completes the proof of Rellich's theorem.

D Vector Analysis

In this Appendix, we collect for reference some of the most useful of the formulas of vector analysis needed in our work. Some of the results are of a sort not usually found in such collections, and are more commonly found in treatments of *differential forms*. The theory of differential forms[1] in differential geometry contains many useful results generalizing those of traditional vector analysis in an elegant fashion. Though we will not use this apparatus here, we will obtain special cases of some of these results, proving them in the more familiar Gibbs notation. We will denote the unit vector in a given direction d by $\mathbf{u}_d$.

D.1 VECTOR IDENTITIES

The usual rules of vector algebra apply. A few nontrivial identities are useful in practice, notably:

$$\mathbf{A} \cdot (\mathbf{B} \times \mathbf{C}) = \mathbf{B} \cdot (\mathbf{C} \times \mathbf{A}) = \mathbf{C} \cdot (\mathbf{A} \times \mathbf{B}) \tag{D.1}$$

$$\mathbf{A} \times (\mathbf{B} \times \mathbf{C}) = (\mathbf{A} \cdot \mathbf{C})\mathbf{B} - (\mathbf{A} \cdot \mathbf{B})\mathbf{C} \tag{D.2}$$

The following vector identities are often found useful; those in the first group hold equally well if

$$\nabla = \mathbf{u}_x \frac{\partial}{\partial x} + \mathbf{u}_y \frac{\partial}{\partial y} + \mathbf{u}_z \frac{\partial}{\partial z}$$

---

[1]See:

F. D. Murnaghan, *Phys. Rev.*, vol. 17, pp. 73-88, 1921.

R. C. Buck, *Advanced Calculus.* New York: McGraw-Hill, 1965, Chapter 7.

H. Flanders, in *Studies in Global Geometry and Analysis*, S. S. Chern, ed. Englewood Cliffs, NJ: Prentice-Hall, 1967, pp. 57-95.

N. V. Balasubramanian, J. W. Lynn, and D. P. Sen Gupta, *Differential Forms on Electromagnetic Networks.* London: Butterworth, 1970.

W. B. Gordon, *J. Math. Phys.*, vol. 16, pp. 448-454, 1975.

P. Hermann, *Arch. Math.*, vol. 29, pp. 598-607, 1977.

G. A. Deschamps, *Proc. IEEE*, vol. 69, pp. 676-696, 1981.

D. Lovelock and H. Rund, *Tensors, Differential Forms, and Variational Principles.* New York: Dover, 1988.

A. Bossavit, *IEE Proc. pt. A*, vol. 135, pp. 179-187, 1988.

P. Bamberg and S. Sternberg, *A Course in Mathematics for Students of Physics*, vols. 1 and 2. Cambridge: Cambridge University Press, 1988.

H. Flanders, *Differential Forms with Applications to the Physical Sciences.* New York: Dover, 1989.

L. Knockaert, *Electromagnetics*, vol. 11, pp. 269-280, 1991.

A. Bossavit, *Computational Electromagnetism.* San Diego: Academic Press, 1998.

S. L. Yap, *Amer. Math. Monthly*, vol. 116, pp. 261-267, 2009.

is replaced by

$$\nabla_t = \mathbf{u}_x \frac{\partial}{\partial x} + \mathbf{u}_y \frac{\partial}{\partial y}$$

and $\mathbf{A}$, $\mathbf{B}$ by transverse vectors $\mathbf{A}_t$ and $\mathbf{B}_t$ in the appropriate formula.

$$\nabla(fg) = f\nabla g + g\nabla f \qquad (D.3)$$

$$\nabla \cdot (f\mathbf{A}) = f\nabla \cdot \mathbf{A} + \mathbf{A} \cdot \nabla f \qquad (D.4)$$

$$\nabla \times (f\mathbf{A}) = f\nabla \times \mathbf{A} - \mathbf{A} \times \nabla f \qquad (D.5)$$

$$\nabla(\mathbf{A} \cdot \mathbf{B}) = (\mathbf{A} \cdot \nabla)\mathbf{B} + (\mathbf{B} \cdot \nabla)\mathbf{A} + \mathbf{A} \times (\nabla \times \mathbf{B}) + \mathbf{B} \times (\nabla \times \mathbf{A}) \qquad (D.6)$$

$$\nabla \times (\mathbf{A} \times \mathbf{B}) = \mathbf{A}\nabla \cdot \mathbf{B} - \mathbf{B}\nabla \cdot \mathbf{A} + (\mathbf{B} \cdot \nabla)\mathbf{A} - (\mathbf{A} \cdot \nabla)\mathbf{B} \qquad (D.7)$$

$$\nabla \times \nabla f = 0 \qquad (D.8)$$

Other such identities that do not specialize to the transverse case in the above way are:

$$\nabla \cdot (\mathbf{A} \times \mathbf{B}) = \mathbf{B} \cdot \nabla \times \mathbf{A} - \mathbf{A} \cdot \nabla \times \mathbf{B} \qquad (D.9)$$

which specializes to

$$\nabla_t \cdot [\mathbf{A}_t \times (\mathbf{u}_z f)] = f\mathbf{u}_z \cdot \nabla \times \mathbf{A}_t - \mathbf{A}_t \cdot \nabla_t \times (\mathbf{u}_z f) \qquad (D.10)$$

and

$$\nabla \cdot (\nabla \times \mathbf{A}) = 0 \qquad (D.11)$$

which becomes

$$\nabla_t \cdot [\nabla_t \times (\mathbf{u}_z f)] = 0 \qquad (D.12)$$

Also of use are the special results

$$\nabla^2 \mathbf{F} = \nabla \nabla \cdot \mathbf{F} - \nabla \times \nabla \times \mathbf{F} \qquad (D.13)$$

$$\nabla \times \mathbf{r} = 0 \qquad (D.14)$$

$$\nabla_t \times \boldsymbol{\rho} = 0 \qquad (D.15)$$

$$\nabla \cdot \mathbf{r} = 3 \qquad (D.16)$$

$$\nabla_t \cdot \boldsymbol{\rho} = 2 \qquad (D.17)$$

$$(\mathbf{A} \cdot \nabla)\mathbf{r} = \mathbf{A} \qquad (D.18)$$

$$(\mathbf{A}_t \cdot \nabla_t)\boldsymbol{\rho} = \mathbf{A}_t \qquad (D.19)$$

where $\mathbf{r} = x\mathbf{u}_x + y\mathbf{u}_y + z\mathbf{u}_z$ is the position vector and $\boldsymbol{\rho} = x\mathbf{u}_x + y\mathbf{u}_y$.

If V is a volume bounded by a closed surface S and $\mathbf{u}_n$ is the outward unit normal to S, then

$$\int_V (\nabla \cdot \mathbf{A}) \, dV = \oint_S \mathbf{u}_n \cdot \mathbf{A} \, dS \qquad (D.20)$$

(the divergence theorem)

$$\int_V (\nabla f)\, dV = \oint_S f \mathbf{u}_n\, dS \qquad (\text{D.21})$$

$$\int_V (\nabla \times \mathbf{A})\, dV = \oint_S (\mathbf{u}_n \times \mathbf{A})\, dS \qquad (\text{D.22})$$

We have also

$$\nabla \times \mathbf{u}_n = 0 \qquad (\text{D.23})$$

$$\nabla \cdot \mathbf{u}_n = -2\mathrm{H} \qquad (\text{D.24})$$

where H is the so-called mean curvature of the surface S at a given point; it is zero for a plane surface.

If S is an open surface whose unit normal is $\mathbf{u}_n$ and whose closed boundary is the contour C, then

$$\int_S (\nabla \times \mathbf{A}) \cdot \mathbf{u}_n\, dS = \oint_C \mathbf{A} \cdot d\mathbf{l} \qquad (\text{D.25})$$

(Stokes' theorem)

$$\int_S (\mathbf{u}_n \times \nabla f)\, dS = \oint_C f\, d\mathbf{l} \qquad (\text{D.26})$$

where the direction of the line element $d\mathbf{l}$ is related to $\mathbf{u}_n$ by the right-hand rule. If S is a surface lying in a plane $z = \text{constant}$, C is its closed boundary, and $\mathbf{u}_n$ is an outward unit normal to C lying in the plane of S, then

$$\int_S (\nabla_t \cdot \mathbf{A}_t)\, dS = \oint_C \mathbf{A}_t \cdot \mathbf{a}_n\, dl \qquad (\text{D.27})$$

$$\int_S (\nabla_t f)\, dS = \oint_C f \mathbf{u}_n\, dl \qquad (\text{D.28})$$

where $\mathbf{A}_t = \mathbf{u}_x A_x + \mathbf{u}_y A_y$ is a vector always perpendicular to the z-axis. The following miscellaneous identities also hold:

$$\mathbf{u}_z \times (\mathbf{u}_z \times \mathbf{A}_t) = -\mathbf{A}_t \qquad (\text{D.29})$$

$$\mathbf{u}_z \times (\nabla \times \mathbf{A}) = \nabla A_z - \frac{\partial \mathbf{A}}{\partial z} \qquad (\text{D.30})$$

$$\nabla_t \times (\mathbf{u}_z \times \mathbf{A}_t) = \mathbf{u}_z (\nabla_t \cdot \mathbf{A}_t) \qquad (\text{D.31})$$

D.2   VECTOR DIFFERENTIATION IN VARIOUS COORDINATE SYSTEMS

Here we present expressions for vector differential operations in the three major coordinate systems. In all these formulas, f represents an arbitrary scalar function, and $\mathbf{A}$ an arbitrary vector function. Note in particular that the expressions for the Laplacian in non-Cartesian coordinates do *not* apply to components of vectors in those systems; e.g., $\mathbf{u}_r \cdot \nabla^2 \mathbf{A} \neq \nabla^2 A_r$ (because the unit vector $\mathbf{u}_r$ is not a constant). The Laplacian of a vector function must in general be obtained from (D.13).

D.2.1 RECTANGULAR (CARTESIAN) COORDINATES

$$\nabla f = \mathbf{u}_x \frac{\partial f}{\partial x} + \mathbf{u}_y \frac{\partial f}{\partial y} + \mathbf{u}_z \frac{\partial f}{\partial z} \tag{D.32}$$

$$\nabla \cdot \mathbf{A} = \frac{\partial A_x}{\partial x} + \frac{\partial A_y}{\partial y} + \frac{\partial A_z}{\partial z} \tag{D.33}$$

$$\nabla \times \mathbf{A} = \mathbf{u}_x \left(\frac{\partial A_z}{\partial y} - \frac{\partial A_y}{\partial z} \right) + \mathbf{u}_y \left(\frac{\partial A_x}{\partial z} - \frac{\partial A_z}{\partial x} \right) + \mathbf{u}_z \left(\frac{\partial A_y}{\partial x} - \frac{\partial A_x}{\partial y} \right) \tag{D.34}$$

$$\nabla^2 f = \frac{\partial^2 f}{\partial x^2} + \frac{\partial^2 f}{\partial y^2} + \frac{\partial^2 f}{\partial z^2} \tag{D.35}$$

D.2.2 CIRCULAR CYLINDRICAL COORDINATES

$$\nabla f = \mathbf{u}_\rho \frac{\partial f}{\partial \rho} + \mathbf{u}_\phi \frac{1}{\rho} \frac{\partial f}{\partial \phi} + \mathbf{u}_z \frac{\partial f}{\partial z} \tag{D.36}$$

$$\nabla \cdot \mathbf{A} = \frac{1}{\rho} \frac{\partial (\rho A_\rho)}{\partial \rho} + \frac{1}{\rho} \frac{\partial A_\phi}{\partial \phi} + \frac{\partial A_z}{\partial z} \tag{D.37}$$

$$\nabla \times \mathbf{A} = \mathbf{u}_\rho \left(\frac{1}{\rho} \frac{\partial A_z}{\partial \phi} - \frac{\partial A_\phi}{\partial z} \right) + \mathbf{u}_\phi \left(\frac{\partial A_\rho}{\partial z} - \frac{\partial A_z}{\partial \rho} \right) + \mathbf{u}_z \left(\frac{1}{\rho} \frac{\partial (\rho A_\phi)}{\partial \rho} - \frac{1}{\rho} \frac{\partial A_\rho}{\partial \phi} \right) \tag{D.38}$$

$$\nabla^2 f = \frac{1}{\rho} \frac{\partial}{\partial \rho} \left(\rho \frac{\partial f}{\partial \rho} \right) + \frac{1}{\rho^2} \frac{\partial^2 f}{\partial \phi^2} + \frac{\partial^2 f}{\partial z^2} \tag{D.39}$$

D.2.3 SPHERICAL COORDINATES

$$\nabla f = \mathbf{u}_r \frac{\partial f}{\partial r} + \mathbf{u}_\theta \frac{1}{r} \frac{\partial f}{\partial \theta} + \mathbf{u}_\phi \frac{1}{r \sin \theta} \frac{\partial f}{\partial \phi} \tag{D.40}$$

$$\nabla \cdot \mathbf{A} = \frac{1}{r^2} \frac{\partial (r^2 A_r)}{\partial r} + \frac{1}{r \sin \theta} \frac{\partial (\sin \theta A_\theta)}{\partial \theta} + \frac{1}{r \sin \theta} \frac{\partial A_\phi}{\partial \phi} \tag{D.41}$$

$$\nabla \times \mathbf{A} = \mathbf{u}_r \frac{1}{r \sin \theta} \left(\frac{\partial (\sin \theta A_\phi)}{\partial \theta} - \frac{\partial A_\theta}{\partial \phi} \right) + \mathbf{u}_\theta \frac{1}{r} \left(\frac{1}{\sin \theta} \frac{\partial A_r}{\partial \phi} - \frac{\partial (r A_\phi)}{\partial r} \right)$$
$$+ \mathbf{u}_\phi \frac{1}{r} \left(\frac{\partial (r A_\theta)}{\partial r} - \frac{\partial A_r}{\partial \theta} \right) \tag{D.42}$$

$$\nabla^2 f = \frac{1}{r^2} \frac{\partial}{\partial r} \left(r^2 \frac{\partial f}{\partial r} \right) + \frac{1}{r^2 \sin \theta} \frac{\partial}{\partial \theta} \left(\sin \theta \frac{\partial f}{\partial \theta} \right) + \frac{1}{r^2 \sin^2 \theta} \frac{\partial^2 f}{\partial \phi^2} \tag{D.43}$$

D.3   POINCARE'S LEMMA

The identities (D.8), (D.11), and (D.12) are special cases of a result known in differential geometry as Poincaré's lemma. In this section, we present several lemmas that are special cases of the converse of Poincaré's lemma. Lemmas 2 and 3 (under more restrictive conditions) are also derivable from Helmholtz's theorem, which is proved in the next section.

Lemma 1:   Any scalar function $f(\mathbf{r})$ can be expressed as the divergence of some vector function $\mathbf{F}(\mathbf{r})$ (possibly in the sense of generalized functions). In a region where f is identically zero, it is possible to choose $\mathbf{F}$ to be zero as well. This vector function is not unique, as the curl of any vector function may be added to a valid $\mathbf{F}$ without changing its divergence because of (D.11).

The proof is by construction:

$$
\begin{aligned}
f(\mathbf{r}) &= \int_0^1 \frac{d}{dt} \left[t^3 f(\mathbf{r}t) \right] \, dt \\
&= \int_0^1 \left[t^3 \frac{d}{dt} f(\mathbf{r}t) + 3t^2 f(\mathbf{r}t) \right] \, dt \\
&= \int_0^1 \left[t^2 (\mathbf{r} \cdot \nabla) f(\mathbf{r}t) + 3t^2 f(\mathbf{r}t) \right] \, dt \\
&= \nabla \cdot \left[\mathbf{r} \int_0^1 t^2 f(\mathbf{r}t) \, dt \right] \qquad\qquad (D.44)
\end{aligned}
$$

so that we can take $\mathbf{F}(\mathbf{r}) = \mathbf{r} \int_0^1 t^2 f(\mathbf{r}t) \, dt$. We have used the fact that

$$
\begin{aligned}
t \frac{d}{dt} f(\mathbf{r}t) &= t \frac{d}{dt} f(xt, yt, zt) \\
&= t \left(x \frac{\partial}{\partial(xt)} + y \frac{\partial}{\partial(yt)} + z \frac{\partial}{\partial(zt)} \right) f(xt, yt, zt) \\
&= (\mathbf{r} \cdot \nabla) f(\mathbf{r}t) \qquad\qquad (D.45)
\end{aligned}
$$

It should be noted that the choice of origin $\mathbf{r} = 0$ in this proof is arbitrary.

Lemma 2:   Let $\mathbf{F}(\mathbf{r})$ be a vector field which (possibly in the sense of generalized functions) satisfies the condition

$$
\nabla \times \mathbf{F} = 0 \qquad\qquad (D.46)
$$

in all space (if $\mathbf{F}$ is defined, and/or (D.46) holds, only in some parts of space, we can set it equal to zero at all other points, understanding that appropriate step functions now enter into its definition). Then there exists some scalar function f such that $\mathbf{F} = \nabla f$. The function f is arbitrary only to within an additive constant.

The proof is similar to that of Lemma 1, and is by construction:

$$
\begin{aligned}
\mathbf{F}(\mathbf{r}) &= \int_0^1 \frac{d}{dt} \left[t\mathbf{F}(\mathbf{r}t) \right] dt \\
&= \int_0^1 \left[\mathbf{F}(\mathbf{r}t) + (\mathbf{r} \cdot \nabla)\mathbf{F}(\mathbf{r}t) \right] dt \\
&= \nabla \int_0^1 \mathbf{r} \cdot \mathbf{F}(\mathbf{r}t) \, dt
\end{aligned}
\tag{D.47}
$$

by (D.4), (D.6), (D.14), and (D.18). We thus take $f(\mathbf{r}) = \int_0^1 \mathbf{r} \cdot \mathbf{F}(\mathbf{r}t) \, dt$.

Lemma 3: Let $\mathbf{F}(\mathbf{r})$ be a vector field which (possibly in the sense of generalized functions) satisfies the condition

$$
\nabla \cdot \mathbf{F} = 0
\tag{D.48}
$$

in all space (if $\mathbf{F}$ is defined, and/or (D.48) holds, only in some parts of space, we can set it equal to zero at all other points, understanding that appropriate step functions now enter into its definition). Then there exists some vector field $\mathbf{A}$ such that $\mathbf{F} = \nabla \times \mathbf{A}$. Clearly $\mathbf{A}$ is not unique, since the gradient of any scalar function can be added to a valid $\mathbf{A}$ without changing its curl.

The proof can be done along the lines of the proofs of the first two lemmas, resulting in

$$
\mathbf{A}(\mathbf{r}) = -\mathbf{r} \times \int_0^1 t\mathbf{F}(\mathbf{r}t) \, dt
\tag{D.49}
$$

and is left as an exercise. An instructive alternative proof uses a different construction as follows.

Suppose there does exist some $\mathbf{A} = \mathbf{A}^{(0)}$ as specified above. Then any other $\mathbf{A} = \mathbf{A}^{(0)} + \nabla f$ is also a possible solution, and in particular the choice of

$$
f = - \int_0^z A_z^{(0)}(x, y, z') \, dz' + \text{function of } (x, y) \text{ only}
$$

results in a vector $\mathbf{A} = \mathbf{A}_t$ that has only x and y components. Then,

$$
\mathbf{F}_t = \frac{\partial \mathbf{Q}}{\partial z}
$$

and

$$F_z = -\nabla_t \cdot \mathbf{Q}$$

where $\mathbf{Q} = \mathbf{u}_z \times \mathbf{A}_t$. Then

$$\mathbf{Q} = \int_0^z \mathbf{F}_t(x, y, z') \, dz' + \mathbf{Q}_0(x, y)$$

where $\mathbf{Q}_0(x, y) = \mathbf{Q}(x, y, 0)$ must be found to satisfy

$$F_z(x, y, 0) = -\nabla_t \cdot \mathbf{Q}_0(x, y)$$

We have reduced a set of equations in three unknowns to one with two unknowns. The observation that a term $\nabla \times [\mathbf{u}_z g(x, y)]$ can be added to any legitimate $\mathbf{Q}_0$ allows us as before to say that if such a vector function exists it must be possible to choose it with only one (say x) component, and thus to construct an explicit solution for it by quadrature. The result is

$$\mathbf{A}(x, y, z) = -\mathbf{u}_z \times \int_0^z \mathbf{F}_t(x, y, z') \, dz' - \mathbf{u}_x \int_0^y F_z(x, y', 0) \, dy'$$

Lemmas 1 and 2 also hold in two-dimensional (xy) versions nearly identical to the three-dimensional versions above. In the case of Lemma 3, we find that a vector $\mathbf{F}_t$ for which $\nabla_t \cdot \mathbf{F}_t = 0$ can always be represented as the curl of a vector with only a z-component: $\mathbf{F}_t = \nabla_t \times (\mathbf{u}_z f)$. The proofs are all similar to those above, and are left as an exercise.

D.4 HELMHOLTZ'S THEOREM

Helmholtz's theorem[2] states that: *A vector function* $\mathbf{F}(\mathbf{r})$ *defined in all of space can be represented as the sum of the curl of some vector function* $\mathbf{A}$ *and the gradient of some scalar function* a:

$$\mathbf{F} = \nabla \times \mathbf{A} + \nabla a \tag{D.50}$$

The divergence of $\mathbf{A}$ *can be specified to be any scalar function* g.

---

[2]We do not address fully all aspects of Helmholtz's theorem for general regions here. For other results and proofs with more technical detail, see:

O. Blumenthal, *Math. Annalen*, vol. 61, pp. 235-250, 1905.

J. Van Bladel, *J. Franklin Inst.*, vol. 269, pp. 445-462, 1960.

M. E. Gurtin, *Arch. Rat. Mech. Anal.*, vol. 9, pp. 225-233, 1962.

E. Martensen, *Potentialtheorie*. Stuttgart: B. G. Teubner, 1968, pp. 97-100, 143-144.

E. Martensen, *Zeits. Angew. Mat. Mech.*, vol. 52, pp. T17-T24, 1972.

J. Van Bladel, *Electromagnetics*, vol. 13, pp. 95-110, 1993.

We prove this first for the case when $\mathbf{F}(\mathbf{r})$ vanishes as $O(1/r)$ and $\partial\mathbf{F}/\partial r$ vanishes as $O(1/r^2)$ as $r \to \infty$. In (7.10), let $k = 0$, so that G is the static free space Green's function

$$G = \frac{1}{4\pi|\mathbf{r} - \mathbf{r}'|} \tag{D.51}$$

and let f be any of the Cartesian components of $\mathbf{F}$, so that s_f is the corresponding Cartesian component of $-\nabla^2\mathbf{F}$. Combining these three scalar equations into one vector equation, we get

$$\mathbf{F}(\mathbf{r}) = -\int G(\mathbf{r}, \mathbf{r}')\nabla'^2\mathbf{F}(\mathbf{r}')\, dV' \tag{D.52}$$

where the surface integral in (7.10) goes to zero as we let S recede to infinity because of the decay of G, $\mathbf{F}$, and their derivatives as $r \to \infty$. Now from (D.13), (D.52) becomes

$$\mathbf{F}(\mathbf{r}) = -\int G(\mathbf{r}, \mathbf{r}')\nabla'\nabla'\cdot\mathbf{F}(\mathbf{r}')\, dV' + \int G(\mathbf{r}, \mathbf{r}')\nabla'\times\nabla'\times\mathbf{F}(\mathbf{r}')\, dV' \tag{D.53}$$

Each of these terms can now be manipulated by means of tricks that are fairly standard in applications of Green's functions. We have for example

$$-\int G(\mathbf{r}, \mathbf{r}')\nabla'\nabla'\cdot\mathbf{F}(\mathbf{r}')\, dV'$$

$$= -\int \{\nabla'[G(\mathbf{r}, \mathbf{r}')\nabla'\cdot\mathbf{F}(\mathbf{r}')] - \nabla'G(\mathbf{r}, \mathbf{r}')\nabla'\cdot\mathbf{F}(\mathbf{r}')\}\, dV'$$

$$= -\nabla\int G(\mathbf{r}, \mathbf{r}')\nabla'\cdot\mathbf{F}(\mathbf{r}')\, dV' \tag{D.54}$$

where (D.21) has been used, together with the decay properties of the integrand at infinity, the fact that functions of primed variables are constants so far as ∇ is concerned, and (7.8). Likewise,

$$\int G(\mathbf{r}, \mathbf{r}')\nabla'\times\nabla'\times\mathbf{F}(\mathbf{r}')\, dV' = \nabla\times\int G(\mathbf{r}, \mathbf{r}')\nabla'\times\mathbf{F}(\mathbf{r}')\, dV' \tag{D.55}$$

where we have used (D.22). Thus, by construction we have (D.50) with

$$a = -\int G(\mathbf{r}, \mathbf{r}')\nabla'\cdot\mathbf{F}(\mathbf{r}')\, dV' \tag{D.56}$$

and

$$\mathbf{A} = \int G(\mathbf{r}, \mathbf{r}')\nabla'\times\mathbf{F}(\mathbf{r}')\, dV' \tag{D.57}$$

Direct evaluation of $\nabla\cdot\mathbf{A}$ using similar manipulations to those above shows that $\nabla\cdot\mathbf{A} = 0$. Since the gradient of an arbitrary scalar function f can be added to $\mathbf{A}$ without affecting (D.50), and since this would give a function $\mathbf{A}$ whose divergence was $\nabla^2 f$, we can achieve a desired value for $\nabla \cdot \mathbf{A}$ using a solution of

$$\nabla^2 f = g$$

which is found using (7.12) as above with $k = 0$.

If the vector field $\mathbf{F}$ does not have the requisite decay properties at infinity, or if it is not defined in all of space, we can modify the above proof by setting $\mathbf{F} = 0$ where it is undefined, or outside a sufficiently large sphere. This has the effect of introducing step functions into $\mathbf{F}$, and we must understand the above formulas in the sense of generalized functions (which means that certain surface integrals appear in the equations for $\mathbf{A}$ and a in addition to the volume integrals given above). Helmholtz's theorem thus remains valid when $\mathbf{r}$ lies in a restricted region of space V bounded by a closed surface S if we put

$$a = -\int G(\mathbf{r}, \mathbf{r}')\nabla' \cdot \mathbf{F}(\mathbf{r}') \, dV' + \oint_S G(\mathbf{r}, \mathbf{r}')\mathbf{F}(\mathbf{r}') \cdot \mathbf{u}'_n \, dS' \qquad (\text{D.58})$$

and

$$\mathbf{A} = \int G(\mathbf{r}, \mathbf{r}')\nabla' \times \mathbf{F}(\mathbf{r}') \, dV' - \oint_S G(\mathbf{r}, \mathbf{r}')\mathbf{u}'_n \times \mathbf{F}(\mathbf{r}') \, dS' \qquad (\text{D.59})$$

where $\mathbf{u}_n$ is the outward unit normal vector to S. In general, then, we can represent $\mathbf{F} = \nabla \times \mathbf{A}$ only if both $\nabla \cdot \mathbf{F} = 0$ in V and $\mathbf{F} \cdot \mathbf{u}_n = 0$ on S. Likewise, $\mathbf{F} = \nabla a$ only if both $\nabla \times \mathbf{F} = 0$ in V and $\mathbf{u}_n \times \mathbf{F} = 0$ on S.[3]

A two-dimensional version of Helmholtz's theorem also exists: *A vector field* $\mathbf{A}_t(\boldsymbol{\rho})$ *having only x and y components and depending only on* $\boldsymbol{\rho} = x\mathbf{u}_x + y\mathbf{u}_y$ *can be expressed as*

$$\mathbf{A}_t = \nabla_t f_1 + \nabla_t \times (\mathbf{u}_z f_2) = \nabla_t f_1 - \mathbf{u}_z \times \nabla_t f_2 \qquad (\text{D.60})$$

where f_1 and f_2 are scalar functions of $\boldsymbol{\rho}$ and $\nabla_t = \mathbf{u}_x \partial/\partial x + \mathbf{u}_y \partial/\partial y$ is the transverse (to z) del operator. Proofs and specification of detailed conditions for validity are left as an exercise.

D.5 GENERALIZED LEIBNITZ RULE

In one-dimensional calculus, consider an integral wherein both the integrand and the limits of integration depend on a parameter t. Then its derivative

---

[3]See

E. Martensen, *loc. cit.*,

P. M. Morse and H. Feshbach, *Methods of Theoretical Physics*, vol. 1. New York: McGraw-Hill, 1953, pp. 52-54.

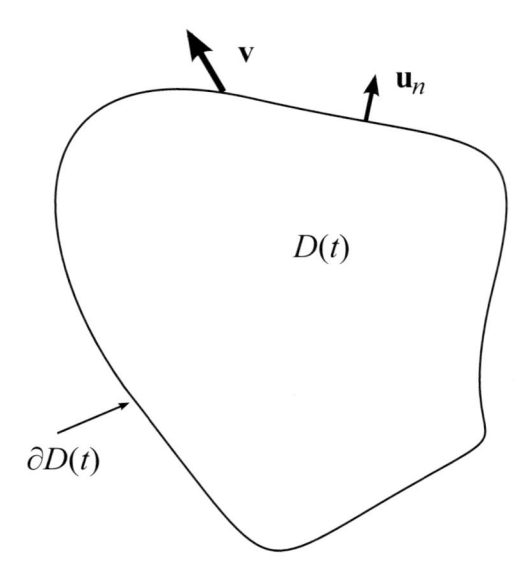

Figure D.1   A time-dependent volume $D(t)$ and its boundary surface $\partial D(t)$.

with respect to t is calculated via *Leibnitz's rule*:

$$\frac{d}{dt}\left(\int_{g(t)}^{h(t)} F(x,t)\,dx\right) = \int_{g(t)}^{h(t)} \frac{\partial F(x,t)}{\partial t}\,dx + \{F[h(t),t]h'(t) - F[g(t),t]g'(t)\}$$

(D.61)

In this section, we discuss the extension of this rule to multidimensional integrals,[4] and some related issues.

The basic trick we use is to remove the variable t from the "limits" of integration, and replace its effect by a suitable step function, which we can then differentiate under the integral sign. Thus, for example, suppose that $D(t)$ is a t-dependent volume in three-dimensional space (see Figure D.1). We write:

$$\int_{D(t)} F(\mathbf{r},t)\,dV = \int F(\mathbf{r},t)\vartheta[-f_D(\mathbf{r},t)]\,dV$$

(D.62)

where the integral sign without subscript implies integration over all of three-dimensional space, ϑ is the Heaviside unit step function, and f_D is a function describing the volume $D(t)$ such that $f_D > 0$ for points $\mathbf{r}$ outside $D(t)$, $f_D < 0$ for points inside $D(t)$, and $f_D = 0$ describes points on the boundary $\partial D(t)$

---

[4]See:

H. Flanders, *Amer. Math. Monthly*, vol. 80, pp. 615-627, 1973.
M. Silberstein, *Radio Science*, vol. 26, pp. 183-190, 1991.

of $D(t)$. It will be convenient in what follows to assume that we have chosen f_D to coincide with the outward normal coordinate (distance) n of $\mathbf{r}$ from $\partial D(t)$, at least in a sufficiently close neighborhood of that boundary where the normal coordinate can be unambiguously defined. The vector $\nabla f_D = \mathbf{u}_n$ is normal to $\partial D(t)$ at any time t. This gradient points toward the side of $\partial D(t)$ on which $f_D > 0$—that is, to the outside of $D(t)$.

We can say a few things about the "motion" of the surface $\partial D(t)$ if t is regarded as a time variable. The "velocity" $\mathbf{v}$ of a point $\mathbf{r}$ on $\partial D(t)$ at time t can be defined by considering additionally the point $\mathbf{r} + \Delta\mathbf{r}$ on $\partial D(t + \Delta t)$ at a later time $t + \Delta t$. Then

$$\mathbf{v} = \lim_{\Delta t \to 0} \frac{\Delta\mathbf{r}}{\Delta t} \tag{D.63}$$

From the definitions of $\partial D(t)$ and $\partial D(t + \Delta t)$, we have

$$0 = f_D(\mathbf{r}, t) \tag{D.64}$$

and

$$0 = f_D(\mathbf{r} + \Delta\mathbf{r}, t + \Delta t) = f_D(\mathbf{r}, t) + \Delta\mathbf{r} \cdot \nabla f_D(\mathbf{r}, t) + \Delta t \frac{\partial f_D(\mathbf{r}, t)}{\partial t} + \cdots \tag{D.65}$$

Then from (D.63)-(D.65), keeping only the lowest-order terms in Δt and $\Delta\mathbf{r}$, we obtain the normal component of the velocity as:

$$\mathbf{v} \cdot \mathbf{u}_n = -\frac{\partial f_D}{\partial t} \tag{D.66}$$

as might have been guessed on an intuitive basis. Observe that we get no information on components of $\mathbf{v}$ tangential to the surface from this procedure.

We are now in a position to differentiate (D.62) with respect to t:

$$\frac{d}{dt} \int_{D(t)} F(\mathbf{r}, t)\, dV = \int \left\{ \frac{\partial F(\mathbf{r}, t)}{\partial t} \vartheta[-f_D(\mathbf{r}, t)] + \frac{\partial \vartheta[-f_D(\mathbf{r}, t)]}{\partial t} F(\mathbf{r}, t) \right\} dV$$

$$= \int \left\{ \frac{\partial F(\mathbf{r}, t)}{\partial t} \vartheta[-f_D(\mathbf{r}, t)] - F(\mathbf{r}, t)\delta[-f_D(\mathbf{r}, t)] \frac{\partial f_D(\mathbf{r}, t)}{\partial t} \right\} dV$$

$$= \int_{D(t)} \frac{\partial F(\mathbf{r}, t)}{\partial t}\, dV + \int \left[\int_{S(n)} F(\mathbf{r}, t)\delta(-n)(\mathbf{v} \cdot \mathbf{u}_n)\, dS \right] dn$$

$$= \int_{D(t)} \frac{\partial F(\mathbf{r}, t)}{\partial t}\, dV + \int_{\partial D(t)} F(\mathbf{r}, t)(\mathbf{v} \cdot \mathbf{u}_n)\, dS \tag{D.67}$$

where $S(n)$ denotes the surface $n = $ const. This is the three-dimensional generalization of (D.61). Using the divergence theorem, (D.67) could also be expressed in the alternate form

$$\frac{d}{dt} \int_{D(t)} F(\mathbf{r}, t)\, dV = \int_{D(t)} \left\{ \frac{\partial F(\mathbf{r}, t)}{\partial t} + \nabla \cdot [F(\mathbf{r}, t)\mathbf{v}] \right\} dV \tag{D.68}$$

provided that $\mathbf{v}$ can be suitably defined for all points in $D(t)$, and not just on its boundary. Many similar results can be obtained using the technique of this section, and are left as exercises for the reader.

D.6 DYADICS[5]

In linear algebra, one often has occasion to transform an ordered set of numbers into another; this kind of transformation is called a *matrix*, and the ordered sets of numbers are called vectors. When we deal with vectors such as electromagnetic fields, currents, etc., however, it is common practice to refer to linear transformations of them not as matrices, but as *dyadics* (or sometimes second-rank *tensors*). Such dyadics are defined to be sums of one or more terms called *dyads*. A dyad is an association of two vectors—not a scalar or vector product—which can operate on ordinary vectors according to the rules we lay out below. We write a dyad $\overleftrightarrow{\mathbf{D}}$ of two vectors $\mathbf{A}$ and $\mathbf{B}$ symbolically as

$$\overleftrightarrow{\mathbf{D}} = \mathbf{A}\mathbf{B} \tag{D.69}$$

without the use of a dot or cross-product sign. The vector $\mathbf{A}$ is called the *anterior element* of $\overleftrightarrow{\mathbf{D}}$, while $\mathbf{B}$ is called the *posterior* element. We can now define right and left dot products of $\overleftrightarrow{\mathbf{D}}$ with a third vector $\mathbf{F}$ as

$$\begin{aligned} \mathbf{F} \cdot \overleftrightarrow{\mathbf{D}} &= (\mathbf{F} \cdot \mathbf{A})\mathbf{B} \\ \overleftrightarrow{\mathbf{D}} \cdot \mathbf{F} &= \mathbf{A}(\mathbf{B} \cdot \mathbf{F}) \end{aligned} \tag{D.70}$$

In this sense, we can think of $\overleftrightarrow{\mathbf{D}}$ as if it were the 3×3 matrix:

$$\overleftrightarrow{\mathbf{D}} \to \begin{bmatrix} A_x B_x & A_x B_y & A_x B_z \\ A_y B_x & A_y B_y & A_y B_z \\ A_z B_x & A_z B_y & A_z B_z \end{bmatrix} \tag{D.71}$$

The identity dyadic is defined by

$$\overleftrightarrow{\mathbf{I}} = \mathbf{u}_x\mathbf{u}_x + \mathbf{u}_y\mathbf{u}_y + \mathbf{u}_z\mathbf{u}_z \tag{D.72}$$

which possesses the property

$$\overleftrightarrow{\mathbf{I}} \cdot \mathbf{F} = \mathbf{F} \cdot \overleftrightarrow{\mathbf{I}} = \mathbf{F} \tag{D.73}$$

An operation that is a little less commonly encountered with matrices is the cross product, but it can readily be defined for dyads as:

$$\begin{aligned} \mathbf{F} \times \overleftrightarrow{\mathbf{D}} &= (\mathbf{F} \times \mathbf{A})\mathbf{B} \\ \overleftrightarrow{\mathbf{D}} \times \mathbf{F} &= \mathbf{A}(\mathbf{B} \times \mathbf{F}) \end{aligned} \tag{D.74}$$

---

[5] J. Van Bladel, *Electromagnetic Fields*, second ed. Hoboken, NJ: Wiley, 2007.
     I. V. Lindell, *Methods for Electromagnetic Field Analysis*. Piscataway, NJ: IEEE Press, 1995.

This definition is readily extended to include any dyadic. Notice that the result of taking the cross product of a dyadic with a vector is once again a dyadic, while the dot product of a dyadic with a vector is again a vector. There are two types of standard representations of a dyadic; one in terms of its anterior components

$$\overset{\leftrightarrow}{\mathbf{D}} = \mathbf{D}_x \mathbf{u}_x + \mathbf{D}_y \mathbf{u}_y + \mathbf{D}_z \mathbf{u}_z \tag{D.75}$$

and the other in terms of its posterior components

$$\overset{\leftrightarrow}{\mathbf{D}} = \mathbf{u}_x \mathbf{D}^x + \mathbf{u}_y \mathbf{D}^y + \mathbf{u}_z \mathbf{D}^z \tag{D.76}$$

The *transpose* $\overset{\leftrightarrow}{\mathbf{D}}^T$ of a dyadic $\overset{\leftrightarrow}{\mathbf{D}}$ is defined as

$$\overset{\leftrightarrow}{\mathbf{D}}^T = \mathbf{D}^x \mathbf{u}_x + \mathbf{D}^y \mathbf{u}_y + \mathbf{D}^z \mathbf{u}_z = \mathbf{u}_x \mathbf{D}_x + \mathbf{u}_y \mathbf{D}_y + \mathbf{u}_z \mathbf{D}_z \tag{D.77}$$

We then have

$$\overset{\leftrightarrow}{\mathbf{D}}^T \cdot \mathbf{F} = \mathbf{F} \cdot \overset{\leftrightarrow}{\mathbf{D}} \tag{D.78}$$

and a host of similar properties. The vector triple product formula extends to certain combinations of vectors and dyadics; for example,

$$\overset{\leftrightarrow}{\mathbf{D}} \times \mathbf{E} \cdot \mathbf{F} = \overset{\leftrightarrow}{\mathbf{D}} \cdot (\mathbf{E} \times \mathbf{F}) \tag{D.79}$$

$$\mathbf{E} \cdot \mathbf{F} \times \overset{\leftrightarrow}{\mathbf{D}} = (\mathbf{E} \times \mathbf{F}) \cdot \overset{\leftrightarrow}{\mathbf{D}} \tag{D.80}$$

Various differential operations can be carried out on dyadics; the meanings of some of them are given below:

$$\nabla \mathbf{A} = \mathbf{u}_x \frac{\partial \mathbf{A}}{\partial x} + \mathbf{u}_y \frac{\partial \mathbf{A}}{\partial y} + \mathbf{u}_z \frac{\partial \mathbf{A}}{\partial z} \tag{D.81}$$

$$\nabla \cdot \overset{\leftrightarrow}{\mathbf{D}} = \nabla \cdot \mathbf{D}_x \mathbf{u}_x + \nabla \cdot \mathbf{D}_y \mathbf{u}_y + \nabla \cdot \mathbf{D}_z \mathbf{u}_z = \frac{\partial \mathbf{D}^x}{\partial x} + \frac{\partial \mathbf{D}^y}{\partial y} + \frac{\partial \mathbf{D}^z}{\partial z} \tag{D.82}$$

$$\nabla \times \overset{\leftrightarrow}{\mathbf{D}} = \nabla \times \mathbf{D}_x \mathbf{u}_x + \nabla \times \mathbf{D}_y \mathbf{u}_y + \nabla \times \mathbf{D}_z \mathbf{u}_z \tag{D.83}$$

$$= \mathbf{a}_x \left(\frac{\partial \mathbf{D}^z}{\partial y} - \frac{\partial \mathbf{D}^y}{\partial z} \right) + \mathbf{u}_y \left(\frac{\partial \mathbf{D}^x}{\partial z} - \frac{\partial \mathbf{D}^z}{\partial x} \right) + \mathbf{u}_z \left(\frac{\partial \mathbf{D}^y}{\partial x} - \frac{\partial \mathbf{D}^x}{\partial y} \right)$$

If $\mathbf{c}$ is a constant vector, then

$$\nabla \nabla f \cdot \mathbf{c} = \nabla \nabla \cdot (f \mathbf{c}) \tag{D.84}$$

D.7 PROBLEMS

D–1 Prove (D.23); that is, show that $\nabla \times \mathbf{u}_n = 0$, where $\mathbf{u}_n$ is a unit vector normal to the surface defined by $n = $ const, and n is an orthogonal curvilinear coordinate that is the distance from a surface S $(n = 0)$ measured along a line perpendicular to S.

D–2 Prove that

$$
\nabla \times [\nabla \times (\mathbf{u}_r f)] = \mathbf{u}_r \left[\frac{1}{r^2} \frac{\partial}{\partial r} \left(r^2 \frac{\partial f}{\partial r} \right) - \nabla^2 f \right] + \nabla_t \frac{\partial f}{\partial r}
$$

for any scalar function f, where

$$
\nabla_t f = -\mathbf{u}_r \times (\mathbf{u}_r \times \nabla f)
$$

is the portion of the gradient perpendicular to the r-direction.

D–3 State and prove the two-dimensional variants of the inverses of Poincaré's Lemmas 1, 2, and 3.

D–4 Prove (D.49) using a technique analogous to (D.44)-(D.45) or (D.47).

D–5 Prove (D.58) and (D.59).

D–6 Prove (D.60) and give conditions under which it is valid.

D–7 Prove that

$$
\nabla \times \left(\overleftrightarrow{\mathbf{I}} f \right) = \nabla f \times \overleftrightarrow{\mathbf{I}}
$$

for any scalar function f.

D–8 Prove that

$$
\left(\overleftrightarrow{\mathbf{I}} \times \mathbf{A} \right) \cdot \mathbf{B} = \mathbf{A} \cdot \left(\overleftrightarrow{\mathbf{I}} \times \mathbf{B} \right) = \mathbf{A} \times \mathbf{B}
$$

for any vectors $\mathbf{A}$ and $\mathbf{B}$.

E Formulation of Some Special Electromagnetic Boundary Problems

The most common scattering problems of electromagnetics were stated in Section 3.4. Some of the more specialized problems are given in this Appendix.

E.1 LINEAR CYLINDRICAL (WIRE) ANTENNAS

Because they can serve in either a transmitting or receiving mode, antenna problems are a rather special mixture of radiation and scattering type problems.[1] We will assume for simplicity that the antenna structure is embedded in otherwise infinite, uniform free space.

E.1.1 TRANSMITTING MODE

We consider first an idealized transmitting antenna as shown in Figure E.1. The antenna itself is a thin tubular perfect conductor of radius a and length $2l$ driven by an ideal voltage source of strength V_t inserted at the point $z = z_0$. The source region is shown in detail in Figure E.2. The voltage appears across a gap of width g that is small compared to a wavelength so that the electric field in the gap region can be treated as approximately static. The voltage

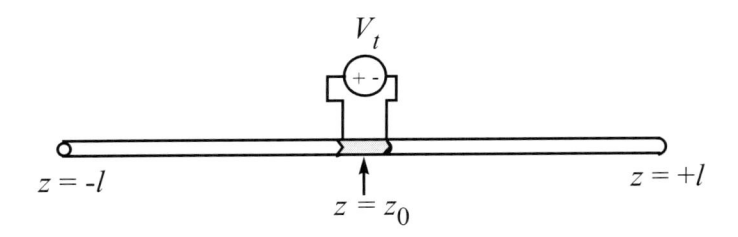

Figure E.1 Transmitting wire antenna.

[1]See, for example,

A. F. Stevenson, *Quart. Appl. Math.*, vol. 5, pp. 369-384, 1948.
Van Bladel, Section 8.7.

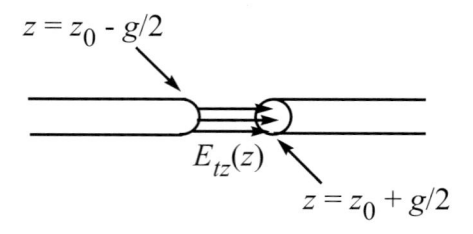

Figure E.2   Detail of voltage source.

corresponds to an electric field $E_{tz}(z)$ in the gap region $|z - z_0| < g/2$ that obeys

$$V_t = -\int_{z_0-g/2}^{z_0+g/2} E_{tz}(z) \, dz \qquad \text{(E.1)}$$

The detailed nature of this field distribution is dependent on the local geometry at the feed point, but for most purposes is not important and could be thought of as having the constant value V_t/g. This is a standard aperture radiation problem, and we denote the solution for its surface current density by $\mathbf{J}_S^{\text{tr}}$. By linearity, it is proportional to V_t, and we write

$$\mathbf{J}_S^{\text{tr}} = V_t \mathbf{J}_2 \qquad \text{(E.2)}$$

The total current passing through the voltage source can be used to evaluate the transmitting impedance of the antenna:

$$Z_{\text{tr}} = -\frac{V_t}{I_{\text{tr}}(z_0)} = -\frac{1}{I_2(z_0)} \qquad \text{(E.3)}$$

where the minus sign appears because of the choices of polarity for V and I at the gap. The total current corresponding to a current density at z_0 is given by integrating $\mathbf{J}_S$ around the circumference of the cylinder:

$$I(z_0) = \oint_C J_{Sz}(a, \phi, z_0) a \, d\phi \simeq 2\pi a \, J_{Sz}|_{\rho=a} \qquad \text{(E.4)}$$

since for a cylinder of small radius, the current is nearly uniformly distributed around the circumference of the wire (cf. Section 5.2).

We designate a suitable point in or on the antenna structure as the origin $\mathbf{r} = 0$. For r large enough, the field radiated by a transmitting antenna has the form of an outgoing *spherical wave* (3.31), described in spherical coordinates as:

$$\mathbf{E} \simeq \boldsymbol{\mathcal{E}}(\theta, \phi) \frac{e^{-jk_0 r}}{r} \quad , \quad \mathbf{H} \simeq \boldsymbol{\mathcal{H}}(\theta, \phi) \frac{e^{-jkr}}{r} \qquad \text{(E.5)}$$

where $k_0 = \omega\sqrt{\mu_0\epsilon_0}$ is the propagation constant of a plane wave in free space, $\boldsymbol{\mathcal{E}}(\theta, \phi)$ is the directional pattern of the electric field in the far-field region, and $\mathcal{H}(\theta, \phi)$ is that of the magnetic field. The wave propagates outward in the radial (r) direction, analogously to the propagation of a plane wave, but there is an additional amplitude decay proportional to $1/r$. The amplitude and polarization of the field may vary with the angular direction (θ, ϕ), but we always have:

$$\mathbf{u}_r \times \mathcal{E}(\theta, \phi) = \zeta_0 \mathcal{H}(\theta, \phi) \tag{E.6}$$

where $\zeta_0 = \sqrt{\mu_0/\epsilon_0} \simeq 377 \ \Omega$ is the wave impedance of free space, and $\mathbf{u}_r$ is the unit vector pointing in the outward radial direction. In other words, the electric and magnetic field vectors are always perpendicular to each other and to the direction of propagation. In this sense, the spherical wave behaves locally like a plane wave, so that the known properties of plane waves can be taken to be approximately true of the far field of a transmitting antenna.

E.1.2 RECEIVING MODE

Suppose on the other hand that the voltage source is replaced by an impedance surface on which

$$\mathbf{E}_{\text{tan}}\big|_{\rho=a} = Z_S \mathbf{J}_S \tag{E.7}$$

and the excitation is now provided by an incident wave $\mathbf{E}^i$, $\mathbf{H}^i$. This is a receiving antenna in which a different voltage V_r appears across the gap, proportional to the strength of the incident field, as is the received current $I_r(z_0)$ appearing at the gap location z_0. Integrating (E.7) over the gap shows that

$$V_r = Z_L I_r(z_0) \tag{E.8}$$

where

$$Z_L \simeq \frac{g Z_S}{2\pi a} \tag{E.9}$$

is the load impedance.

But by decomposing the boundary condition (3.61) into two parts and using superposition, the total field in the receiving case can be expressed as the sum of the field due to the incident wave scattering from a perfect conductor (i.e., the gap is metallized over), and that of a transmitting antenna with source voltage $Z_L I_r(z_0)$ appearing across the gap. Calling $I_1(z)$ the current for the first of these contributions—the *short-circuit receiving antenna current*—we have that the total loaded receiving antenna current must be

$$I_r(z) = I_1(z) + Z_L I_r(z_0) I_2(z) \tag{E.10}$$

and thus, putting $z = z_0$ and using (E.3), we get

$$I_r(z_0) = I_1(z_0) - \frac{Z_L}{Z_{\text{tr}}} I_r(z_0)$$

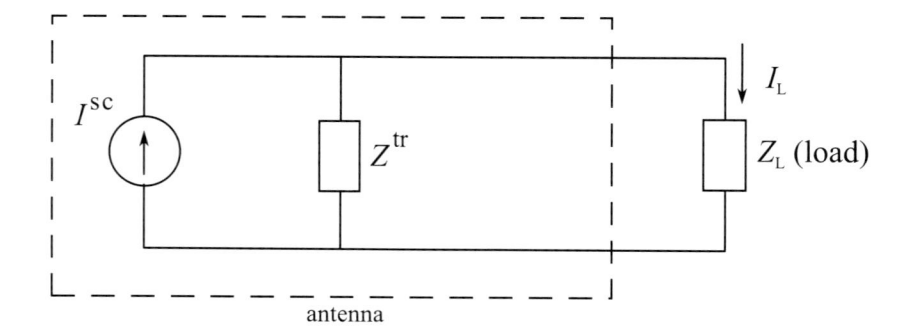

Figure E.3 Norton equivalent circuit for a receiving antenna.

or,

$$I_r(z_0) = \frac{Z_{\mathrm{tr}}}{Z_L + Z_{\mathrm{tr}}} I_1(z_0) \tag{E.11}$$

This equation can be interpreted by means of the Norton equivalent circuit shown in Figure E.3. The current $I_{\mathrm{sc}} = I_1(z_0)$ is the current that would be induced at z_0 by the receiving antenna if its load point were short-circuited. The current $I_L = I_r(z_0)$ actually delivered to a load impedance Z_L connected at z_0 is the result of a current divider having Z_L in parallel with the transmitting antenna impedance Z_{tr}. Maximum power is delivered to the load if there is a conjugate impedance match: $Z_L = Z_{\mathrm{tr}}^*$. It is clear then that Z_{tr} is more than simply a transmitting antenna impedance; it has important impact on the behavior of the receiving antenna as well.

Also of interest is the way in which the short-circuit current $I_1(z_0)$ depends upon the incident field. In particular, if the incident field is the far field of a second antenna, we wish to find the dependence of $I_1(z_0)$ on the direction of propagation and polarization of the incident wave. Use of the Lorentz reciprocity theorem on the fields of the transmitting and receiving modes shows that there is a relationship between this dependence and the far-field pattern of the transmitting mode. Details are left as an exercise.

E.2   STATIC PROBLEMS

Problems involving static electric or magnetic fields must be handled slightly differently than the time-harmonic ones dealt with in the foregoing subsections. Because $\omega = 0$, the derivations of the radiation condition as well as the uniqueness theorems in Chapter 3 no longer hold, and we find in fact that conditions that guarantee uniqueness for a time-harmonic field will not do so for a static field. In this section we will consider only static fields in the presence of ideal conductors. Static field problems with dielectric or magnetic bodies can also be formulated in a similar fashion, using the ideas of this section. Details are left as an exercise.

E.2.1   ELECTROSTATIC PROBLEMS

Consider first an electrostatic problem. An "incident" field $\mathbf{E}^i = -\nabla\Phi^i$ is produced by a given distribution of impressed charges ρ_{ext} in empty space. A perfectly conducting body whose surface is S and whose interior volume is V_{in} is placed in this incident field, resulting (as in the time-harmonic problem) in an additional scattered field $\mathbf{E}^s$, such that the total field $\mathbf{E}^{tot} = \mathbf{E}^i + \mathbf{E}^s$ satisfies the electrostatic Maxwell equations

$$\nabla \times \mathbf{E} = 0 \qquad \nabla \cdot \mathbf{D} = \rho_{ext} \tag{E.12}$$

as well as the boundary condition that the tangential component of $\mathbf{E}^{tot}$ be zero on S. It is assumed that the impressed sources of the incident field are located completely in the region V_e *exterior* to S. As we will see, these conditions are not enough to uniquely determine the resulting electric field.

To understand the reason for this, suppose that $\mathbf{E}_d$ is the difference between two static fields produced by the same sources in the presence of the same conducting obstacle. For simplicity, we assume that the obstacle is connected— that is, "in one piece." From (2.3), we can write

$$\mathbf{E}_d = -\nabla\Phi_d \tag{E.13}$$

and thus the nonnegative integral representing the stored electric energy (assumed finite) in the difference field can be written as

$$
\begin{aligned}
\frac{1}{2}\int_{V_e} \mathbf{D}_d \cdot \mathbf{E}_d\, dV &= -\frac{1}{2}\int_{V_e} \mathbf{D}_d \cdot \nabla\Phi_d\, dV \\
&= \frac{1}{2}\int_{V_e} [\Phi_d \nabla \cdot \mathbf{D}_d - \nabla \cdot (\mathbf{D}_d \Phi_d)]\, dV \\
&= \frac{1}{2}\oint_S \mathbf{D}_d \cdot \mathbf{u}_n \Phi_d\, dS \\
&= \frac{1}{2}\Phi_{Sd} q_{Sd}
\end{aligned}
\tag{E.14}
$$

where Φ_{Sd} is the constant value of the difference potential on S, while q_{Sd} is the total "difference" charge on S. We have assumed without loss of generality that an arbitrary additive constant in the difference potential Φ_d has been chosen so that it approaches zero as $r \to \infty$ (cf. (E.18) below). If the total charge *or* the potential on S were fixed, the charge or potential for the difference field would vanish and the value of (E.14) would be zero. We could then conclude that $\mathbf{E}_d$ was identically zero, and uniqueness would follow. Thus one of these two additional quantities needs to be specified in an electrostatic problem. If the conductor S is the union of more than one disjoint pieces, then the charge or potential on each of the pieces needs to be specified in order to render the field uniquely determined. We also note that the energy in the actual electrostatic field can be found by a derivation similar to that which led to (E.14):

$$\frac{1}{2} \int_{V_e} \mathbf{D} \cdot \mathbf{E} \, dV = \frac{1}{2} \Phi_S q_S \tag{E.15}$$

where Φ_S is the constant value of the total potential on S, while q_S is the total charge on S.

E.2.2 THE CAPACITANCE PROBLEM

In the capacitance problem, we stipulate that the potential at the conducting body (which we assume here to be of one connected piece), $\Phi_S \equiv \Phi_{Sc}$, has been specified, and additionally that no externally impressed sources ρ_{ext} are present. The solution of the boundary problem, as shown above, is now unique, its fields, charges and potentials indicated with the subscript $_c$. The total charge on the conductor is $Q_c = \oint_S \mathbf{D}_c \cdot \mathbf{u}_n \, dS$, from which we obtain the capacitance C, which is defined as:

$$C = \frac{Q_c}{\Phi_{Sc}} \tag{E.16}$$

We can also express C in terms of the stored energy in the field (see (E.15)):

$$C = \frac{1}{\Phi_{Sc}^2} \int_{V_e} \mathbf{D}_c \cdot \mathbf{E}_c \, dV \tag{E.17}$$

The capacitance problem can also be characterized alternatively by its far field. From (4.79), we can obtain an expression for the far field of a charged conductor of any shape. Let the charge density ρ be independent of time t for this static problem, and let $r \to \infty$ so that $|\mathbf{r} - \mathbf{r}'| \simeq r$. Then (4.79) reduces to

$$\Phi \simeq \frac{Q}{4\pi\epsilon_0 r}$$

for large enough r, which is the potential of a point charge whose strength is

$$Q = \int \rho \, dV$$

Making the appropriate adaptations for the capacitor problem, we have:

$$\Phi_c(\mathbf{r}) \sim \frac{Q_c}{4\pi\epsilon r} = \Phi_{Sc}\frac{C}{4\pi\epsilon r} \tag{E.18}$$

For the simplest three-dimensional geometry (S is a sphere of radius a), the capacitance problem has a well-known solution obtained by Gauss' law and symmetry arguments:

$$\Phi_c = \Phi_{Sc}\frac{a}{r}; \qquad Q_c = 4\pi\epsilon a \Phi_{Sc} \tag{E.19}$$

We can characterize the capacitance of *any* three-dimensional conductor in terms of that of a sphere of an *equivalent radius* a_{eq}:

$$C = 4\pi\epsilon a_{eq} \quad \text{or} \quad a_{eq} = \frac{C}{4\pi\epsilon} \qquad \text{(E.20)}$$

so that the far field of the charged conductor in the absence of an incident field is

$$\Phi_c(\mathbf{r}) \sim \Phi_{Sc}\frac{a_{eq}}{r} \qquad \text{(E.21)}$$

as $r \to \infty$. Thus a_{eq} can serve as an alternative parameter to C for characterizing the capacitance problem.

E.2.3   THE ELECTRIC POLARIZABILITY PROBLEM

The electric polarizability problem, on the other hand, is characterized by the presence of a nonzero, *uniform* incident field $\mathbf{E}^i = -\nabla\Phi^i$, where $\Phi^i = -\mathbf{r}\cdot\mathbf{E}^i$. Moreover, it is stipulated that the total charge on the conductor is zero. We think of the situation thus: an initially uncharged conductor is introduced into the impressed field without changing the total charge on the conductor. In this case, the potential Φ_{Sp} at the conductor, though still a constant, is not known. Nevertheless, the prescription of zero total charge on the conductor leads as described above to a unique solution for the fields, which will be indicated with the subscript $_p$. The total field is the sum of the incident field and a scattered field: $\mathbf{E}_p = \mathbf{E}^i + \mathbf{E}_p^s$.

To find the value of Φ_{Sp}, suppose the solution to the capacitance problem above is known. Moreover, suppose that the solution to the scattering problem with the same given incident potential Φ^i, *and a given conductor potential* Φ_{Sg} (but unspecified total charge) is also known, it being labelled with the subscript $_g$.

Then the solution to the polarizability problem can differ from Φ_g only by a solution to the static field equations that has zero incident field and constant potential at S—that is, by a constant multiple of the solution Φ_c of the capacitance problem. Because the total charge $Q_p = 0$, we find easily that

$$\Phi_{Sp} = \Phi_{Sg} - \frac{Q_g}{C} \qquad \text{(E.22)}$$

The polarizability problem is succinctly characterized in terms of the dipole moment $\mathbf{p}$ induced by the action of the incident field:

$$\mathbf{p} = \oint_S \mathbf{r}\rho_S\, dS \qquad \text{(E.23)}$$

It is evidently proportional to the strength of the incident field, a relationship that is used to define the electric polarizability of the scatterer:

$$\mathbf{p} = \epsilon\overset{\leftrightarrow}{\boldsymbol{\alpha}}_E \cdot \mathbf{E}^i \qquad \text{(E.24)}$$

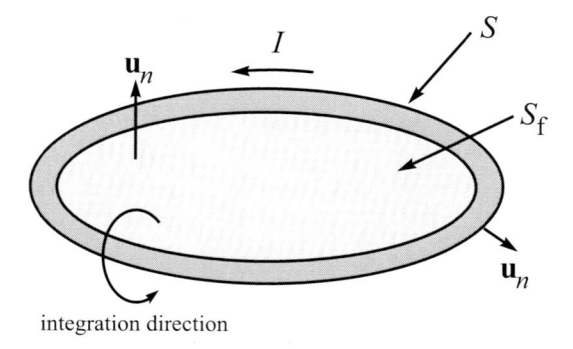

integration direction

Figure E.4 Multiply connected conductor.

for any possible uniform incident field $\mathbf{E}^i$. The energy stored in the scattered part of the field in the polarizability problem can be shown to be

$$\frac{1}{2}\int_{V_e} \mathbf{D}_p^s \cdot \mathbf{E}_p^s \, dV = \frac{1}{2}\mathbf{E}^i \cdot \mathbf{p} - \frac{1}{2}V_{\text{in}}\epsilon\mathbf{E}^i \cdot \mathbf{E}^i \qquad (\text{E.25})$$

where the notation V_{in} is used here to denote the volume of the region contained inside S. We see that the scattered field energy is $\mathbf{p} \cdot \mathbf{E}^i/2$ minus the energy of the incident field located inside S that is no longer present when the conductor is placed into the incident field. Since both V_{in} and the left side of (E.25) are nonnegative, it is clear that the polarizability thus defined is nonnegative, in the sense that $\epsilon\mathbf{E}^i \cdot \overset{\leftrightarrow}{\boldsymbol{\alpha}}_E \cdot \mathbf{E}^i \geq 0$ for any incident field $\mathbf{E}^i$.

E.2.4 MAGNETOSTATIC PROBLEMS

The boundary condition for a magnetostatic field at a perfect conductor is that the normal component of $\mathbf{B}$ should be zero. Our discussion of uniqueness for this problem parallels that of the electrostatic case to some extent, but must take account of the fact that the dual scalar potential Ψ introduced in (2.16) may not always be at once continuous *and* single-valued. For example, a conducting loop as shown in Figure E.4 must have the closed line integral $\oint \mathbf{H} \cdot dl$ equal to the current I flowing in the loop if the path of integration encircles the conductor once in the sense indicated. But if $\mathbf{H} = -\nabla\Psi$, this line integral must be zero unless the potential Ψ has a discontinuity at some point on the contour of integration and the closed contour of integration is interrupted at this point. We must therefore introduce a surface of discontinuity for Ψ that covers every "hole" in the conductor topology[2] like a soap film, and the amount of the discontinuity must be I.

---

[2]The region exterior to an object with at least one hole is said to be multiply connected; that exterior to an object with exactly one hole is said to be doubly connected.

Now consider the difference $\mathbf{H}_d$ between two static magnetic fields produced by the same sources in the presence of the same conducting obstacle. We choose the arbitrary additive constant in the potential Ψ_d so that $\Psi_d \to 0$ at infinity. The nonnegative integral representing the stored magnetic energy can be written as

$$
\begin{aligned}
\frac{1}{2} \int_{V_e} \mathbf{B}_d \cdot \mathbf{H}_d \, dV &= -\frac{1}{2} \int_{V_e} \mathbf{B}_d \cdot \nabla \Psi_d \, dV \\
&= \frac{1}{2} \int_{V_e} [\Psi_d \nabla \cdot \mathbf{B}_d - \nabla \cdot (\mathbf{B}_d \Psi_d)] \, dV \\
&= \frac{1}{2} \oint_{S_0} \mathbf{B}_d \cdot \mathbf{u}_n \Psi_d \, dS
\end{aligned}
\tag{E.26}
$$

where S_0 is not the conductor surface only, but a simply connected surface formed by adding to S both sides of each "soap film" surface S_f across which Ψ_d is discontinuous. For a conductor with only one such "hole," the integral in (E.26) reduces to

$$
\frac{1}{2} \int_{V_e} \mathbf{B}_d \cdot \mathbf{H}_d \, dV = \frac{1}{2} I_d \psi_{m,d}
\tag{E.27}
$$

where $\psi_{m,d}$ is the difference magnetic flux $\int_{S_f} \mathbf{B}_d \cdot \mathbf{u}_n \, dS$ through the hole and $I_d = I_1 - I_2$ is the difference between the currents corresponding to the two possible fields. In order to ensure that the magnetic field in this problem is unique, we must specify either the magnetic flux through each hole of the conductor, or the current in the closed loop around each hole.

The use of a composite surface S_0 for the magnetostatic problem with a multiply-connected conductor is the classical approach. We can avoid its introduction by treating the fields and potential in the context of generalized functions. Consider the scalar potential defined above. Suppose we were to follow a path that repeatedly looped through the surface S_f as shown, and that this path is described by a monotonic increase in some coordinate ϖ that takes the value of some multiple of 2π each time the path passes through S_f (ϖ is analogous to one of the coordinates used in toroidal coordinate systems). A continuous potential Ψ^c could be defined that is single-valued as a function of ϖ but multi-valued as a function of $\mathbf{r}$, because of the fact that an infinite number of values of ϖ will correspond to the same spatial point. The *single-valued* but discontinuous function $\Psi(\mathbf{r})$ defined above by the introduction of a discontinuity at S_f is related to Ψ^c by

$$
\Psi = \Psi^c + I \Upsilon_{S_f}
$$

where Υ_{S_f} is a "staircase" function of ϖ given by:

$$
\Upsilon_{S_f} = \sum_{p=-\infty}^{\infty} \vartheta_{S_f}(\varpi - 2p\pi)
$$

Since $\varpi = 2p\pi$ coincides with the surface $n = 0$, the function Υ_{S_f} behaves near S_f as $\vartheta_{S_f}(n) + \text{constant}$. Now the actual field is given by $\mathbf{H} = -\nabla\Psi^c$, while the hypothetical field

$$\mathbf{H}_0 = -\nabla\Psi = \mathbf{H} - I\mathbf{u}_n\delta_{S_f}(n)$$

contains a delta-function term in addition to the actual field. The field $\mathbf{H}_0$ now includes behavior *at* $n = 0$ that had been excluded before. Therefore, we can write, in terms of the discontinuous potential Ψ:

$$\mathbf{H} = -\nabla\Psi + I\mathbf{u}_n\delta_{S_f}(n) \tag{E.28}$$

The only difference between $\mathbf{H}$ and $\mathbf{H}_0$ is a delta-function term at S_f. In terms of the stream potential concept of Section 1.4.2, we can put $\mathbf{T}_m = 0$ and

$$\mathbf{T}_e = I\mathbf{u}_n\delta_{S_f}(n)$$

so that $\mathbf{J}_{\text{eq}} = \mathbf{J}_{\text{ext}}$ are the actual impressed current sources (if present) and $\mathbf{M}_{\text{eq}} = j\omega\mu\mathbf{T}_e \to 0$ for a static field problem. Thus, by (1.102),

$$\rho_{m,\text{eq}} = -I\nabla \cdot \left(\mu\mathbf{u}_n\delta_{S_f}(n)\right)$$

We interpret the extra term in $\mathbf{H}_0$ as the field at a double layer of magnetic charge:

$$\rho_{m,\text{eq}} = -I\left[\mu\delta'_{S_f}(n) + \delta_{S_f}(n)\nabla \cdot (\mu\mathbf{u}_n)\right]$$

Note that the closed line integral of $\mathbf{H}_0$ around any loop (whether current is enclosed by this loop or not) is zero (in the sense of generalized functions), and the only contribution to $\oint \mathbf{H} \cdot d\mathbf{l}$ is from the second term of (E.28). The evaluation of the magnetostatic energy now proceeds as:

$$
\begin{aligned}
\frac{1}{2}\int_{V_e} \mathbf{B} \cdot \mathbf{H}\, dS &= -\frac{1}{2}\int_{V_e} \mathbf{B} \cdot \nabla\Psi\, dV + \frac{1}{2}I\int_{S_f} \mathbf{B} \cdot \mathbf{u}_n\, dS \\
&= \frac{1}{2}I\psi_m \tag{E.29}
\end{aligned}
$$

analogous to what we found in (E.27) for the difference fields, having treated all these integrals in the sense of generalized functions.

E.2.5   THE INDUCTANCE PROBLEM

If there is no external incident field, we have the *inductance problem*,[3] wherein a circulating current I flows on the conductor around the hole, resulting in a net magnetic flux

$$\psi_{m,\text{ind}} = \int_{S_f} \mathbf{B}_{\text{ind}} \cdot \mathbf{u}_n\, dS \tag{E.30}$$

---

[3]Our approach uses concepts from the treatments found in:

P. Werner, *J. Math. Anal. Appl.*, vol. 14, pp. 445-462, 1966.

E. Arvas and R. Harrington, *IEEE Trans. Ant. Prop.*, vol. 31, pp. 719-725, 1983.

through the hole (see Figure E.4). The inductance of the conductor is then defined to be

$$L = \frac{\psi_{m,\text{ind}}}{I} \tag{E.31}$$

in the usual fashion. Alternatively, the inductance is expressible in terms of the stored energy:

$$L = \frac{1}{I^2} \int_{V_e} \mu \mathbf{H}_{\text{ind}} \cdot \mathbf{H}_{\text{ind}} \, dS \tag{E.32}$$

If the inductance problem is to be solved by means of a magnetic scalar potential, we must face the difficulty that Ψ_{ind} may be multi-valued as was discussed above. We will circumvent this difficulty by using a scalar magnetic potential to represent only a portion of the total magnetic field. Let $\tilde{\mathbf{J}}$ be some current distribution contained entirely in V_{in}, and such that the total current through a surface bounded by any path surrounding the conductor and passing through its hole is equal to I, the same as for the actual loop. Now let $\mathbf{T}_e$ be a stream potential chosen so that $\nabla \times \mathbf{T}_e = \mathbf{J}'$. This stream potential does not necessarily satisfy the boundary condition that $\mu \mathbf{u}_n \cdot \mathbf{T}_e = 0$ on the surface of the conductor, nor that $\nabla \cdot (\mu \mathbf{T}_e) = 0$ in V_e. A line current loop I located completely in V_{in} and surrounding its hole produces an $\mathbf{H}$-field that would serve as $\mathbf{T}_e$. We then put

$$\mathbf{H} = \mathbf{T}_e + \mathbf{H}_0 \tag{E.33}$$

where $\mathbf{H}_0$ is an auxiliary field expressible as the gradient of a single-valued scalar function:

$$\mathbf{H}_0 = -\nabla \Psi_0 \tag{E.34}$$

so that Ψ_0 obeys:

$$\nabla \cdot (\mu \nabla \Psi_0) = -\rho_{m,\text{eq}} = \nabla \cdot (\mu \mathbf{T}_e) \qquad \text{in } V_e \tag{E.35}$$

and

$$\mu \frac{\partial \Psi_0}{\partial n} = -\rho_{mS,\text{eq}} = \mu \mathbf{u}_n \cdot \mathbf{T}_e \qquad \text{on } S \tag{E.36}$$

When this problem is solved for Ψ_0, either the magnetic field is reconstituted and the inductance is computed from (E.31), or we use the stored energy expression, which can be shown to reduce to:

$$L = \frac{1}{I^2} \int_{V_e} \mu \left[\mathbf{T}_e \cdot \mathbf{T}_e - (\nabla \Psi_0)^2 \right] dV \tag{E.37}$$

P. W. Karlsson, *Arch. Elektrotech.*, vol. 67, pp. 29-33, 1984.

V. P. Kazantsev, *Sov. Phys. J.*, vol. 30, pp. 624-628, 1987.

J. C. Verité, *IEEE Trans. Magnetics*, vol. 23, pp. 1881-1887, 1987.

M. R. Krakowski, *Arch. Elektrotechnik*, vol. 74, pp. 329-334, 1991.

O. A. Zolotov and V. P. Kazantsev, *Sov. Phys. J.*, vol. 34, pp. 824-829 and 829-835, 1991.

The first term in the integrand of (E.37) is due to the stored energy in a field corresponding to a certain approximate current distribution within the loop, and is often used as an approximation to the inductance, particularly when the loop conductor is thin. The second term is thus a correction to this (small, it is hoped), and from its form we observe that the use of the first term only would give an approximate inductance that is larger than the true inductance.

E.2.6   THE MAGNETIC POLARIZABILITY PROBLEM

The other magnetostatic problem is the polarizability problem,[4] wherein the conductor is placed in a locally uniform incident magnetostatic field described by a magnetic scalar potential $\mathbf{H}^i = -\nabla \Psi^i$ where $\Psi^i = -\mathbf{r} \cdot \mathbf{H}^i$ in the neighborhood of S, and the constraint of zero current flow around any hole in the conductor is enforced. Because of this, we are entitled to use a single-valued scalar potential to represent globally the scattered field:

$$\mathbf{H}_p = \mathbf{H}^i + \mathbf{H}_p^s = \mathbf{H}^i - \nabla \Psi_p^s \qquad (E.38)$$

The induced surface current on S produces a magnetic dipole moment $\mathbf{m}$, which by (1.21) is:

$$\mathbf{m} = \frac{1}{2} \oint_S \mathbf{r} \times \mathbf{J}_S(\mathbf{r})\, dS = \frac{1}{2} \oint_S [(\mathbf{r} \cdot \mathbf{H}_p)\mathbf{u}_n - (\mathbf{r} \cdot \mathbf{u}_n)\mathbf{H}_p]\, dS \qquad (E.39)$$

We split $\mathbf{H}_p$ in (E.39) into incident and scattered parts using (E.38), and deal with each separately.

First we treat (E.39) with $\mathbf{H}_p$ replaced by $\mathbf{H}_p^s$. The first integral on the right side can be rewritten through liberal use of the identities of vector calculus:

$$\oint_S \mathbf{u}_n(\mathbf{r} \cdot \mathbf{H}_p^s)\, dS = -\int_{V_e} \nabla(\mathbf{r} \cdot \mathbf{H}_p^s)\, dV = -\int_{V_e} [(\mathbf{r} \cdot \nabla)\mathbf{H}_p^s + \mathbf{H}_p^s]\, dV \quad (E.40)$$

where V_e is the region *exterior* to S, and we have assumed sufficient decay of the scattered field as $r \to \infty$. The second integral on the right side of (E.39) is dealt with by taking a typical component (in the x-direction, say):

$$\mathbf{u}_x \cdot \oint_S (\mathbf{r} \cdot \mathbf{u}_n)\mathbf{H}_p^s\, dS = -\int_{V_e} \nabla \cdot (\mathbf{r} H_{px}^s)\, dV = -\int_{V_e} [3 H_{px}^s + (\mathbf{r} \cdot \nabla)H_{px}^s]\, dV$$
$$(E.41)$$

Doing likewise with the other Cartesian components, we get:

$$\oint_S (\mathbf{r} \cdot \mathbf{u}_n)\mathbf{H}_p^s\, dS = -\int_{V_e} [3\mathbf{H}_p^s + (\mathbf{r} \cdot \nabla)\mathbf{H}_p^s]\, dV \qquad (E.42)$$

---

[4]J. B. Keller, R. E. Kleinman, and T. B. A. Senior, *J. Inst. Math. Appl.*, vol. 9, pp. 14-22, 1972.

Combining (E.42) and (E.40) with (E.39) gives:

$$\frac{1}{2} \oint_S \left[(\mathbf{r} \cdot \mathbf{H}_p^s) \mathbf{u}_n - (\mathbf{r} \cdot \mathbf{u}_n) \mathbf{H}_p^s \right] dS = \int_{V_e} \mathbf{H}_p^s \, dV = - \int_{V_e} \nabla \Psi_p^s \, dV$$

$$= \oint_S \Psi_p^s \mathbf{u}_n \, dS \qquad (E.43)$$

The rewriting of the part of (E.39) coming from $\mathbf{H}^i$ is similar, but we must now transform to volume integrals *inside* S (that is, over V_{in}) rather than V_e, because (a) $\mathbf{H}^i$ does not decay as $r \to \infty$, and (b) the incident field is not necessarily expressible from a scalar potential over all of V_e. The result is:

$$\frac{1}{2} \oint_S \left[(\mathbf{r} \cdot \mathbf{H}^i) \mathbf{u}_n - (\mathbf{r} \cdot \mathbf{u}_n) \mathbf{H}^i \right] dS = - \int_{V_{in}} \mathbf{H}^i \, dV = \int_{V_{in}} \nabla \Psi^i \, dV$$

$$= \oint_S \Psi^i \mathbf{u}_n \, dS \qquad (E.44)$$

so that

$$\mathbf{m} = \oint_S \Psi_p \mathbf{u}_n \, dS \qquad (E.45)$$

where $\Psi_p = \Psi^i + \Psi_p^s$.

The stored energy in the scattered part of the magnetic field can be related to the incident field and the induced dipole moment, similarly to the electrostatic polarizability problem (compare (E.25)). The result is:

$$\frac{1}{2} \int_{V_e} \mathbf{B}_p^s \cdot \mathbf{H}_p^s \, dV = -\frac{1}{2} \mathbf{B}^i \cdot \mathbf{m} - \frac{1}{2} V_{in} \mathbf{B}^i \cdot \mathbf{H}^i \qquad (E.46)$$

Since the volume $V_{in} \geq 0$, and the left side of (E.46) is also nonnegative, it is clear that the induced magnetic dipole moment is in a direction *opposite* to that of the incident magnetic field. It is of course also related to the strength of the incident field: in this linear problem, $\mathbf{m} \propto \mathbf{H}^i$. With $\mathbf{H}^i$ uniform at S, we use

$$\mathbf{m} = \overset{\leftrightarrow}{\boldsymbol{\alpha}}_M \cdot \mathbf{H}^i \qquad (E.47)$$

to define the magnetic polarizability dyadic $\overset{\leftrightarrow}{\boldsymbol{\alpha}}_M$.[5]

---

[5] A minus sign is sometimes used in the definition of magnetic polarizability in (E.47) in order that the polarizability of a perfectly conducting scatterer be nonnegative (in the extended sense that $\mu \mathbf{H}^i \cdot \overset{\leftrightarrow}{\boldsymbol{\alpha}}_M \cdot \mathbf{H}^i \geq 0$ for any incident field). The reader is warned to check carefully the definition used in other literature.

E.3   PROBLEMS

E–1   Provide the details of the derivation of (E.25).

E–2   Derive (E.37).

E–3   A dielectric body of permittivity ϵ is placed in an incident static electric field $\mathbf{E}^i$ that would exist in free space in the absence of the body.

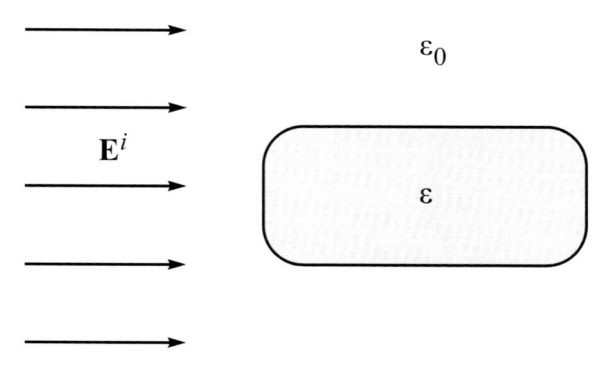

Formulate the conditions to be satisfied for this static scattering problem to have a unique solution.

E–4   A thin loop of ideally conducting (superconducting) wire lying along a curve C in a homogeneous isotropic medium with permeability μ has inductance L. Obtain an explicit expression for its magnetic polarizability dyadic, using the condition that the wire is thin as an approximation.

Index